T0137811

Advances in Intelligent Systems and Computing

Volume 925

Series editor

Janusz Kacprzyk, Systems Research Institute, Polish Academy of Sciences, Warsaw, Poland

The series "Advances in Intelligent Systems and Computing" contains publications on theory, applications, and design methods of Intelligent Systems and Intelligent Computing. Virtually all disciplines such as engineering, natural sciences, computer and information science, ICT, economics, business, e-commerce, environment, healthcare, life science are covered. The list of topics spans all the areas of modern intelligent systems and computing such as: computational intelligence, soft computing including neural networks, fuzzy systems, evolutionary computing and the fusion of these paradigms, social intelligence, ambient intelligence, computational neuroscience, artificial life, virtual worlds and society, cognitive science and systems, Perception and Vision, DNA and immune based systems, self-organizing and adaptive systems, e-Learning and teaching, human-centered and human-centric computing, recommender systems, intelligent control, robotics and mechatronics including human-machine teaming, knowledge-based paradigms, learning paradigms, machine ethics, intelligent data analysis, knowledge management, intelligent agents, intelligent decision making and support, intelligent network security, trust management, interactive entertainment, Web intelligence and multimedia.

The publications within "Advances in Intelligent Systems and Computing" are primarily proceedings of important conferences, symposia and congresses. They cover significant recent developments in the field, both of a foundational and applicable character. An important characteristic feature of the series is the short publication time and world-wide distribution. This permits a rapid and broad dissemination of research results.

** Indexing: The books of this series are submitted to ISI Proceedings, EI-Compendex, DBLP, SCOPUS, Google Scholar and Springerlink **

Advisory Editors

Nikhil R. Pal, Indian Statistical Institute, Kolkata, India

Rafael Bello Perez, Faculty of Mathematics, Physics and Computing, Universidad Central de Las Villas, Santa Clara, Cuba

Emilio S. Corchado, University of Salamanca, Salamanca, Spain

Hani Hagras, Electronic Engineering, University of Essex, Colchester, UK

László T. Kóczy, Department of Automation, Széchenyi István University, Gyor, Hungary

Vladik Kreinovich, Department of Computer Science, University of Texas at El Paso, EL PASO, TX, USA

Chin-Teng Lin, Department of Electrical Engineering, National Chiao Tung University, Hsinchu, Taiwan

Jie Lu, Faculty of Engineering and Information Technology, University of Technology Sydney, Sydney, NSW, Australia

Patricia Melin, Graduate Program of Computer Science, Tijuana Institute of Technology, Tijuana, Mexico

Nadia Nedjah, Department of Electronics Engineering, University of Rio de Janeiro, Rio de Janeiro, Brazil

Ngoc Thanh Nguyen, Faculty of Computer Science and Management, Wrocław University of Technology, Wrocław, Poland

Jun Wang, Department of Mechanical and Automation Engineering, The Chinese University of Hong Kong, Shatin, Hong Kong

More information about this series at http://www.springer.com/series/11156

Paolo Ciancarini · Manuel Mazzara ·
Angelo Messina · Alberto Sillitti ·
Giancarlo Succi
Editors

Proceedings of 6th International Conference in Software Engineering for Defence Applications

SEDA 2018

 Springer

Editors
Paolo Ciancarini
University of Bologna
Bologna, Italy

Manuel Mazzara
Innopolis University
Innopolis, Russia

Angelo Messina
Innopolis University
Innopolis, Russia

Alberto Sillitti
Innopolis University
Innopolis, Russia

Giancarlo Succi
Innopolis University
Innopolis, Russia

ISSN 2194-5357   ISSN 2194-5365 (electronic)
Advances in Intelligent Systems and Computing
ISBN 978-3-030-14686-3   ISBN 978-3-030-14687-0 (eBook)
https://doi.org/10.1007/978-3-030-14687-0

Library of Congress Control Number: 2019933335

This Springer imprint is published by the registered company Springer Nature Switzerland AG
The registered company address is: Gewerbestrasse 11, 6330 Cham, Switzerland

Preface

Not only the military world but the whole defence and security community has always demonstrated interest and expectation in the evolution of software application and in the way it has been designed and manufactured through the years. The first real standard in the area of software quality was originated by the US DOD (2167A and 498) to demonstrate the need for this particular user, the US Department of Defence, to implement repeatable and controllable processes to produce software to be used in high-reliability applications. Military systems, security systems, and most of the mission critical systems rely more and more on software than older systems did.

Security in airports and train stations rely more and more on the correct functioning of specific software applications. This reliance on software and its reliability is now the most important aspect of military systems, and the same will happen in the close related which share the same "mission criticality". SEDA has, of course, a special attention for the military area which is seen as one of the most challenging and then as a benchmark for software reliability. The military-specific area of application includes mission data systems, radars/sensors, flight/engine controls, communications, mission planning/execution, weapons deployment, test infrastructure, programme life cycle management systems, software integration laboratories, battle laboratories, and centres of excellence. Even if it is slightly less significant, the same scenario applies to the land component of the armed forces. Software is now embedded in all the platforms used in operations, starting from the wearable computers of the dismounted soldier up to various levels of command and control, and every detail of modern operations relies on the correct behaviour of some software product. Many of the mentioned criticalities are shared with other public security sectors such as the police, the firefighters, and the public health system.

The rising awareness of the critical aspects of the described software diffusion convinced the Italian Army General Staff that a moment of reflection and discussion was needed and with the help of the universities, the SEDA conference cycle was started. For the third conference SEDA 2014, it was decided to shift the focus of the event slightly away from the traditional approach to look at innovative software engineering. Considering the title: software engineering for defence application, this time, the emphasis was deliberately put on the "defence application" part. For the first time, papers not strictly connected to the "pure" concept of software engineering were accepted together with others that went deep into the heart of this science. The reasons for this change were, first of all, the need for this event to evolve and widen its horizon and secondly the need to find more opportunities for the evolution of military capabilities. In a moment of economic difficulty, it is of paramount importance to find new ways to acquire capabilities at a lower level of funding using innovation as a facilitator. It was deemed very important, in a period of scarce resources to look ahead and leverage from dual use and commercial technologies. Software is, as said, a very pervasive entity and is almost everywhere, even in those areas where it is not explicitly quoted. A mention was made to the changes in the area of software engineering experienced in the Italian Army and the starting of a new methodology which would then become "Italian Army Agile" and then DSSEA® iAgile.

SEDA 2015 pointed out that in the commercial world, "Agile" software production methods have emerged as the industry's preferred choice for innovative software manufacturing as pointed out in the various Chaos reports by the Standish Group. Agile practices in the mission critical and military arena seem to have received a strong motivation to be adopted in line with the objectives the USA DoD is trying to achieve with the reforms directed by Congress and DoD Acquisition Executives. DoD Instruction 5000.02 (December 2013) heavily emphasizes tailoring programme structures and acquisition processes to the programme characteristics. At the same time, in May 2013, the Italian Army started an effort to solve the problem of the volatility of the user requirement that is at the base of the software development process. The Army project is: **LC2Evo.** The results and outcome of the SEDA 2015 conference are very well presented in the post-proceedings that can be found online.

The LC2Evo results and analysis marked the pace of the 5th SEDA 2016 conference, the first one under coordination of DSSEA. The acronym stands for Land Command & Control Evolution, and this is a successful effort the Italian Army General Staff made to device a features and technology demonstrator that could help identifying a way ahead for the future of the Command & Control support software. The main scope, related to the software engineering paradigm change in the effort, was to demonstrate that a credible, innovative, and effective software development methodology could be applied to complex user domains even in the case of rapidly changing user requirements. The software project was embedded in a more ambitious and global effort in the frame of the Italian defence procurement innovation process aimed at implementing the Concept Development & Experimentation (NATO CD&E) which was initially started by the "Centro Innovazione Difesa (CID)".

The military operations in Iraq and Afghanistan had clearly demonstrated that the operating scenario was changing in a few months cycle and the most required characteristic for a C2 system by the user was flexibility. The possibility of adapting the software functions to an asymmetric dynamically changing environment seemed to be largely incompatible with the linear development life cycle normally used for mission critical software in the defence and security area. The major features needed for a rapid deployment software prototype are:

- Responding rapidly to changes in operations, technology, and budgets;
- Actively involving users throughout development to ensure high operational value;
- Focusing on small, frequent capability releases;
- Valuing working software over comprehensive documentation.

Agile practices such as SCRUM include planning, design, development, and testing into an iterative production cycle (Sprint) able to deliver working software at short intervals (3–4 weeks). The development teams can deliver interim capabilities (at demo level) to users and stakeholders monthly. These fast iterations with user community give a tangible and effective measure of product progress meanwhile reducing technical and programmatic risk. Response to feedback and changes stimulated by users is far quicker than using traditional methods. The user/stakeholder community in the Army is very articulated, including operational units, main area commands, and schools. The first step we had to take was the establishment of a governance body which could effectively and univocally define the "Mission Threads" from which the support functions are derived.

The first LC2Evo Scrum team (including members from industry) was established in March 2014. In the framework of a paramount coordination effort led by The Italian Army COFORDOT (three star level command in charge, among other things of the Army Operational Doctrine), the Army General Staff Logistic Department got full delegation to lead, with the help of Finmeccanica (now Leonardo), a software development project using agile methodology (initially Scrum, then ITA Army Agile, and finally DSSEA® iAgile) aimed at the production of a technology demonstrator capable of implementing some of the functional area services of a typical C2 Software.

Strictly speaking software engineering, one of the key issues was providing the users with a common graphical interface on any available device in garrison (static office operation) in national operations (i.e. "Strade sicure") or international operations. The device type could vary from desktop computers to mobile phones.

The development was supposed to last 6 to 8 months at the Army premises to facilitate the build-up of a user community network and to maximize the availability of user domain experts, both key features of the new agile approach. In the second phase, the initial team was supposed to move to the contractor premises and serve as an incubator to generate more teams to work in parallel. The first team outcome was so surprisingly good, and the contractor software analysts and engineers developed such an excellent mix with the Army ones that both parts agreed to continue phase two (multiple teams) still at the Army premises. The effort reached the peak activity after 18 months from start when 5 teams were active at the same time operating in parallel (the first synchronized "scrum of scrum like" reported in the mission critical software area).

As per the results presented at SEDA 2016, more than 30 basic production cycles (Sprints of 4–5 weeks) were performed; all of them delivered a working software increment valuable for the user. The delivered FAS software tested in real exercises, and some components deployed in operations. One of the initial tests was performed during a NATO CWIX exercises and concerned cyber security. The product, still in a very initial status, was able to resist more than 48 hours to the penetration attempt by a very good team of "NATO hackers". More than a million equivalent line of software was developed at a unit cost of less than 10 Euros, with an overall cost reduction of 90% reaching a customer satisfaction exceeding 90%. One of developed FAS is still deployed in Afghanistan at the multinational command.

The preparatory work for SEDA 2016 made it clear that the delivered working software and the impressive cost reduction were not the most important achievements of the Italian Army experiment. The most important result was the understanding of what is needed to set up a software development environment which is effective for a very complex and articulated set of user requirements and involves relevant mission critical and high-risk components.

After a year into the experience, the LC2Evo project and the collateral methodological building efforts had already substantially involved a community much wider than the Italian Army and the Italian MoD, including experts from universities, defence industry, and small enterprises, making it clear that there was an urgent need to preserve the just born improved agile culture oriented to the mission critical and high-reliability applications. The community of interests build around these efforts, identified four key areas called "Pillars" (explained through the conference sessions) on which any innovative agile software development process for mission critical applications should invest and build. Surprisingly (may be not) the collected indications mostly concern the human component and the organization of the work, even if there are clear issues on the technical elements as well.

To act as a "custodian" of the new born methodology, the no-profit association DSSEA® took the lead of the methodology development, now DSSEA® iAgile, and of the SEDA conference cycle organization. As a result, the methodology and the conference are available to developers and researchers for free. In the area of innovation and towards building a new software engineering paradigm, DSSEA® iAgile constitutes a real breakthrough and for this NCI Agency (NATO Communication and Information Agency) organized two different workshops aimed at devising a strategy to introduce this methodology into the NATO procurement cycle.

The DSSEA coordination in the preparation and execution of SEDA 2016 has initiated a series of collateral discussion and elaboration processes which resulted in many continuous methodological and technical efforts mainly at the Italian Directorate of Armaments Agency: DAT, at NATO NCI and at some universities: Innopolis University (Russian Federation) being the most active, University of Bologna (Italy), University of Regina (Canada), and University of Roma 1 Sapienza (Italy). It appears that this DSSEA coordination activity is capable of generating a year-round production of technical papers as a spin-off of any SEDA event. For this reason, it was decided to decouple the post-proceedings publication date and the conference date keeping as the only requirement to publish before the date of the next conference.

SEDA 2018 was held at the premises of the Italian Defence General Secretariat (Segretariato Generale della Difesa) and the main military body actively supporting the conference was the Land Armament Directorate (Direzione Armamenti Terrestri). This is an important evolution of the Italian SEDA military community, as it shows that the interest for the SEDA debate is now shared at a "joint" level including all the Italian services: Army, Navy, Air Force, and Carabinieri.

At the same time, the awareness of the threat posed to the modern society through software is raised and the perception there is more to do, and more problems are coming is clear. Together with the major SEDA debate streamline on software engineering, tools and defects control, more have appeared and become very vital. For the first time a thorough discussion on the complexity of users requirement has started with the presentation of the specific needs of a real operation mission. This includes the description of the political and international scenario, giving a full picture of the complexity.

Very fruitful exchanges have been ignited by the two full-day event with an unprecedented production of papers, most of which are included in this post-proceedings. The debate is still alive, and new areas are being outlined such as the one on "teaching" innovative software engineering methodology in various contexts.

Acknowledgements

The SEDA committees and DSSEA wish to express the warmest appreciation and gratitude to:

- The Segretariato Generale della Difesa and in particular to Lt. Gen Castrataro and Col. Cotugno for the incredible support including a perfect conference facility and the related logistics.
- The Officers of the Italian Army Corp of Engineers who attended in large number the conference adding an interesting dynamic to the event.
- Innopolis University, and in particular the Chairman of the Board of Trustee, Nikolay Nikiforov, the CEO, Kirill Semenikhin, and the Viceprovost for Academic Affairs, Sergey Masyagin, for generously supporting the fruitful research and the rich discussion that have permeated the whole conference.
- All the universities from Italy and abroad for the great contribution to the conference.

January 2019   Paolo Ciancarini
                                                                         Manuel Mazzara
                                                                         Angelo Messina
                                                                         Alberto Sillitti
                                                                         Giancarlo Succi

Organization

Programme Committee

Ilya Afanasyev	Innopolis University, Russia
Luigi Benedicenti	University of Regina, Canada
Marcello M. Bersani	Politecnico di Milano, Italy
Joseph Alexander Brown	Innopolis University, Russia
Jean-Michel Bruel	IRIT, France
Paolo Ciancarini	University of Bologna, Italy
Michele Ciavotta	Università di Milano-Bicocca, Italy
Franco Raffaele Cotugno	MOD, Italy
Nicola Dragoni	Technical University of Denmark, Denmark
Mohamed Elwakil	Innopolis University, Russia
Luca Galantini	Università Cattolica del S.Cuore Milano, Italy
Mansur Khazeev	Innopolis University, Russia
Vivek Kumar	NUST-MISIS, Russia
Jooyoung Lee	Innopolis University, Russia
Nikita Lozhnikov	Innopolis University, Russia
Manuel Mazzara	Innopolis University, Russia
Angelo Messina	Innopolis University, Russia
Alexandr Naumchev	Innopolis University, Russia
Francesco Poggi	University of Bologna, Italy
Nafees Qamar	Vanderbilt University, USA
Victor Rivera	Innopolis University, Russia
Francesco Rogo	Leonardo SpA, Italy
Davide Rossi	University of Bologna, Italy
Stefano Russo	University of Naples Federico II, Italy
Salah Sadou	IRISA, University of South Brittany, France
Larisa Safina	Innopolis University, Russia
Alberto Sillitti	Innopolis University, Russia
Giancarlo Succi	Innopolis University, Russia

Additional Reviewers

Ashlock, Daniel
Aslam, Hamna
De Donno, Michele
Giaretta, Alberto
Kavaja, Juxhino

Lee, Colin
McHugh, Michael
Ohlke, C.
Qamar, Nafees
Wintjes, Jorit

Contents

Bi-lingual Intent Classification of Twitter Posts: A Roadmap

Akinlolu Solomon Adekotujo[1,2,3(✉)], JooYoung Lee[2],
Ayokunle Oluwatoyin Enikuomehin[1], Manuel Mazzara[2],
and Segun Benjamin Aribisala[1]

[1] Department of Computer Science, Lagos State University, Lagos, Nigeria
adekotujoakinlolu@gmail.com,
toyinenikuomehin@gmail.com,
benjamin.aribisala@gmail.com
[2] Innopolis University, Innopolis, Russia
{j.lee,m.mazzara}@innopolis.ru
[3] Computer, Information and Management Studies Department,
The Administrative Staff College of Nigeria, Badagry, Nigeria

Abstract. A core advantage of social media platforms is the freedom that comes with the way users express their opinions and share information as they deem fit, in line with the subject of discussion. Advances in text analytics have allowed researchers to adequately classify information expressed in natural language text, which emanates in millions per minute, under well-defined categories like "hate" or "radicalized" content which provide further insight into intent of the sender. This analysis is important for social media intelligence and information security. Commercial intent classifications have witnessed several research attentions. However, social intent classification of topics in line with hate, radicalized posts, have witnessed little research effort. The focus of this study is to develop a roadmap of a model for automatic bilingual intent classification of hate speech. This empirical model will involve the use of bi-gram words for intent classification. The feature extraction will include expected cross entropy, while topic modeling will use supervised context-based n-gram approach. Classification will be done using ensemble-based approach which will include the use of Naïve Bayes and Support Vector Machine. This study will also discuss the differences between the concept of fake news, stance and intent identification. We anticipate that the proposed roadmap, if implemented, will be useful in the classification of intent as it relates to hate speech in bilingual twitter post. The proposed model has the potential to improve intent classification and that could be useful in hate speech detection, which can avert social or security problems.

Keywords: Intent classification · Hate speech · Machine learning classifier

1 Introduction

Personal expressions or comments on social media are largely about users' emotions, sentiments or goals (intents) which are particularly valuable, for instance, for monitoring activities to ensure security of lives and properties [1]. Understanding user's

P. Ciancarini et al. (Eds.): SEDA 2018, AISC 925, pp. 1–9, 2020.
https://doi.org/10.1007/978-3-030-14687-0_1

intent for hate from speech or text in general, is a natural language problem which is usually difficult to solve, as we are now confronted with much more short texts and news every day. There are three commonly confused terminologies related to personal expressions, opinions, sentiments and emotions, these are fake news, stance and intent. Fake news are intentional false information [2] and stance are the stand of a person on a topic [3] while intent are every day or futuristic behavior or goals. Fake news is commonly motivated by financial and ideological reasons [4]. On the other hand, stance are motivated by sentiments powered by prior knowledge and analytical skills [3], while intent are motivated by individual desires. Intent can help identify actionable information [5, 6].

1.1 Fake News Detection

Fake news detection has recently attracted growing interest in research due to increasing misinformation essentially on social media feeds. Until recently, detection of fake news relied on satire based news system and fact finding news portals like politifact (https://www.politifact.com/). Fake news can be detected by first formalizing the news as an input to a fake news detection equation (Eq. 1) whose output is either 1 or 0 (True or False respectively). Given the social news engagements $\mathcal{E}$ among n users for news article a, the task of fake news detection is to predict whether the news article a is a fake news piece or not,

i.e., $\mathcal{F}: \mathcal{E} \rightarrow \{0,1\}$

such that,

$$\mathcal{F}(a) = \begin{cases} 1, \text{ if } a \text{ is a piece of fake news,} \\ 0, \text{ otherwise.} \end{cases} \tag{1}$$

where $\mathcal{F}$ is the prediction function.

Note that we define fake news detection as a binary classification problem because fake news is essentially a distortion on information released by the publisher [4].

1.2 Stance Detection

Stance detection are also becoming common in research environment. The main aim of stance detection is to identify if the author of a piece of text is supportive of a given target or against it. For clarity, stance should be seen as a subtask of opinion mining. Stance can detection can be represented using Eq. 2, which shows that the output of stance detection model is either Agree (0), Disagree (1), Discuss (2) or Unrelated (3) [7, 8]. Given a set of social media posts D related to the K target entities $T_1,\ldots, T_k$ then, the goal is to determine the value of mapping S. The task of stance detection is to identify the cognitive position of an author of individual posts in his reaction, towards a given statement or claim.

i.e., S: $T_1 \times ... \times T_k \times D \rightarrow$ {Agree, Disagree, Discuss, Unrelated} k for each post $d \in D$

such that,

$$S(d) = \begin{cases} 0, & \text{if agreed to a claim in D,} \\ 1, & \text{if disagree to a claim in D,} \\ 2, & \text{if a claim in D is discussed} \\ 3, & \text{if a claim in D is unrelated.} \end{cases} \quad (2)$$

where S is the claim function.

Stance classification systems normally require identification of a claim as belonging to any of 4 [7, 9–11] categories namely supporting, denying, querying, and commenting. Supporting means that the claim is supported, while denying means that the claim is not supported, but disagreed. Querying implies that the claim is being discussed or questions related to the claim are being raised, while commenting implies that reactions to the claim are unrelated to the claim [3]. Table 1 gives an example of a claim and the reactions related to these four categories [7, 10].

Table 1. A sample claim and reactions showing the four categories of stance

Claim: Robert Plant Ripped up $800M Led Zeppelin Reunion Contract	
Snippet	Stance
Led Zeppelin's Robert Plant turned down £500m to reform supergroup…	Agree
Robert Plant's publicist has described as "rubbish" a Daily Mirror report that he rejected a £500m Led Zeppelin reunion…	Disagree
Robert Plant reportedly tore up an $800 million Led Zeppelin reunion deal…	Discuss
Richard Branson's Virgin Galactic is set to launch SpaceShipTwo today….	Unrelated

1.3   Intent Detection

Intent(ion) detection and analysis are very important aspect of social media modelling. This is largely because intents are hidden and execution of same could unrest or social misbehaviour. Intent detection is commonly used alongside Slot Filling. Using slot filling, the conservational flow within an intent can be determined could serve as a means of determining the validity of the detected intent.

Intent can be detected using Eq. 3. The output of an intent detection algorithm is either True (1) or False (0). Given the social news engagements $\mathcal{E}$ among n users for news article or twitter post a, the task of intent detection is to predict whether the news article or twitter post a has intent of hate speech or not,

i.e., $I: \mathcal{E} \rightarrow \left[0,1 \right]$

such that,

$$I(a) = \begin{cases} 1, & \text{if } a \text{ contains hate intent or speech,} \\ 0, & \text{otherwise.} \end{cases} \tag{3}$$

where I is the prediction function.

A variety of factors like personal interest, religion and social influence, affect an individual's expression of intentionality or post [12–15]. Intentions that have multiple potential often complicates natural language clarification in short-text documents. In a bid to make the hate intent or goal mining problem computationally manageable, we need to first mine specific intent classes with corresponding hate speech, and therefore define a multiclass intent classification problem.

The task intent analysis is defined [6] as to approximate the unknown function in Eq. (4)

Given that: $S_i \in d_i$, $d_i \subset D$,

$$f: S \times C^I \rightarrow \{True, \ False\}, \tag{4}$$

where $C^I = \{c_1^I, c_2^I \ldots c_n^I\}$ is a set of predefined intent categories from document, where D is a domain that consist of text documents and each reviewed document d_i contains a sequence of sentences $S = \{s_1, s_2 \ldots s_{|S|}\}$.

Intent classification, is known to focus on futuristic action, it is a form of text classification, whereas sentiment and opinion analysis that compute subjectivity text classification, focuses on the current state of affairs [1, 6]. For instance, in a message "I wanna visit the magnificent Walt Disney animal hotel", topic classification focuses on the noun, the hotel 'Walt Disney Animal Hotel'; sentiment and emotion classification focused on the positive feeling of the author's message expressed with the adjective 'magnificent'. In contrast, intent classification concerns the author's intended future action or goal, i.e. going to visit the hotel. Given that the focus of this study is on intent classification, the next section and subsequent ones will focus on intent classification.

2   Related Work to Intent Classification

This section presents the review of existing works on intent classification. The review was divided into commercial intent classification and social intent classification for easy understanding [12].

2.1   Commercial Intent Classification

Commercial intents are intent that are related to buying and selling or marketing, they can be identified by verbs in a post. Hollerit et al. [16] proposed an automatic method for classifying commercial intent in tweets based on Bayes Complement Naïve Bayes classifier in the textual domain, and a linear logistic regression classifier. Benczúr et al. [17] proposed features for web spam filtering based on the presence of keywords with high advertisement value, same spam filtering power was demonstrated on online commercial intention value.

Lewandowski et al. [18] concluded that neither crowdsourcing approach nor questionnaire approach lead to satisfying result during the survey on a commercial search engine's portal. Guo et al. [19] explore client-side instrumentation for inferring personalized commercial intent of user's, by investigating whether mouse movement over search result can provide clues into the user's intent using clich-through on ads, the result came out that mouse movement analysis can provide such clues.

Lewandowski [20] used retrieval effectiveness test design on three prominent web search engines by classification of the result in relation to their commercial intent, the result showed that Google more significant commercial intent.

Purohit et al. [5] present a hybrid feature approach of combining top-down processing using a bag-of-tokens model to address the problem of multiclass classification of intent on twitter for crisis events dataset and address the problem of ambiguity and sparsity in order to classify the intent of narrative.

2.2   Social Intent Classification

Social intent are intents that do not have verbs representing buying and selling, but that are related to social activities or events. Hate intent is a good example of social intent and the focus of this review is on hate intent.

Ben-David et al. [21] using longitudinal multimodal and network analysis, claimed that hate speech and discriminatory engagements are not only explicated with user's motivations and actions, but also included the formation of a network of ties of platform policy, technological availability, and the manners of communication of its users. The study affirmed that the platform encourages discrimination, which the users eventually exhibits. Wang et al. [22] affirmed that word order and phrases are important in giving understanding to text in many texts mining job, the study proposed a topical n-gram topic model that is able to identify topics and topical phrases using probabilistic model.

The problem of identifying hatred videos is proposed, [23] with the implementation of classification algorithm named shark search, the study focused on the creation of a web portal and produced a framework that will resolve the problem of finding out hatred videos. Agarwal et al. [24], claimed that just a keyword spotting based techniques cannot accurately identify the intent of a post, the study developed a mono-lingual cascaded ensemble learning classifier for identifying the posts having racist or radicalized intent with the use of open sources API's for feature extractions. It was also reaffirmed [25–28], while presenting an unsupervised method for polarity classification in twitter that little work has been noticed in the area of multilingual domain, the method [25] is based on the expansion of the concepts expressed in the tweets through the application of Page Ranking to Wordnet. Additionally, Gomes et al. and Agarwal et al. [24, 29] emphasized the high performance of ensemble learning for data stream classification of intent above one single classifier. Gomes et al. [29] proposed taxonomy for data stream ensemble learning and listed popular open source tools, while Agarwal et al. [24], used cascaded ensemble learning classifier for only mono lingual post.

Sanfilippo et al. [30] developed a violet intent modeling software (VIM) [30], using a Wikibased expert knowledge management approach and content extraction and content extraction, which describe a framework that implemented a multidisciplinary approach in the emergence of radicalization leading to violet intent based on English language only.

3   Problem Definition

The increase in the emerging trend of hate groups, and their presence online in recent years call for concern. In 2014, a survey conducted by the New York Times with the Police Executive Research Forum reported that rightwing extremism is the primary source of "ideological violence" in America and that Europe is dazed by the rise of far-right extremist groups [31]. From literature review, there is no single study that has investigated the methods of identifying hate speech in multilingual languages, in terms of word order and n-gram concept. With features concept analysis, as against a key word technique, there is need to apply an improved feature model for extraction, and a context-based technique that combines bi-grams to accurately classified intent, hence the computational research problem as stated below:

In this study, hate detection will be modeled for bi-lingual posts. We assume that we are interested in classifying a dataset D of twitter feeds t, represented in Eq. 5.

$$D^l = t_i\{1| \leq i \leq n\} \quad \text{where } l \in \{l_1, l_2\}. \tag{5}$$

D_1^l and D_2^l are two distinct documents with two distinct different languages l_1 and l_2.

The first is to preprocess D with n-gram concept, using Topical n-gram (TNG) model to identify a set of topics, C, for each tweet, where C = {c_1, c_2, ..., c_k}. Then we assign the mixture of topics to n-gram phrases of words to get a set of predefined hate terms, S. Finally, we classify vectored terms. Our goal is to define f which can classify the set of terms to either True or False.

$$f: S \rightarrow \{True, False\} \text{ where } f \text{ is the unknown intent function.}$$

3.1   Proposed Methodology

Here we propose a framework for the model for classification of hate intent of multi-lingual twitter posts. The framework is represented in Fig. 1.

The proposed frame work in Fig. 1, will take two different twitter dataset that is bilingual, t_i^1 and t_j^2. Each dataset will be subjected to pre-processing by cleaning the bilingual tweets of non-textual information and not relevant topics. Next, stop-word removal as well as tokenization, stemming, Part of Speech tagging, and lemmatization, using appropriate lexicons and semantics appropriate for each dataset will be carried out.

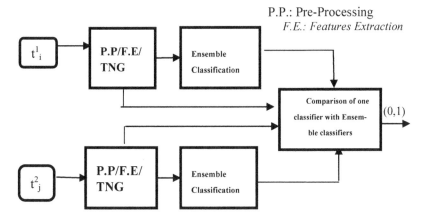

Fig. 1. The structural framework of the proposed solution to the problem statement.

The feature extraction will include Expected Cross Entropy (ECE) and Gibbs Sampling for vectorization, while supervised context-based N-gram topic model (TNG) will be employed for topic modeling. One term keyword spotting based techniques cannot be used to accurately identify the intent of a post, it requires context-based techniques that combines bi-grams to accurately classified intent as this can better defines intent in a post. Feature extraction and topic modeling will be employed for the two datasets separately, so also with the Ensemble Classification using Naïve Bayes, and Support Vector Machine for classification. The performance of each classifier will be assessed using precision, recall and accuracy and performance will also be compared with existing methods.

4   Conclusion

Existing literature have shown that hate and radicalism speech are not tied to only minor communities, but larger communities. Researchers have shown much concern to commercial intent analysis, while very few attentions are shown towards social hate intent classification in multilingual natural language text. While Sentiment serves as a rationale for emotional expressions, hate intent can characterize a person's goal and thus provides additional information about the person itself. One term keyword-based techniques cannot accurately identify the intent of a post. Hence, the focus of this study was to develop a framework for a cascaded ensemble learning based classifier for hate intent classification in multilingual languages using topical n-gram model.

We have proposed a framework for intent identification in bilingual twitter post, using context-based technique. If implemented, the algorithm will help in ensuring national, social, and information security.

References

1. Kröll M, Strohmaier M (2015) Associating intent with sentiment in weblogs. In: International conference on applications of natural language to information systems. Springer, Cham
2. Albright J (2016) The# Election2016 micro-propaganda machine. https://medium.com/@d1gi/the-election2016-micro-propaganda-machine-383449cc1fba#.idanl6i8z. Accessed 15 Jan 2017
3. Lozhnikov N, Derczynski L, Mazzara M (2018) Stance prediction for Russian: data and analysis
4. Shu K et al (2017) Fake news detection on social media: a data mining perspective, 19 (1):22–36
5. Purohit H et al (2015) Intent classification of short-text on social media. In: 2015 IEEE international conference on smart city/SocialCom/SustainCom (SmartCity). IEEE
6. Kröll M, Strohmaier M (2009) Analyzing human intentions in natural language text. In: Proceedings of the fifth international conference on knowledge capture. ACM
7. Mohtarami M et al (2018) Automatic stance detection using end-to-end memory networks
8. Zubiaga A et al (2018) Detection and resolution of rumours in social media: a survey, 51 (2):32
9. Lozhnikov N, Derczynski L, Mazzara M (2018) Stance prediction for russian: data and analysis. arXiv preprint arXiv:1809.01574
10. Dai HK et al (2006) Detecting online commercial intention (OCI). In: Proceedings of the 15th international conference on world wide web. ACM
11. Kirsh D (1990) When is information explicitly represented? Information, language and cognition - the Vancouver studies in cognitive science. UBC Press, pp 340–365
12. Ajzen I (1991) The theory of planned behavior, 50(2):179–211
13. Malle BF, Knobe J (1997) The folk concept of intentionality, 33(2):101–121
14. Sloman SA et al (2012) A causal model of intentionality judgment, 27(2):154–180
15. Melnikov A et al (2018) Towards dynamic interaction-based reputation models. In: 2018 IEEE 32nd international conference on advanced information networking and applications (AINA). IEEE
16. Hollerit B, Kröll M, Strohmaier M (2013) Towards linking buyers and sellers: detecting commercial intent on Twitter. In: Proceedings of the 22nd international conference on world wide web. ACM
17. Benczúr A et al (2007) Web spam detection via commercial intent analysis. In: Proceedings of the 3rd international workshop on adversarial information retrieval on the web. ACM, pp 89–92
18. Lewandowski D, Drechsler J, Von Mach S (2012) Deriving query intents from web search engine queries. J Am Soc Inform Sci Technol 63(9):1773–1788
19. Guo Q, Agichtein E, Clarke CL, Ashkan A (2008) Understanding "abandoned" ads: towards personalized commercial intent inference via mouse movement analysis. Inf Retr Advert IRA 2008:27–30
20. Lewandowski D (2011) The influence of commercial intent of search results on their perceived relevance. In: Proceedings of the 2011 iConference. ACM, pp 452–458
21. Ben-David A, Matamoros-Fernandez A (2016) Hate speech and covert discrimination on social media: monitoring the Facebook pages of extreme-right political parties in Spain. Int J Commun 10:1167–1193
22. Wang X, McCallum A, Wei X: Topical n-grams: phrase and topic discovery, with an application to information retrieval. In: ICDM. IEEE, pp 697–702

23. Chavhan RN (2016) Solutions to detect and analyze online radicalization, 1(4)
24. Agarwal S, Sureka A (2017) Characterizing linguistic attributes for automatic classification of intent based racist/radicalized posts on Tumblr micro-blogging website
25. Montejo-Ráez A et al (2014) A knowledge-based approach for polarity classification in Twitter, 65(2):414–425
26. Balahur A, Perea-Ortega JM (2015) Sentiment analysis system adaptation for multilingual processing: the case of tweets, 51(4):547–556
27. Montoyo A, MartíNez-Barco P, Balahur A (2012) Subjectivity and sentiment analysis: an overview of the current state of the area and envisaged developments. Elsevier (2012)
28. Vilares D et al (2017) Supervised sentiment analysis in multilingual environments, 53 (3):595–607
29. Gomes HM et al (2017) A survey on ensemble learning for data stream classification, 50 (2):23
30. Sanfilippo A et al (2009) VIM: a platform for violent intent modeling. In: Social computing and behavioral modeling. Springer, Heidelberg, pp 1–11
31. Ben-David A, Matamoros-Fernandez A (2016) Hate speech and covert discrimination on social media: monitoring the Facebook pages of extreme-right political parties in Spain, 10:1167–1193

Evolving Diverse Cellular Automata
Based Level Maps

Daniel Ashlock$^{(\boxtimes)}$ ⓘ and Matthew Kreitzer

University of Guelph, Guelph, ON N1G 2W1, Canada
{dashlock,mkreitze}@uoguelph.ca

Abstract. This chapter generalizes a technique for creating terrain maps using a generative fashion based cellular automata representation. The original technique, using fashion based cellular automata, generated terrain maps that exhibit a consistent texture throughout. The generalization presented here co-evolves rules to permit a spatially varying type of map. Pairs of fashion based cellular automata rules are evaluated with objective functions that require connectivity within the terrain and encourage other qualities such as entropic diversity of terrain type, separation of the rule types, and a specified fraction of clear terrain pixels. These three encouraged properties are independently switchable yielding eight different possible fitness functions which are tested and compared. Use of the entropic diversity reward is found to strongly encourage good results while rewarding separation of the two rules without the entropic diversity reward was found to yield bad results with an excess of empty space. The matrix encoding of cellular automata rules yields a discrete granular space encoded with real parameters. Some properties of this space are provided.

Keywords: Automatic content generation ·
Evolutionary computation · Cellular automata

1 Introduction

The problem of automatically generating terrain maps is an area within the field of *Procedural Content Generation* (PCG). The goal of PCG is to provide a map for a game scenario. This chapter will use an evolutionary algorithm [1] to generate diverse examples of terrain maps that can easily be sub-selected and scaled. A key point is that the representation or encoding used to find the terrain maps is unusual in that it creates scalable maps. Changing the representation used by an evolutionary algorithm, even while retaining many of the other algorithmic details, can radically change the type of content that is generated [2]. The evolution of maps is an example of *search-based procedural content generation* (SBPCG), a variant of PCG that incorporates search rather than generating

The authors would like to thank the University of Guelph for its support of this work.

ⓒ Springer Nature Switzerland AG 2020
P. Ciancarini et al. (Eds.): SEDA 2018, AISC 925, pp. 10–23, 2020.
https://doi.org/10.1007/978-3-030-14687-0_2

acceptable content in a single pass. SBPCG is typically used when a single pass will not suffice to locate content with the desired qualities. A survey and the beginnings of a taxonomy of SBPCG can be found in [3].

Though automated map generation in games can arguably be traced back to the roguelikes of the 1980s (Rogue, Hack, NetHack, . . .), the task has recently received some interest from the research community. In [4] levels for 2D sidescroller and top-down 2D adventure games are automatically generated using a two population feasible/infeasible evolutionary algorithm. In [5] multi-objective optimization is applied to the task of search-based procedural content generation for real time strategy maps. In [6] cellular automata are used to generate, in real time, cave like levels for use in a roguelike adventure game. The work presented here is part of a project extending this idea. Earlier publications in this project include [7–11].

2   Background

This study evolves a morphable cellular automata rule to create a more diverse type of map that those in earlier studies. We must review both digital evolution and cellular automata before we can plunge into the generalization. While we use the term *evolutionary algorithm* this style of computation has been discovered several times and the names *genetic algorithms*, *evolutionary programming*, and *evolution strategies* are also common in the literature.

2.1   Evolutionary Algorithms

An evolutionary algorithm is any algorithm that uses some version of the biological theory of evolution [12]. The pseudocode for evolutionary computation is given below, as Algorithm 1.

Algorithm 1: basic evolutionary algorithm.

Generate an initial collection of solutions
Evaluate the quality of the solutions
Repeat
    Select parent solutions to reproduce
    Copy parents and apply variation operators to the copies
    Evaluate the quality of the resulting children
    Conditionally replace population members with children
Until(Problem solved or out of time)

A key issue within evolutionary algorithms is *representation*. The representation for an evolutionary algorithm encompasses the way that solutions to the problem are encoded and their variations are generated. The representation used here is based on $n \times n$ matrices that specify the updating rule for a cellular automata. In order to make morphable rules, two separate matrices are evolved together. The details of the representation for locating scalable terrain maps, including the variation operators, are in given in Sect. 3.

2.2   Cellular Automata

Cellular automata instantiate discrete models of computation. A cellular automaton has three components:

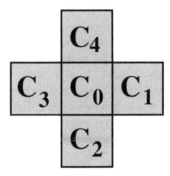

Fig. 1. The cells forming the neighborhood of cell C_0 in the cellular automata used in this study.

1. A collection of cells. In this work a rectangular grid of cells that make up the terrain map being created,
2. A designation, for each cell, of which set of cells form the neighborhood of the cell. In this work this will be the cell itself and the four cells closest to it in a grid, as shown in Fig. 1,
3. A set of states that cells can take on. In this work, 0 implies an empty cell while every other state represents a different type of full cell. The automata has n states, one empty, $n - 1$ full.
4. An updating rule that maps the states of a cell's neighborhood in the current time step to a new state for the cell in the next time step. The rule is encoded as an $n \times n$ matrix, a square matrix with dimension equal to the number of states in the automata.

Cellular automata are a type of discrete dynamical system that exhibits self-organizing behavior. When the cells of the automata are updated according to local transition rules, complex patterns form. The updating may be synchronous, as it is here, or asynchronous. Cellular automata are potentially valuable models for complex natural systems that contain large numbers of identical components experiencing local interactions [13, 14].

Examples of cellular automata based cavern maps from earlier studies are shown in Fig. 2. The automata used in this chapter are called *fashion-based* cellular automata because the updating rule may be thought of as following the current fashion within each neighborhood. The locality property of cellular automata turns out to be quite valuable as it both enables re-use and permits scaling of the generated map.

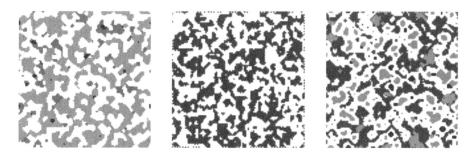

Fig. 2. Examples of renderings of three different single-rule fashion based cellular automata.

The rule for updating the cellular automata is encoded as a matrix that is the focus of evolutionary search. Different matrices encode different automata that, in turn, yield different maps. Here is how the encoding operates.

A state of the automata, at any one time, consists of an assignment of the available state values $\{0, 1, \ldots, n-1\}$ to all the positions in the grid the map is to occupy. These assignments are updated with the automata rule. The matrix entry $M_{i,j}$ specifies the score that a cell of type i obtains from having a neighbor of type j. The neighborhood given in Fig. 1 specifies who the neighbors of each cell are; the grid is toroidal, wrapping at the edges. The updating rule first computes the score of each cell, based on its neighbors, and then changes the cell state to the type of the cell in the neighborhood with the highest score, breaking ties in favor of the current cell state.

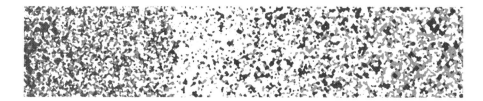

Fig. 3. An example of a morphed automata, created by a smooth lateral morph from one matrix specified rule to another, from left to right.

The cellular automata rule described thus far is a single rule that specifies an entire map by starting with random cell states and then updating them a fixed number of times to obtain the map which is evaluated for fitness. In this study *two* rules are evolved simultaneously. Examining the examples in Fig. 2, the maps have similar character at all points in the map. Figure 3 shows a map generated using two matrices M_1 and M_2 in the following fashion. If cell c in the grid is a fraction γ of the way across the map then the matrix

$$(1 - \gamma)M_1 + \gamma M_2$$

is used to generate the updating score for c. This has the effect of changing the rule smoothly across the map. The left-to-right morph is used for quality assessment during evolution, but the morphing parameter γ can be assigned in a variety of different geometric fashions.

3   Experimental Design

The goal of the evolutionary algorithm designed in this study is to achieve a pair of $n \times n$ matrices that can generate a spatially varying morphed cellular automata. We must specified how the matrices are stored for evolution. In order to evolve a population of paired matrices, we must give a numerical quality metric that captures the notion of "good terrain map" for the map generated by a morphed rendering based on the matrices.

For testing purposes we used six cellular automata states with 0 encoding empty space and 1–5 encoding different types of filled space. Thus the two matrices evolved for the morph are 6×6. We store the data structure as a vector of 72 real numbers in the range $0 \leq M_{i,j} \leq 1$. The first thirty-six members are the entries of the first matrix, read row-wise. The second thirty-six entries designate the second matrix, read in the same fashion.

When two pairs of matrices are selected for reproduction, they are copied into child structures and a middle segment of the 72-number vectors is selected, uniformly at random, and exchanged. This is called *two point crossover* because there are two positions where the material in the children changes where it originates. Crossover is analogous to chromosomal recombination in biology.

After crossover, from 1–7 positions are chosen in each child and those entries replaced with new numbers from the interval $[0, 1]$. The number of positions used it, itself, chosen uniformly at random. This process is analogous to biological mutation and is also called *mutation*. The number of positions changes in each child are chosen independently and uniformly at random. These variation operators are adopted from [11] and work well; a parameter study might improve performance.

Selection of parents and population members to be replaced is performed by picking seven population members. Within the group of seven, the two most fit are chosen as parents and the two least fit are replaced with the children.

We say that the two matrices specified by a single member of the evolving population of map-specifiers are *co-evolved* because they undergo joint fitness evaluation and so must adapt to one another in a fashion that permits all the intermediate rules to achieve a high score according to the quality measure.

3.1   The Quality Measure

The basis of the quality measure is to maximize the total path length from a white area in the middle of the left face of the grid of cells to similar areas in the middle of the other three faces. A fixed initial random state, used in all fitness evaluation, is updated 20 times using the cellular automata to create the map that undergoes quality evaluation. Figure 4 shows approximate paths between the areas in the middle of the left face and the middle of the other three faces of each map. A simple dynamic programming algorithm [15] is used to compute the shortest open path between the centers of the left face of the grid and the centers of the other three faces. The sum of these three distances is the base fitness value. If the center of a face is inaccessible, the map is awarded a base fitness of zero. The secondary rewards used are all multiplicative, and so this zero fitness is fatal.

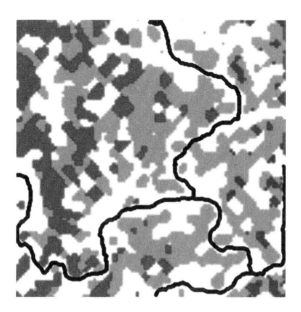

Fig. 4. The map generated from two co-evolved matrices with the paths between grid-faces shown in black.

The basic fitness measure ensures that a fairly complex and winding map is generated, unless it is counteracted by one of the secondary rewards. In order to create consistent quality evaluations, the random state information used is generated at the beginning of a run of the evolutionary code and retained throughout. This might be problematic if the action of a cellular automata rule was not purely local. A set of automata rules that generate good maps for one initialization of the cell array will generate very similar morphs for other initialization. This is demonstrated in some detail in [10].

A collection of eight fitness functions is derived by incorporating any of three multipliers that modify the fitness to bring out other qualities in a map. The first modifier is the *density modifier*. The user specifies a desired fraction α of non-empty cells in the final map. The actual fraction β is computed and then the basic fitness is multiplied by

$$R_d = \frac{1}{(\alpha - \beta)^2 + 0.4} \tag{1}$$

The number 0.4 was chosen to make the maximum reward 2.5, when $\alpha = \beta$, in order to give it a similar scale to the other rewards.

The second modifier, the *diversity modifier* multiplies the basic fitness by the entropy of the relative density of cell states that appear in the map. The number of cells of each type are computed and normalized to one to give a rate of empirical occurrence p_i of cell type i. The entropy is then

$$R_e = -\sum_{i=0}^{n-1} p_i \cdot log_2(p_i) \tag{2}$$

The maximum reward is $log_2(n)$ where n is then number of states the automata possesses. The reward R_e is greatest when the distribution among state types is even. For the six cell states available in the system used for testing, this reward has a maximum value of $Log_2(6) \cong 2.58$.

The third modifier is the *angular modifier*. Notice that the matrices M_1 and $c \cdot M_1$, for a positive constant c, encode the same cellular automata updating rule: all the updating rules look only at relative score. This means that there is a line of matrices that are scalar multiples of one another that all encode the same updating rule. A pair of matrices along one of these lines would encode a single rule and so fail to produce the diversity of appearance desired.

Table 1. Listed are the reward patterns used for each of the experiments.

Experiment	R_d	R_e	R_θ
1	No	No	No
2	$\alpha = 0.3$	No	No
3	No	Yes	no
4	$\alpha = 0.3$	Yes	No
5	No	No	Yes
6	$\alpha = 0.3$	No	Yes
7	No	Yes	Yes
8	$\alpha = 0.3$	Yes	Yes

If we regard a matrix as a vector, then we can reward being far from lying on the same line if we reward having the largest possible angle between the vectors. Recall that if v_1 and v_2 are non-zero vectors and θ is the angle between then

$$cos(\theta) = \frac{v_1 \cdot v_2}{||v_1|| \cdot ||v_2||} \tag{3}$$

The cosine is largest when the angles are most similar to the angular reward is given as the reciprocal of the cosine. In terms of the matrix entries $(M_1)_{i,j}$ and

$(M_2)_{i,j}$ this is:

$$R_\theta = \frac{\left(\left[\sum_{i=1}^{n}\sum_{j=1}^{n}(M_1)_{i,j}^2\right] \cdot \left[\sum_{i=1}^{n}\sum_{j=1}^{n}(M_2)_{i,j}^2\right]\right)^{1/2}}{\sum_{i=1}^{n}\sum_{j=1}^{n}(M_1)_{i,j} \cdot (M_2)_{i,j}} \qquad (4)$$

Since this reward is the reciprocal of the cosine of an angle, it has a potential maximum value of infinity, as the angle approaches a right angle, but this does not occur in practice. The maximum reward observed was approximately 53, in runs not using the diversity reward, which produced very bad maps; the maximum angular reward in runs that produced acceptable maps was 2.43.

The eight fitness functions tested arise from all possible combinations of added rewards to the base fitness function. The units of the base fitness function are distance, measured in grids. When an added reward is used, the units change, and each reward changes the units in a different fashion. The fitness values generated by different versions of the fitness function are thus not comparable to one another.

All three added rewards are all designed to increase fitness when the special condition they reward is realized more strongly and an effort has been made to scale the maximum possible added reward to be as similar as possible. All versions of the fitness functions are tested using a population size of 360. The experiments reward usage is given in Table 1

4   Experimental Results

Figures 5 and 6 show renderings of the thirty morphed automata produced for each of the eight experiments performed. The experiments are segregated by use of the diversity reward function with Fig. 5 showing the results without the diversity reward while Fig. 6 showing results that use the diversity reward. The following conclusions are supported by these displays.

1. Maps evolved without the diversity reward often display quite a similar character across the width of the morph between the two encoded rules. The diversity reward is needed to discourage this outcome.
2. When the reward for the angle between the matrices is used *without* the diversity reward, maps with a large amount of white space seem to be common. While there may be situations where this is a desirable property, it was one that the current work was attempting to avoid; there are much easier ways to leave a part of the map blank.
3. When the diversity reward is used, the impact of the other two rewards seems substantially reduced. The reward for being 30% full seems not to have much impact - it varies from left to right in the figures.

An important point to raise is that the morphable automata rules that got high fitness from run-away angular rewards, in Experiments 5 and 6, are still present in the search space of Experiments 7 and 8. They may even be higher

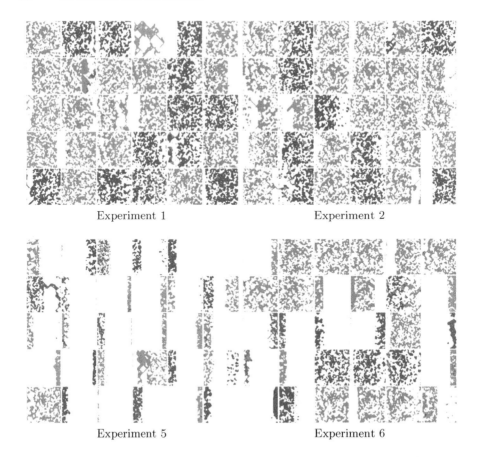

Experiment 1   Experiment 2

Experiment 5   Experiment 6

Fig. 5. Shown are the terrain maps resulting from thirty runs of the evolutionary algorithm for the four experiments that do not use the reward for entropic diversity of states. The right two panels are from experiment that reward 30% density and the bottom two reward angular separation of the two matrices.

fitness than the more aesthetically satisfying rules located. Evolution, many claims in the literature to the contrary, is *not* a global optimizer unless run for infinite time with well-chosen crossover and mutation operators. The presence of the diversity reward probably *deflected* evolution away from the white-in-the-middle maps because the diversity reward was easier to earn early in evolution. The remarkably high rewards, up to 53, for high angular values in Experiments 5 and 6 were in a part of the rule search space ignored in Experiments 7 and 8.

The results using both the diversity reward and the angular reward seem to yield the best qualitative results, overall. We use these results, from Experiment 7, for larger displays of the terrain morphing that occurs.

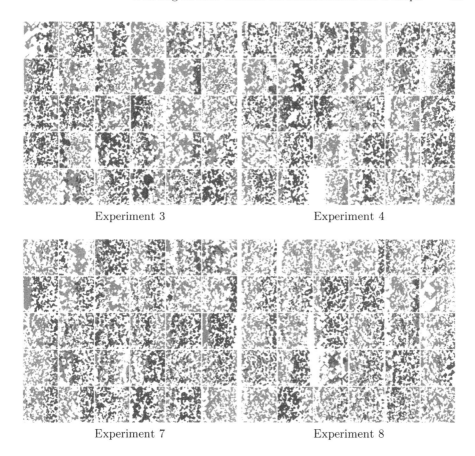

Experiment 3   Experiment 4

Experiment 7   Experiment 8

Fig. 6. Shown are the terrain maps resulting from thirty runs of the evolutionary algorithm for the four experiments that use the reward for entropic diversity of states. The right two panels are from runs that reward 30% density and the bottom two reward angular separation of the two matrices.

5   Discussion

The work presented here permits the automatic generation of a wide variety of terrain maps. This permits a programmer designing a war game or a video game to replace hand design with a simple procedural test for desired qualities in a map. This might be as simple as having particular regions of the map, where game-critical objects will go, empty and connected to one another or it might ask for more stringent conditions such as proximity goals on those empty regions (nearness, fairness). The ability to generate maps rapidly permits a scenario to be rerun with a different, but tactically equivalent map, rapidly and at low cost.

Figure 7 shows seven extended morphs for pairs of rules evolved in Experiment 7. Several of these rules, the first, fifth, and seventh, have areas that are too dense for passage of agents in a simulation and so are not good maps. Portions

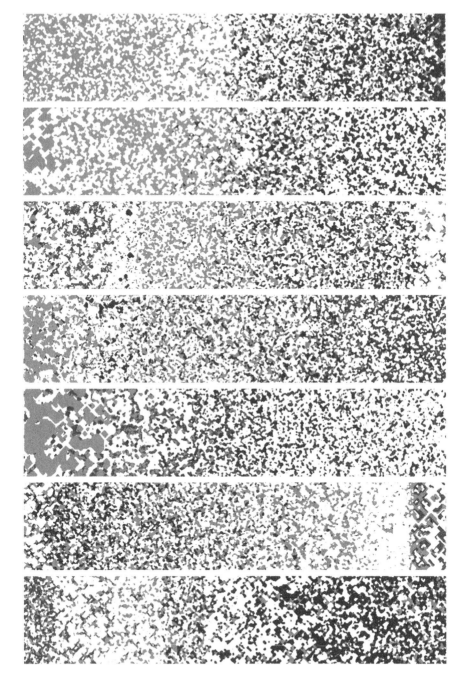

Fig. 7. Shown are seven selected renderings of extended morphs of evolved pairs of matrices from Experiment 7.

of these renderings span acceptable terrain types and there is a simple repair possible: restrict the morphing parameter to the usable portions of the maps. It is practical, also, to simply evolve a large number of maps and select the ones that are usable across their span.

During fitness evaluation, the denser portions of the terrain morphs shown in Fig. 7 had a single passage through or around them. This issue may be solvable by re-writing the quality measure in a fairly simple way, e.g. using more check points than the centers of the four faces of the map testing area.

In the earlier parts of this study, the scalability of cellular automata maps was alluded to. The extended morphs use both different initial conditions (the random starting point is different) and a much larger area. Many of the co-evolved rules still provide adequate terrain maps over this modified domain. Part of the problem arises from testing on relatively small 51 × 51 maps. Running evolution with larger maps would slow the software in direct proportion to the increase in the area of the testing map, but would mean that the behavior of the intermediate rules in the morph were evaluated more completely.

5.1 The Cellular Automata Rule Space

Figure 8 shows a pair of morphs between pairs of co-evolved rules. The morphing parameter still moves between the matrices by going from zero to one, but a much smaller increment is used to permit the behavior of the automata to be more visible.

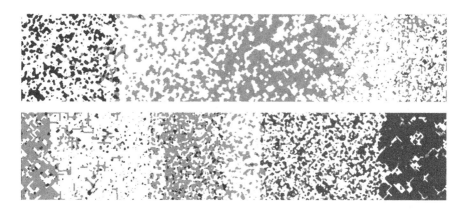

Fig. 8. Shown are two extended morphs of rules from Experiment 7. These rules display the granular character of the rule space fairly clearly. Note how discrete behavioral regions exist in different lateral chunks.

The updating of the automata depends on the comparison of sums of four scores from the matrix encoding the updating rule. This means that its behavior is discrete and so making small changes in the entries of the matrix specifying the rule will often not change the automata's behavior. This implies that the space of matrices is *granular* in the sense that contiguous regions of matrix space encode the same automata behavior. Examining Fig. 8 the top morph displays five or six behavioral regions that are large enough to see, while the bottom morph displays seven or eight behaviors. The size of the grains is not uniform and it seems likely, given the high dimensionality of the space, that there are very small regions that intersect large numbers of grains.

The morphs used in both fitness evaluation and to display the extended behavioral portraits of the sort in Fig. 8 are lines transecting 36-dimensional matrix space. They suffice to demonstrate granularity, but do not begin to discover the complexity of behavioral matrix space. This complexity was a primary indicator for choosing an evolutionary algorithm to perform search for good pairs of matrices that encode a good line of automata rules.

5.2 Extensions of the Work

There is substantial room to contrive additional fitness modifiers to control the character of the maps discovered. The three rewards presented here, touching on density, diversity of cell state usage, and angular separation of rules, were made to have roughly equal scale. It is not clear that choosing other scales, e.g. rewarding correct density more strongly, would be a bad idea. That reward appeared to have little impact on the maps located via evolution.

While the maps were evolved using a left-to-right morph, there is no reason that other morphs could not be used. In [11] radial morphs, of pair of rules not co-evolved with one another, was tried and while the maps were seldom high-quality, the radial morphing worked perfectly well. A more radical approach to try in the future is to have an underlying height map with height being taken as specifying the morphing parameter. This would permit a single pair of cellular automata rules to correctly decorate a terrain map.

In service of this notion of terrain maps, it might be worth fixing a moderately sparse "lowland" matrix and then evolving a collection of co-evolved partners. Assigning these partner rules to hills in a height map controlling choice of automata rule would permit the creation of highly diverse landscapes.

There is also an entire mathematical theory of the space of rules encoded by matrices. We already have the obvious result that scaling a matrix by a positive constant yields a matrix that encodes exactly the same rule. It is also clear, from the results, that there are many rules that lead to a completely empty map. Characterizing the space, the distribution of sizes of behaviorally uniform regions would require sophisticated mathematical tools, but would yield a higher level of control over the terrain maps generated.

References

1. Ashlock D (2006) Evolutionary computation for opimization and modeling. Springer, New York
2. Ashlock D (2018) Exporing representation in evolutionary level design. Morgan and Claypool, San Rafel

3. Togelius J, Yannakakis G, Stanley K, Browne C (2010) Search-based procedural content generation. In: Applications of evolutionary computation. Lecture notes in computer science, vol 6024. Springer, Heidelberg, pp 141–150

4. Sorenson N, Pasquier P (2010) Towards a generic framework for automated video game level creation. In: Proceedings of the European conference on applications of evolutionary computation (EvoApplications). LNCS, vol 6024. Springer, pp 130–139

5. Togelius J, Preuss M, Yannakakis GN (2010) Towards multiobjective procedural map generation. In: Proceedings of the 2010 workshop on procedural content generation in games, PCGames 2010. ACM, New York, pp 1–8

6. Johnson L, Yannakakis GN, Togelius J (2010) Cellular automata for real-time generation of infinite cave levels. In: Proceedings of the 2010 workshop on procedural content generation in games, PCGames 2010. ACM, New York

7. Ashlock D, Lee C, McGuinness C (2011) Search-based procedural generation of maze-like levels. IEEE Trans Comput Intell AI Games 3(3):260–273

8. Ashlock D, Lee C, McGuinness C (2011) Simultaneous dual level creation for games. Comput Intell Mag 2(6):26–37

9. Ashlock D, McNicholas S (2013) Fitness landscapes of evolved cellular automata. IEEE Trans Evol Comput 17(2):198–212

10. Ashlock D (2015) Evolvable fashion based cellular automata for generating cavern systems. In: Proceedings of the 2015 IEEE conference on computational intelligence in games. IEEE Press, Piscataway, pp 306–313

11. Ashlock D, Bickley L (2016) Rescalable, replayable maps generated with evolved cellular automata. Acta Phys Pol (B) Proc Suppl 9(1):13–24

12. Darwin C (1859) On the origin of species by means of natural selection. Murray, London

13. Wolfram S (1984) Universality and complexity in cellular automata. Phys D: Nonlinear Phenom 10(1–2):1–35

14. Sapin E, Bailleux O, Chabrier J (1997) Research of complexity in cellular automata through evolutionary algorithms. Complex Syst 11:1

15. Ashlock D, McGuinness C (2016) Graph-based search for game design. Game Puzzle Des 2(2):68–75

Affordance Theory Applied to Agile Development: A Case Study of LC2EVO

Hamna Aslam[1], Joseph Alexander Brown[1(✉)], and Angelo Messina[2]

[1] Artifical Intelligence in Games Development Lab, Innopolis University, Innopolis, Republic of Tatarstan 420500, Russia
{h.aslam,j.brown}@innopolis.ru
[2] Innopolis University, Innopolis, Republic of Tatarstan 420500, Russia

Abstract. The Theory of Affordances considers the interaction between the people and their environment based on their mental models of the environment. This study analyses a software development process on the theory of affordances. The LC2EVO software has been developed in Italy. There were two onsite team groups located five kilometers apart in separate buildings who were using the agile framework in their development. The product was delivered successfully under this framework. Team members in the case study were allowed inside of their environment to arrange their workspaces. By utilizing the works of Gibson and Norman on affordances, we found that arrangement of the workspace broadly met with both ideas of affordances. The examined case study demonstrates that a collaborative process on the design of a work space inside of an agile methodology increases opportunities for collaboration and communication among team members. The design of the work space affect the application of the agile methodology. Worker's affordances with the environment, both through their perception and interaction is fundamental to consider for the optimal and productive environment setup.

1 Introduction

The motivation behind our work is to analyze the interactions between a worker and the workspace based on the idea of affordances. The two most discussed ideas are from Gibson and Norman. Gibson specifies affordance as all possible interactions between the object and the environment. Norman focusses on the design aspects and emphasizes the need of using signifiers and visual clues. The signifiers and clues embedded in the design lead user towards the correct usability of the design without cognitive overload.

The term affordance was introduced by J.J. Gibson. Gibson defines affordances as all possible interactions between the individual and the environment. In the book *An Ecological Approach to Visual Perception*, Gibson states:

> The affordances of the environment are what it offers the animal, what it provides or furnishes, either for good or ill [11].
> According to Gibson, affordances are all action possibilities concealed in the environment irrespective of the individual's ability to recognize them.

© Springer Nature Switzerland AG 2020
P. Ciancarini et al. (Eds.): SEDA 2018, AISC 925, pp. 24–35, 2020.
https://doi.org/10.1007/978-3-030-14687-0_3

Therefore, an affordance of an object or an environment might possibly differ from one individual to another. The concept of affordance has been studied by D. Norman in the realm of Human-Computer Interaction and Graphic Design. Norman, in the book *The Design of Everyday Things*, illustrates [15], "Affordances are the possible interactions between the people in the environment. Some affordances are perceivable, others are not." Norman further describes that "Perceivable affordances often act as signifiers, but they can be ambiguous. Whereas signifiers signal things, what actions are possible and how they should be done. Signifiers must be perceivable otherwise they fail to function" [15]. Also, signifiers are more important in design than affordances because they communicate how to use the design. Designers can add forcing functions in the environment or in the object to promote the desired behavior or to prevent the incorrect behavior. These are defined by Norman, constraints or forcing functions can be added as *Interlocks*, which forces the function to take place in the correct pattern, *Lock-ins*, which prevent users from prematurely stopping an action and *Lockouts*, preventing someone from entering a space that is unsafe [15]. All these mechanisms in design naturally propel a user toward the correct understanding of the model and force the desired behavior.

Though Norman mentions that a design is a tradeoff among alternatives, the system designer should primarily consider that their design must support the appropriate system model through all components of the design the user will interact with.

The affordances concept has been studied widely and significant research has been carried in different domains grounding on the idea of affordances and its influence upon achieving desired objectives or understanding underlying phenomenon. Hammond [12] discusses the concept of affordances in the field of Information and Communications Technology (ICT) and its contribution towards understanding the use of ICT to support teaching and learning mechanisms. Affordance revolves around the user and the tool a user interacts with. Hammond presents a comparison between different researchers about user and tool interaction. The comparison has been based on different arguments such as, the concept of affordance points to the tool and the person interacting with the tool (Gibson and Norman's affordances aligns with this argument) or, affordances are directly perceived (Gibson's affordance concept agrees with and Norman's concept does not comply with direct perception). Further, the notions from different researchers are presented which are divided into two categories, the one focusing on how opportunities of learning should be perceived and the second which focuses on how opportunities of learning are perceived.

It is indispensable to take into consideration peoples' perceptions about any interaction before expecting a certain behavior in regards to that interaction. Boschi et al. [6] examines peoples' perceptions about oddly shaped dice(which conform to the distribution presented by two common six sided dice). In another example, Sadler [19] has analyzed the behavior of graduate students in context with the library usage. The study examined two questions, what affordances are perceived by the students in the library and to which extent student's perceived affordances differ from the librarian's anticipated affordances for the students. The qualitative analysis of the students and librarian's interviews did not show an enormous difference between the librarians intended and students perceived affordances, however the theory of affordances could point out not just intended (intended by the librarians and perceived by the students) but unintended (unintended by the librarians and perceived by the students) affordances as well as the affordance gap (affordances not clearly recognizable by the students). Similarly, Aslam et al. [1] analyses game design for its intuitiveness in terms of three age groups and what affordances players perceive for different game objects.

With affordance theory spanning over multiple domains, our research examines the work cell layout on affordance concepts. A case study has been done where multiple teams worked together to build a software. Though the case study concerns software development environment, we believe that this analysis can benefit any work environment to be convenient and productive for the workers.

2   Case Study: An Affordance Theory-Based Analysis of Work Cell Design During the Development of LC2EVO Software

The software industry is one of the fastest growing industries in the world and a software development process involves team(s) which work in collaboration to deliver the final product. There are numerous software development methodologies, the most widely adopted are agile [10] and waterfall methodologies [16,18]. Agile is rapidly replacing waterfall and other linear methodologies because it is more suitable to cope with changing and incomplete requirements which are the normality when complex domain applications are developed. In the last few years, it has been demonstrated that a version of the agile methodology can be effectively used in the development of mission-critical applications as well [4]. Agile which is central to our case study, the software teams had to undergo planning, adopt tasks, build an effective communication network and be prepared for adaptability.

To account for effective interaction among teams and a productive work environment, human factors have become indispensable to consider. In this respect, Colonese [7] talks about human factors in agile and points out that agile methodologies are likely to fail if human factors are not considered such as people's preferences and focus when they make decisions and how they like to communicate in a team. The paramount importance of the analyst/programmers skills in the quick delivery development process which is dictated by agile is also documented [3]. A solution to this has been proposed as Myers-Briggs Type Indicator (MBTI) model. It is a self-reported questionnaire that helps to identify a person's strengths, preferred style of working. As agile methodologies incorporate teamwork, team members can be assigned tasks according to their match on MBTI model that will ultimately maximize productivity, will reduce conflicts as well as stressful situations. However, Myers-Briggs model is considered limited because of its statistical structure, validity and also reliability, such as the test results differ over a period of time [17]. Hence, other measures such as OCEAN tests perhaps could be used in the same manner [2].

Moreover, along with understanding the human factors, an environment that supports the desired behaviour of team members and does not hinder their preferred style of working is fundamental otherwise team members will not be able to manifest their full productivity. The work cells must provide a comfortable environment to let people work their full potential in comfortable surroundings. An appealing cell design would not serve the purpose if it does not accommodate the requirements of the workflow. As Marr points out, *Usefulness of a representation depends upon how well suited it is to the purpose for which it is used* [13].

2.1 Land Command and Control EVOlution System (LC2EVO)

Land Command and Control EVOlution System is a strategic management tool developed for use by the Italian army. The main product is a software accompanied by hardware support systems [14]. The Continuous evolution of the Operational Environment generates instability of the Command and Control (C2) systems requirements and obliged the developers to work with unstable and unconsolidated mission needs. North Atlantic Treaty Organization (NATO) new "Resolute Support" (RS) mission in Afghanistan, was focusing on the training and advising to the Afghan National Defence and Security Forces - ANDSF, introducing new dimension that transcends the canonical range of military operations. In 2013, the Italian Army General Staff Logistic Department decided to overcome the problem of the "volatile requirement" transitioning to a completely different software development methodology derived from the commercial area but almost completely new to the "mission critical" software applications: the so-called "agile" methodology. The introduction of "agile" in the development of high-reliability software was not easy and applied the generation of a brand new "agile" methodology called: Italian Army Agile or ITA2. Setting up the LC2EVO (the evolution software of the land C2) applied to be the solution of many problems and the construction of a solid structure based on four "pillars": User Community Governance, Specific "agile" Training, new "agile" CASE Tools and custom "agile" development doctrine [8,9].

Most of the effort performed to generate an adequate production structure for the LC2EVO has been devolved to the setting of an innovative cultural and technical environment. Majority of the difficulties found in the way of this innovation process were "human-based", essentially due to cultural resistance based on consolidated practices. A whole brand-new environment had to be built. The four "pillars" of this "innovative software engineering paradigm" are User Community Governance, Innovative Agile training, Innovative CASE tools, and the High-Reliability Agile doctrine. A huge transformation from the traditional software development working space into a social and human-centred working space was required to accommodate the above-mentioned characteristics. The traditional structure of software factories was mainly based on ICT support and the developers would interact using their own workstation to circulate the artefacts. In the traditional software factory environment, the developers would spend most of their time in their personal office facing their PC screen developing code or architectural views with no or very limited interpersonal communication. For LC2EVO development, there were two onsite working teams. The teams were allocated workspace in two separate buildings which were five kilometres apart. The spaces designed to be work cell were different in terms of the design and placement of structures such as furniture and tools etc. The team working area was an open space and had to provide areas for large dashboards and a separate room for meetings. The dashboard/visualization board was installed to provide to the team and to external visitors a synthetic view on the state of the current production.

2.2   The Visualization of the Work Flow During LC2EVO Development

As is already recognized, the demonstration and visualization of the ongoing work are significant in every project to measure the progress as well as to trace factors that hinder or accelerate the workflow. For LC2EVO development, the team members had the liberty to opt for any arrangement of resources they can find optimal in their workflow. The first workspace for team 1 (consisting of 7 members), consisted of a room of size approximately 5 by 6 m. As shown in Fig. 1, the room consisted of 4 tables with chairs along with visualization boards. The second workspace for team 2 (for convenience, the two groups working together and sharing the space are collectively being referred as team 2) consisted of a larger space and it was in a separate building which was five kilometres apart from the team 1's allotted building. The team 2 work cell was approximately 15 by 20 m, as shown in Fig. 2. The placement of equipment and furniture was entirely decided by the team members. In the beginning, the team members did not put the visualization boards prominently but after a few days, they realized the heavy need to put visualization boards without which they could not adopt agile methodology completely. The boards were placed on the walls of the room, Fig. 3. In the room where there was not much space, walls were perceived by all team members as the best place to install boards, which is the same idea as Gibson suggests, *perception drives action*. The same approach has been adopted by Team 2 who created a soft partition for themselves through the movable boards and further utilized these boards as visualization screens.

The visualization boards were set to fulfill agile requirements, see Fig. 4a-c. The board was divided into six separate columns where each column represents the ongoing state of each task assigned to the team. With this division, the team members could easily move the stickers to the next column to show the flow of work.

As Norman's view of affordance suggests that clues or signifiers must exist to lead the actor towards the correct use of the design. The board setup is an example of this as clues on the board in form of stickers and column headings lead the team members toward adopting agile rules by naturally encouraging the desired behaviour for board usage in the agile development process.

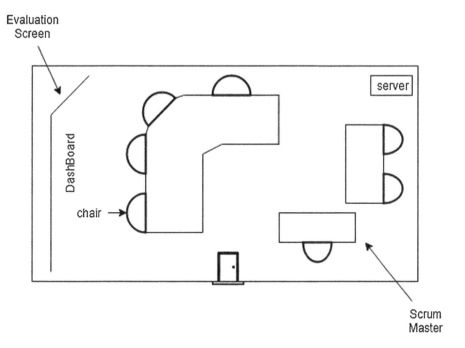

Fig. 1. Team 1 room layout

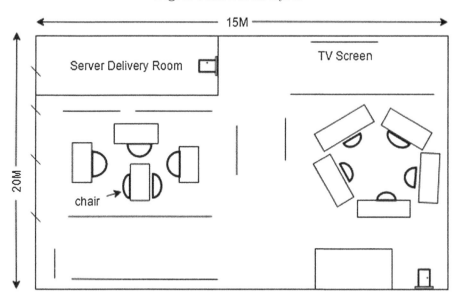

Fig. 2. Team 2 room layout

Fig. 3. LC2EVO development team utilizing visualization board

(a)   (b)   (c)

Fig. 4. The visualization of work progress on visualization board

The analysis of team's actions while setting up visualization boards show that their decisions were based on the environment's affordances. While finding the appropriate space for the visualization board, they analyzed the space from Gibson's view of affordance and considered all setup possibilities latent in the environment. After finding the best possible space for the board setup, the teams instinctively followed Norman's approach for a useful design and created columns as well as utilized different colour stickers that worked as signifiers and propelled all team members towards correct usability of the board for the agile methodology. Moreover, for other members of the project who had to be aware of the progress periodically, the board was satisfactory to describe the state of each task without having to demonstrate workflow verbally. However, to maintain records for a longer period, the data was also stored electrically.

2.3 Work Cell Design and Interpersonal Space

In the agile teams, transparency and sharing are necessary so all the developers share the same working space and may interrupt their main work activity to discuss or explain a solution. Separation of workstations must be the minimum possible to avoid interference but to encourage looking at each other's work.

For LC2EVO, the work cell has to accommodate multiple screens where the outputs of the measurement tools are continuously displayed. In both layouts (Team 1 and Team 2 workspaces), we see evidence of a self-organization of teams towards a circular layout in their method of action. In Fig. 1, this is evidenced that the team has arrayed themselves to be seen and see the scrum master. It is very similar layout to the Panopticon [5] and stimulates an environment of over sight of the scrum leader by the team members. However, the paired teams while having minimal contact with other teams, still have highly effective communications between each other. Note that in this layout individual screens are arrayed to allow for personal privacy between teams and the scrum leader, but are able to be seen by a pair.

In Fig. 2, there are two different arrangements shown and an evolution of the communicative ability is seen. The teams (two groups sharing the same space) all have given up for the most part personal privacy of their screens while still arranging in a circle in order to efficiently allow for computation and use of space. In the group on the left, a middle desk is used with two members in order to allow for a semi-circle, on the right, it is expressly a circle as shown in Fig. 2. There is a measure of allowing for individual work, one focused on a task can look at their screen and not be open to communication, however, the chairs used, allow for movement, swiving, repositioning. One team member can easily 'slide over' to support another team member in a pair programming event, and such actions were evidenced during the sprints.

The space in the second case, Fig. 2, is further arranged in order to have both commutative and workspaces, leading to a conceptual change in the mindset depending upon location. The upper right hand of the space with the TV screen is clearly denoted as a communal space between the teams, however, the arrangement of boards allows for functional separations during the sprints. These spaces are naturally formed as consequences of the space and are not laid down to the teams from a higher design. The self-organization of the space fell into the natural tradeoff between personal action and communication, considering communication as a public action. The arrangements of the room naturally fell into the systems that allow for a high degree of communication between team

Fig. 5. Team 2 in the server room

members. When team members afforded the ability to move an object in this case, the agile methodology was best suited to have high communication because of the arrangement of their objects. They naturally changed the space to fit the agile methodology, as can be seen in Fig. 5.

The agile methodology creates a model that requires communication and the arrangement of the objects in this space was to support that cognitive model, that falls under the Norman definition. However, in this case, the system that was utilized was far more in the ideas of Gibson on affordances because they were able to move the object in the position they wished, and they had all possible options for arrangements available to them.

3   Data Analysis

Although the physical space arrangements were not at the center of a structured research activity, it is fact that Team 1 started software development activities in February 2014 using a standard old style "software factory" set up (single room with separation between individual working spaces) and almost immediately started changing this set up to adapt it to the needs of the new agile development methodology whose details have been already described in this paper and the reference ones. Another fact is the improvement of the team performance in the quality and quantity of the produced software. A good deal of this improvement is due to the better understanding of the domain and a general growth of the analysts skill but a part of the improvement is also linked to the new working environment. Teams 2, which started with the "agile" set up already, experienced a better performance since the very beginning of their activity and a lower increase of their performance through the lifecycle as well.

Considering that the initial phase consisted in eight sprints (time boxed production periods of four weeks each, it is possible to observe an increase of a factor of five in the LOCs delivered per Sprint until the product maturity is reached and the first product release is delivered (S8), Fig. 6. Through S1-S3 (initial three months) the major part of the environment changes were made. A more thorough analysis could be run collecting questionnaires among the team members on the obstacle to agile production that were actually removed, but simply extrapolating from the daily reports (not releasable because marked as Defense administration internal) it is possible to observe that many strictly connected to the producing environment problems were solved during the first three months, such as positioning of the agile dashboard and provision of a common open space.

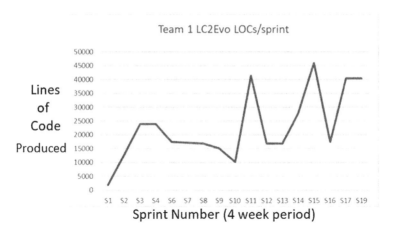

Fig. 6. Team 1 performance graph

4   Discussion and Implications of Research

The case study presented in this paper analyzed the actions and decisions of developers in relation to their environment, working together to develop a software, adopting agile methodology. There was no strict requirement upon the team from the authorities on the space or work cell setup.

By having no strict guidelines, the opportunity was adopted by team members to discover the affordances of their environment through trying out various setup schemes. The furniture was arranged how it was found to be most comfortable in that space. The team members considered the constraints or the requirements of their work plan. Effective communication was crucial. The layout chosen facilitated the effective communication in that space while maintaining their personal space.

The interaction of people with the workspace was also influenced by their perceptions. Initially, the space was freely investigated, analyzing the environment, work requirements, and considering all possible layouts. Final setup was done based on what they perceived to be the optimal option. Through working in their preferred setup, the team successfully delivered the project.

While making decisions for space utilization, team members began with their conceptual model of the space and discovered more interaction possibilities within the environment over time. Based upon these outcomes, we conclude that design of workspaces requires consideration of workers perceived conceptual model of the task. In the absence of predefined space setup guidelines, people's interaction with the environment and their decisions were not limited to their perceived affordances but they discovered more interaction possibilities with the environment and established the optimal setup.

References

1. Aslam H, Brown JA, Reading E (2018) Player age and affordance theory in game design. In: King D, (ed.) Proceedings of the 19th international conference on intelligent games and simulations, GAME-ON 2018. Eurosis, pp 27–34
2. Barrick MR, Mount MK (1991) The big five personality dimensions and job performance: a meta-analysis. Pers Psychol 44(1):1–26
3. Benedicenti L, Cotugno F, Ciancarini P, Messina A, Pedrycz W, Sillitti A, Succi G (2016) Applying scrum to the army: a case study. In: Proceedings of the 38th international conference on software engineering companion. ACM, pp 725–727
4. Benedicenti L, Messina A, Sillitti A (2017) iAgile: mission critical military software development. In: 2017 international conference on high performance computing & simulation (HPCS). IEEE, pp 545–552
5. Bentham J (1995) Panopticon; or the inspection-house: Containing the idea of a new principle of construction applicable to any sort of establishment, in which persons of any description are to be kept under inspection; and in particular to penitentiary-houses, prisons, houses of industry, work-houses, poor-houses, lazarettos, manufactories, hospitals, mad-houses, and schools: With a plan of management adapted to the principle: in a series of letters, written in the year 1787, from Crecheff in white russia to a friend in England. In: Bozovic M (ed) The panopticon writing. Verso, London, pp 29–95
6. Boschi F, Aslam H, Brown JA (2018) Player perceptions of fairness in oddly shaped dice. In: Proceedings of the 13th international conference on the foundations of digital games, no 58. ACM
7. Colonese E (2016) Agile: the human factors as the weakest link in the chain. In: Proceedings of 4th international conference in software engineering for defence applications. Springer, pp 59–73
8. Cotugno F, Messina A (2014) Implementing scrum in the army general staff environment. In: The 3rd international conference in software engineering for defence applications-SEDA, Roma, Italy, pp 22–23
9. Cotugno FR, Messina A (2014) Adapting scrum to the Italian army: methods and (open) tools. In: IFIP international conference on open source systems. Springer, pp 61–69
10. Fertalj K, Katic M (2008) An overview of modern software development methodologies. In: 19th central european conference on information and intelligent systems
11. Gibson JJ (2014) The ecological approach to visual perception: classic edition. Psychology Press, New York/London
12. Hammond M (2010) What is an affordance and can it help us understand the use of ICT in education? Educ Inf Technol 15(3):205–217
13. Marr D (1982) Vision: a computational approach
14. Messina A, Fiore F (2016) The Italian army C2 evolution: from the current SIAC-CON2 land command & control system to the LC2EVO using "agile" software development methodology. In: Military communications and information systems (ICMCIS). IEEE, pp 1–8
15. Norman D (2013) The design of everyday things: revised and expanded edition. Basic Books (AZ), New York City
16. Petersen K, Wohlin C, Baca D (2009) The waterfall model in large-scale development. In: International conference on product-focused software process improvement. Springer, pp 386–400

17. Pittenger DJ (1993) Measuring the MBTI... and coming up short. J Career Plan Employ 54(1):48–52
18. Royce WW (1987) Managing the development of large software systems: concepts and techniques. In: Proceedings of the 9th international conference on software engineering. IEEE Computer Society Press, pp 328–338
19. Sadler E, Given LM (2007) Affordance theory: a framework for graduate students' information behavior. J Doc 63(1):115–141

Multi-level Security Approach on Weapon Systems

Mirko Bastianini🆔, Luca Recchia$^{(\boxtimes)}$, Mario Trinchera,
Emiliano de Paoli, and Christian di Biagio

MBDA Italia S.p.A., Rome, Italy
{mirko.bastianini,luca.recchia,mario.trinchera,
emiliano.de-paoli,christian.di-biagio}@mbda.it

Abstract. Scope of this paper is to show how is possible to implement Bell-LaPadula model using Linux Container on a single host machine; this innovative technology, which will be explained better in next chapters, allows to create segregated and isolated environment into same machine.

The creation of multi-level security system inside a container environment can effectively guarantee the confidentiality of the application software and data, restrict illegal attacker, and prevent attackers from arbitrarily destructing system information; moreover, it can also be easy to configure and manage the system policy.

Keywords: Multi level security · Linux Container · Weapon systems · Segregation

1 Introduction

Military companies and government can hold information at different levels of classification (Confidential, Secret, Top Secret, etc.) and it is important being sure that data can only be read by a person whose level is at least as high as the data's classification; this kind of policy is called *Multi Level Security (MLS)* and the most famous implementation is made by Bell-LaPadula model, developed by David Elliot Bell and Leonard J. LaPadula.

Thanks to three main properties, this model guarantees *"no read up"* and *"no write down"* principles: the former is necessary to avoid a user with low security level to have access to high security level information, while the latter removes the possibility that a user with high security level can write or modify document with low security one; this is necessary to prevent disclosure of information from high to low security level.

This model is often not very easy to implement, as is very restrictive and Unix traditionally does not strongly implement the principle of least privilege and the least common mechanism principle: in fact, most objects in Unix, including the file system, processes, and the network stack are globally visible to all users.

P. Ciancarini et al. (Eds.): SEDA 2018, AISC 925, pp. 36–43, 2020.
https://doi.org/10.1007/978-3-030-14687-0_4

2 Related Works

Multi-Level security is a technique currently adopted by military companies, government, defense and other domains in which information need to be divided into security levels. The idea behind this approach is simple but effective: users at different security levels are provided with information at corresponding security levels. This kind of mechanism is strictly interconnected with the safety of the entire operating system, and this is finally proved to be not practical.

2.1 Data Confidentiality and Access Control

Bell-LaPadula security model focuses on data confidentiality and controls access to classified information. Represented entities are divided into *subjects* and *objects* and entire system can be defined secure only if one secure state actually exists; this condition is verified when each state transition preserves security by moving from one state to another. Security levels on objects are called *Classifications* while security levels on subjects are called *Clearances*.

Briefly, this formal definition can concern to a concept of a state machine with a set of allowable states: the transition from one state to another is defined by **transition functions**. This model defines one *discretionary access control* (DAC) rule and two *mandatory access control* (MAC) rules:

- **Simple Security Property:** a subject with given security level shall not read an object at a higher security level.
- *** (star) Security Property:** a subject with given security level shall not write to any object at a lower security level.
- **Discretionary Security Property:** it is necessary to use an access matrix to specify the discretionary access control.
- Rules above could make the model very difficult to apply in a real scenario and not very flexible at all so this is why there are two other principles:
- **Trusted Subject:** it can transfer information from a document with high security level to document with low security level, because it's not restricted by the Star-property;
- **Tranquility principle:** it states that the classification of a subject or an object does not change while it is being referenced; there are two forms of this principle:
  - *Strong tranquility:* security levels do not change during the normal operation of the system;
  - *Weak tranquility:* security levels can't change in such a way as to violate a defined security policy;

Weak tranquility is always desirable because allows systems to keep the principle of least privilege (Fig. 1).

Giving a practical example, secret researchers can create secret or top-secret files but cannot create public files; on other hand, this kind of researchers can view public or secret files, but cannot view top-secret files.

As said into the introduction, this model is based on *"no read up"* and *"no write down"* principles.

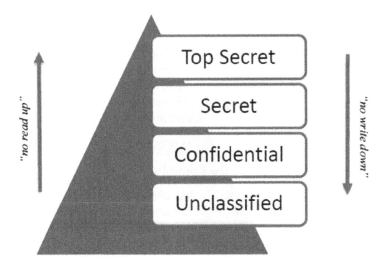

Fig. 1. Bell-LaPadula system applied to military classification

2.2   Container Based Virtualization

Linux Container is an innovative virtualization technology for running multiple seg-regated environments (called *containers*) in complete isolation on a single computer; this "new" virtual machine behaves like a full OS and it is able to run *init, inetd, sshd, syslogd, cron*, and other core services.

Host machine with Container support shares kernel among containers, so this is why there is only one level of resource allocation and scheduling, and thanks to Linux control groups (*cgroups*) feature, resource consumption can be strictly limited. All of this happens directly at Operating System's level; guest processes are executed on the host machine kernel, keeping general execution's overhead near to zero.

As said above, this virtualization mostly depends on two features offered by the kernel:

- **Control Groups:** provide hierarchical grouping of processes in order to limit and/or prioritize resource utilization.
- **Namespaces:** they isolate processes by presenting them an isolated instance of global resources.

While these two features above are probably the most important to implement a container system on Linux, other features can be used to further harden the isolation between processes, for example: Container system uses six different features to ensure that each guest is strictly isolated from the rest of the system and cannot damage host or other guests.

To keep entire architecture reliable and scalable, it is possible to create an image of an O.S. as a single file; output contains application along with all its dependencies, and can be used to create a new container, providing fast deployment (Fig. 2).

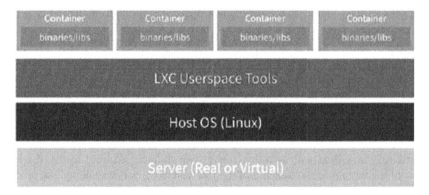

Fig. 2. Architecture of container virtualization.

OS-level virtualization is a feature of the host operating system, so there are no additional hardware dependencies: this means that OS-level virtualization is restricted to only providing support for containers using the same operating system API as the host operating system provides.

Regarding security and segregation, Container virtualization has a big role: host operating system, thanks to a Linux Daemon, ensures that container only interacts with those parts of the system it is allowed to; permission for using specific device and for file system access must be explicitly granted, and containers cannot directly access the contents of other running containers.

3   Implementing MLS Using Linux Container

3.1   No Read Up/No Write Down Rules

Scope of Multi Level Security approach is to divide (segregate) the information into classification levels, and to guarantee flowing of data only in one way: from low security level to high security level.

Here in this paragraph will be defined a formal way to represent this security system, based on Bell-LaPadula requirements; to do this, first it's important to define first some variables and related transition functions:

- **Domain System (DS):** security levels of system under analysis.
- **Address Space (AS):** space in where a container can write or read information.
- **Information Level (IL):** security level of specific domain.
- **level (AS, IL)**: transition function reporting *Information Level* of a specific *Address Space*.
- **write (DS, AS)**: transition function reporting *Address Space* writable from the *Domain System*.
- **read (DS, AS)**: transition function reporting *Address Space* readable by a *Domain System*.

Thanks to these definitions, it is possible to create a rule to formally define a MLS system.

Definition I: $\forall x, y \in DS$ and $level(x) \leq level(y)$ it is possible to say that a system is MLS compliant if:

$$(1)\; write\,(x) \cap read\,(y) \neq \emptyset$$

$$(2)\; read\,(x) \cap write\,(y) \neq \emptyset.$$

First rule says that, with two domains at two different security levels, the intersection of the writable address space of the low security-level domain and the readable address space of high security-level domain must be not empty: this guarantees low security-level information domain to flow to the high security-level domain.

Second rule says that the intersection of the readable address space of the low security - level domain and the writable address space of high security-level domain must be empty, so low security-level domain can't read the information written by a high security-level domain.

These are focal points for having *"no read up"* and *"no write down"* principles.

3.2   Linux Container Implementation

To transform concepts above into a practical implementation, it is necessary to divide available space of host machine into two main parts, called domains:

- **Container Domain (CD):** it represents entire file system of the container under analysis:
  - *C-Private Domain (CPD):* a private part of container file system, accessible only by container itself;
  - *C-Shared Domain (CSD):* a shared part of container file system, accessible with write permission by container itself.
- **Host Domain (HD):** it represents the file system of host machine without consider containers one:
  - *H-Private Domain (HPD):* a private part of host file system, accessible only by the host machine itself;
  - *H-Shared Domain (HSD):* a shared part of host file system, accessible with read permission by containers and with write permission by host machine itself.

These explained rules represent the basis for applying the concept of MLS to Linux containers, and they can be easily understood by looking at the following picture (Fig. 3):

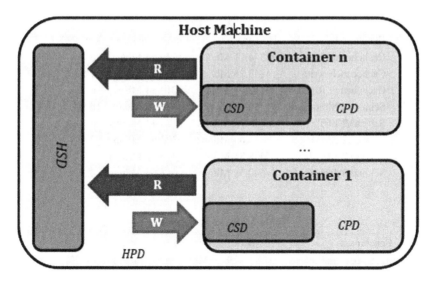

Fig. 3. Architecture of MLS system using Linux container.

As showed in the picture, user inside container has permission to read and write files into his private domain and into his shared domain, but he can only read files published into host shared domain; this will prevent disclosure of information from high security level to low security level (*"no write down"*).

So, by the same logic, user inside host machine has permissions to read and write files into his private domain and into his shared domain, but he can only write file into container shared domain: this will prevent flows of information from low security level to high security level (*"no read up"*).

The above represented scenario is fully compliant with Multi Level Security approach and can be easily realized using Container virtualization technique in combination with some security mechanisms like *unprivileged container* and shared folders between host and guest machine. This approach can be also easily extended with other containers, implementing more than two security levels in the same machine.

3.3   Linux Container Performance

Purpose of this section is to briefly illustrate performance of Linux Container, which have been tested using a benchmark test suite. These tests are used to verified capacities of CPU, RAM and graphics card and results are compared against same tests performed on the host machine; tables show name of benchmark, performances on Host machine and on Linux Container and finally there is a percentile variation between them.

Here follow a summary Table (1, 2 and 3):

Table 1. CPU performance

Benchmark	Host machine	Linux container	Var. (%)
Cachebench read	1344.71 Mb/s	1236.44 Mb/s	8.1%
Cachebench write	4843.25 Mb/s	4771.12 Mb/s	1.5%
Cachebench R/M/W	20142.47 Mb/s	20010.34 Mb/s	0.7%
Scimark2 Montecarlo	79.1 Mflops	70.8 Mflops	10.5%
Scimark2 FFT	78.75 Mflops	73.44 Mflops	6.7%
Scimark2 SMM	310.65 Mflops	302.55 Mflops	2.6%
Scimark2 LU MF	428.81 Mflops	410.69 Mflops	4.2%
Scimark2 Jacobi	640.36 Mflops	613.44 Mflops	4.2%

Table 2. RAM performance

Benchmark	Host machine	Linux container	Var. (%)
Ramspeed: add integer	8829.89 Mb/s	8763.79 Mb/s	0.7%
Ramspeed: copy integer	10637.13 Mb/s	9347.18 Mb/s	12.1%
Ramspeed: scale integer	11927.69 Mb/s	11438.22 Mb/s	4.1%
Ramspeed: triad integer	12451.7 Mb/s	11913.1 Mb/s	4.3%
Ramspeed: average integer	9542.24 Mb/s	9277.62 Mb/s	2.8%
Ramspeed: add floating	8624.48 Mb/s	8566.74 Mb/s	0.7%
Ramspeed: copy floating	7846.63 Mb/s	7427.88 Mb/s	5.3%
Ramspeed: scale floating	6354.21 Mb/s	6118.91 Mb/s	3.7%
Ramspeed: triad floating	10245.8 Mb/s	10087.5 Mb/s	1.5%
Ramspeed: average floating	9546.22 Mb/s	9433.92 Mb/s	1.2%

Table 3. Graphics performance

Benchmark	Host machine	Linux container	Var. (%)
Plot 3D	3830 points	3821 points	0.2%
Furmark	605 points	589 points	2.6%
Triangle	101426 points	100354 points	1.1%
Pixmark Piano	225 points	223 points	0.9%
Pixmark Volplosion	744 points	741 points	0.4%

4   Conclusions

Linux container can be a good technology to implement Multi Level Security mechanism; thanks to segregation offered by containers each userspace does not have visibility of others and the communication between areas at different level of security can be made using shared folders with appropriately read/write permissions. Every single container must be compliant with policies imposed by Bell-LaPadula model, so users inside a container with higher level of security cannot modify documents into containers with lower security level (but they must be able to view them). Conversely, users inside containers with lower security level cannot view documents in containers with higher security level, but they must be able to wrote documents inside them.

A future implementation is the building of a container orchestrator which allows to instantly configure containers corresponding to their levels of security.

References

1. Felter W, Ferreira A, Rajamony R, Rubio J (2015) An updated performance comparison of virtual machines and Linux containers
2. Bacis E, Mutti S, Capelli S, Paraboschi S (2015) DockerPolicyModules: mandatory access control for docker containers
3. Javed A (2017) Linux containers: an emerging cloud technology
4. Grattafiori A (2016) Understanding and hardening Linux container
5. Souppaya M, Morello J, Scarfone K (2017) Application container security guide
6. Li XY, Ji C, Liu G (2017) A container-based trusted multi-level security mechanism

Procedural Generation for Tabletop Games: User Driven Approaches with Restrictions on Computational Resources

Joseph Alexander Brown[1](✉) and Marco Scirea[2]

[1] Artificial Intelligence in Games Development Lab, Innopolis University, Innopolis, Republic of Tatarstan, Russia
j.brown@innopolis.ru
[2] Mærsk Mc-Kinney Moller Institute, University of Southern Denmark, Odense, Denmark
msc@mmmi.sdu.dk

Abstract. Procedural Content Generation has a focus on the development of digital games, this leaves out a number of interesting domains for a generation of wargames and other board game types. These games are based on limited computational resources, using dice for random numbers and creations via tables. This paper presents a historical look at non-digital methods for PCG application and in particular a taxonomy of methods and locations of their use.

Keywords: Procedural content generation · Analog computing · War games · Taxonomy

1 Introduction

The idea of a configurable board which can be generated predates the computer. The training of Prussian officers included the war game Kriegspiel (formal rules in 1812). The table for this game, presented to King Friedrich Wilhelm III, included such features as movable terrain pieces as board tiles. An even earlier example, created in 1780 by Johann Christian Ludwig Hellwig, boasted a board of 1600 color-coded terrain squares as an expansion on *Königsspiel* developed in 1664, with a relatively small five-hundred squares – though it is not known if these terrain squares were preset or somehow configured [7]. These early Procedural Content Generation (PCG) for games far predate the examples as shown by Smith [10]. These titles represented fields, forests, streams, etc. and would be configured or "rolled up", depending upon the wishes of the commander leading the examination. *Little Wars* [16] would further advise the players to use such terrain as what was available to them, which undoubtedly lead to a number of broken objects being swept up by maids as the Victorian wood dowel firing

© Springer Nature Switzerland AG 2020
P. Ciancarini et al. (Eds.): SEDA 2018, AISC 925, pp. 44–54, 2020.
https://doi.org/10.1007/978-3-030-14687-0_5

cannons streaked their ordnance across the gardens and studies of the British stately homes.[1]

Later board games, both of German and Western designs, such as *Carcassonne* and *Zombies!!!*, utilize a configurable board as part of the core game mechanics of play. Neither of these has a much complex application of resources nor delivers a playing experience which truly meets with the requirements of an actionable field-ready military game, like Kriegspiel. Still, they allow for massive combinations of new outcomes in terms of the space in which the game is fought over.

The randomness of war has been remarked on by great historical generals; Tzu [12] as early as the fifth century refers to the high of generals being able to anticipate the enemies plans and thwart them. He advises that the general should act in accordance with their own forces, and in accordance to terrain, and should embrace intelligence gathering and spies in order to better plan. Moreover, Sun Tzu remarks on the randomness inherent in war, "According as circumstances are favorable, one should modify one's plans." Von Clausewitz [15] would echo this sentiment into the Napoleonic era stating that:

> . . . the great uncertainty of all data in War is a peculiar difficulty, because all action must, to a certain extent, be planned in a mere twilight, which in addition not unfrequently [sic] – like the effect of a fog or moonshine – gives to things exaggerated dimensions and an unnatural appearance.
> What this feeble light leaves indistinct to the sight talent must discover, or must be left to chance. It is therefore again talent, or the favor of fortune, on which reliance must be placed, for want of objective knowledge.

In such plans it is obvious that the commander must deal in vagaries for the opponent is not a known entity, nor should the commander allow others to know plans beforehand.

At this time, generation techniques are mainly focused on PCG for computer games and do not consider the user requirements for board games. Looking at reviews of future directions in PCG [11,19], there is no mention of non-digital methods. Tools for the generation of computer driven terrain are costly in terms of the required computation and do not meet with the needs of the average wargamer standing at the table making an arrangement for a session. While the

---

[1] There is a long historical precedence of wives being upset with their wargaming husbands. Catherine the Great of Russia, born Princess Sophie of Anhalt-Zerbst, led a coup d'etat brought about by Peter III's neglect. His predilection to spend his time enjoy wargaming, as opposed to running a nation state. She writes in her autobiography, "the main plaything of the Grand Duke, when in town, was an excessive quantity of small puppets, of soldiers made of wood, lead, starch and wax, which he arrayed on very narrow tables that filled the entire room; one could hardly pass between those tables. He had nailed long strips of brass along the lengths of these tables; these bands had wires attached to them, and when one pulled these, the brass strips made a sound which, according to him, was akin to the rolling fire of muskets." [6].

work of Ashlock and McGuiness [5], Valtchanov and Brown [13], have presented developments which can address the creation of dungeon games such as *D&D* or *Pathfinder*, the aim of this paper is to examine the techniques allowing the layout of such boards which would be within the domain of a board or wargame played without access or need for digital devices. The tools required should not go beyond the requirement of tables, dice, drawing cards, etc. common to the consumer (Fig. 1).

We present a taxonomy of wargames, based on previous research and expanded to include specific characteristics of wargames in Sect. 2. In Sect. 3, we present a survey of the main methods for PCG in boardgames, discussing their benefits and shortcomings and providing examples of commercial games that include such features.

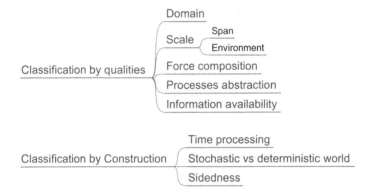

Fig. 1. The proposed taxonomy of Wargames.

2   A Taxonomy of Wargames

Wargames are a subset of games; Van Creveld defines a wargame as a contest of opposing strategies that, while separated from real warfare, simulates some key aspects of real war [14]. Interestingly, in the research related to wargames – especially from the military perspective – there appears to be some disregard for games that simulate non-realistic war, as exemplified by a US army officer's statement: "This is not Dungeons and Dragons we're doing here" (Allen [1]). This sentiment is further supported by the Prussian High Command's remarks in 1824 (but only recorded in 1874[2]) that Kreigsspil was "not an ordinary game, but a school of war" [17]. See [18] for a more through review of Kreigsspil development.

In this paper we do argue that according to Van Creveld's definition all kind of games that represent a conflict with some war-like aspect are indeed wargames, *Dungeons & Dragons* included. Sabin [9] agrees with this, in general

---

[2] We would like to thank the reviewer to brought this to our attention.

stating that, "The key characteristic uniting war and games, and which sets them apart from most human activities, is their competitive and agonistic nature. In games, this conflict is mainly artificial, while war is mainly situational, but the effect is the same." All wargames are abstractions, as noted by Nakamura [8], "correct simulation is always dependent on the player's *image* of the subject being simulated" (author's emphasis retained).

While many works have been written on wargames, not a lot of attention has been given to ways to classify wargames as a separate subset of games. A rare example of such a taxonomy can be found in a 1989 workshop report by Anderson *et al.* sponsored by the Military Operations Research Society (MORS) [2,3]. While Anderson's taxonomy has many interesting qualities, it is based on a definition of wargames that is closer to "military simulations", in fact excluding all wargames that are played outside of a military environment. The following taxonomy aims to allow for categorization of wargames both for personal enjoyment and for more practical uses. The taxonomy is divided into two sub-classifications: by the quality and by construction. The classification by qualities focuses on differentiating between games based on the type of conflict that is being represented, while the classification by construction looks at distinctions of game mechanics.

We hope that this taxonomy will help frame the possible aspects that can be procedurally generated, and moreover give a more solid framework on which to analyze wargames.

2.1 Classification by Qualities

Domain. By domain, we define the type of conflict represented. An incomplete list of examples of domains is land, sea, n-dimensional space, psychological, or economical conflict. *Warhammer* mostly represents land conflict, while *General Quarters* represents sea conflict, and *Crimson skies* air conflict Table 1.

Scale. We define scale as the size of the described conflict since this can have multiple facets – such as the level of control of the player on the units, or the role of the represented conflict – we further divide this category in *Span* and *Force composition*.

Span. While this quality might seem similar to a domain, it serves to express what role does the conflict represented in the game in the bigger context of a war. Possible spans of conflict include global, theater, local, etc.
Of course, a game might represent a single battle of a larger campaign based game - in which a separate set of rules would be adopted in order to represent the campaign level. There are also a number of games with a 'Legacy' effects (*Risk Legacy*), in which the outcome of previous battles impacts on the future set.

Force Composition. The force composition quality deals with what is represented as a "unit" by the wargame. The typical representations are:
   – Division, corps, or in general army-sized units.

- Battalions to divisions.
- Squads, platoons, companies, vehicles.
- Individual soldiers

Note how in many of these, individuals are not directly represented but are an abstract component of the "unit".

Environment. In the context of wargames, we define environment as how the playing field is structured. For example is the construction of the playing field achieved through placement of set pieces (terrain reliefs, fortifications, rivers, etc.), or is the terrain divided in tiles that have to be connected, or is there a static predefined map?

Processes Abstraction. This quality describes the level of abstraction of the actions the units can execute. If we consider the omnipresent "attack" action we could, for example, differentiate between calculating damage shot-by-shot, by average, by damage-per-second (DPS), piece capture, etc.

Information Availability. With this attribute, we want to differentiate between games that present complete or incomplete information to the players. While partial information, especially through the use of the 'fog of war' is very common in digital games, it has also been previously implemented in board games; an example is *Quebec 1759*.

2.2   Classification by Construction

Time Processing. Wargames can be differentiated by how they deal with time, the most obvious distinction being if the progress of the game is based on real-time or turns. At the same time, there are more subtle distinctions, such as the time relevance of orders or interrupt actions. Interrupt actions are special actions that can be either triggered manually by the player in some conditions, or automatic actions outside the player's control. An example of automatic actions can be seen in *Warhammer 40000*'s overwatch action: once a unit has overwatch applied to it, it will perform an automatic attack on the first enemy unit to come into range. Variations of interrupts can be seen in both real-time and turn-based wargames Table 2.

Stochastic vs Deterministic World. Some wargames might present a completely deterministic world – empathising player strategy and fairness – while others might include stochastic elements. Randomness can be used to create interesting and surprising situations, and including a more continuous sense of tension since the result of each move is not completely assured a priori. An example of classic wargames that present such opposing approaches are Chess and Risk: given any state of the board, in Chess, each action has only one possible resolution, while Risk requires the players to roll dice to determine the result of an attack.

Sidedness. Finally, sidedness differentiates between games depending on where the players stand in the conflict. We differentiate games into three categories:

One player side (multiple players) vs the game: in these games the player(s) all stand on the same side of the conflict, while the game itself provides the enemy. Examples of wargames that fall into this category are *Space Alert* and *Dungeons & Dragons*.

1 vs 1: in these games there are two sides, both manned by a human player. A classical example would be most two-player Chess-like games.

1 vs many: these games present more than two sides to the conflict, which could be controlled by human players or by the game itself. *Risk* would fall into this category, as is usually played by more than 2 players. A game that can also present game-controlled sides is the *Game of Thrones boardgame*, as it has a static number of "houses" that can be more than the players' number.

3 Techniques

This section examines the techniques used in wargames to provide a procedural content. In many of these methods, the provider of randomness is a dice roll or a card draw, we do not see dice or cards themselves as a generator, but instead as the analog randomness tool which a generation method utilizes Table 3.

3.1 Lookup Tables

This technique is based on having a table (or something that can be abstracted to a table) of possible outcomes that are selected through a stochastic process. We divide this technique into two categories: where the lookup table is integrated into an analog "random number generator", or where it is external to the object. It is important to note that the use of an integrated lookup table in a game does not preclude the same game from using an external lookup table in a different mechanic. For example, a game may include a deck of cards for selecting and issuing orders (an integrated table) and a dice roll for checking in a rule book the number of casualties if a unit is hit with shot (an external table).

Integrated. This category includes techniques in which the consequence of a "roll" is integrated into the physical object used as a source of randomness. An example would be the dice in *Risk*, which, while being six sided, it only has blue/red colors on the faces to indicate a positive/negative result. In this category also fall examples of randomly shuffled decks (see event cards in *Betrayal at the House on the Hill*), in these cases by drawing a card from the deck a procedural element might be introduced where the consequence of the randomness is explicit on the card itself.

External. In this category fall all other examples of decision making where an external table (often in the game manual) is used to determine a random numerical value. *Dungeon & Dragons* (and in general most pen and paper games) presents many examples of this approach. In fact, it is not only used to determine how successful attacks are in combat but is also used in dialogue, traversal of levels, character creation, etc. It seems likely to us that such a technique is mainly favored by very open-ended games, such as pen-and-paper role-playing games since, given the extreme range of game mechanics, having specialized artifacts to determine the outcome of all of them would be severely impractical.

3.2 Tiles

A very common technique for generating playing boards consists of delegating the creation to the map to the players themselves through the positioning of resources. We will refer to this resources as *tiles*, since very often they are just that, but could take different shapes than, for example, a square or hexagonal tile. We define *tiles* as a representation of resources, in fact while often directly representing some type of terrain (e.g. *Carcassonne, Settlers of Catan*) or dungeon layout (e.g. *Dungeons & Dragons*), it could also represent a resource to place on a predefined map (e.g. *Game of Thrones, the board game*). This approach has many variations, which we have split into two main categories: the ones that encompass the usage of some geometric tile and the ones that use a more free-form representation (and usually placement) of the resources.

Geometric. The geometric approach is based on the usage of tiles that present some geometrical shape which allows the players to connect them together to create the playing field. The most common shapes used for these techniques are squares and hexagons since they allow connecting a large number of tiles in a space fitting. One of the common issues with this approach is that the maps might not look very consistent if the rules for placing adjacent tiles are too strict (a well known problem in *Settlers of Catan*). On the other hand, when rules for connecting pieces are quite strict to encourage consistent and pretty boards the risk is that the initial board generation might take the players a long time. An interesting example to create complex, "natural", and fast maps, while using a geometrical tiled approach can be observed in *Victory: The Blocks of War*. In this game, the movement tiles and the board-generating tiles are decoupled (respectively hexagons and rectangles), and each board-tile contains a number of movement tiles. This means that a quite large playing map can be created by connecting just a small amount of tiles.

The most common approach game designers use to solve this dilemma is to make the placement rules an integral part of the gameplay. An example can be seen in *Carcassonne* and similar games (e.g. *Kingdomino*): the placement of the tiles to generate the map becomes part of the core gameplay, with the players drawing and placing a tile each turn to distribute the resources in the most profitable way. This method often leads to "incomplete" maps (with gaps), since the conflict shifts from a more direct fight to an economic conflict of resource obtainment.

Terrain. This approach allows instead for a more fluid placement of resources, that could be represented by non-geometric or irregular shapes. The historical Kriegspiel already presented this approach, as various tiles (or tokens) could be placed on the playing field to represent terrain characteristics such as mountains, trenches, rivers, etc. Most modern miniature wargames (or table-top wargames) present some similar terrain modification mechanism, examples include H.G. Wells' *Little Wars*, the *Warhammer* games, and *Axis & Allies*.

The placement of these tiles can create bonuses (e.g. cover or high ground) or penalties (e.g. limited movement or visibility) for the units in the vicinity and can have a great impact on the strategy necessary to achieve victory. It is also interesting to note that, while the game publishers often provide tile sets for the players to buy, it's quite common for a hobbyist to create their own set pieces or use common objects to improvise new tiles.

3.3 Control Structures

Control methods place restrictions on the types of orders which may be given or the order in which units gain the initiative. While the common game method is a *You go - I go* approach to turns, control structures may allow for a random unit to gain the initiative or be played during a turn, either by use of a roll of a die, or a card based system. Games such as *Bolt Action* implements placing an order die for each unit on the table for both sides. A player draws from this pool one at a time and the colour of the selected die allows that army to give an order to one of the units. *Star Wars Legion* has each player take turns pulling unit type counters from a personal bag in order to decide on who is allowed to move in a turn by turn order. Other game effects can interrupt this process allowing a unit to move without being drawn from the bag.

Kingdom Death is a recent and interesting example of the level of the control structure that can be implemented inside of a game only using analog generation. It implements a sophisticated analog AI via a series of cards drawn in sequence from an AI deck with conditions of the actions. This is a similar representation to a Finite State Machine or If-Skip-Action list [4] implemented by cards. The cards might also be self referential to the AI, allowing for a reordering of future actions.

Conditions include targeting the closest threat, visible threat, an enemy unit with an affinity or item, and have a default action when those conditions are not met. These defaults are normally searched actions, attempting to repair damage, or activating buffs (bonuses).

Table 1. Taxonomy analysis of war-games **qualities** for the games *Dungeons & Dragons*, *Axis and Allies*, *Kingdom Death*, and *Little Wars*

Game	Domain	Span	Force composition	Environment	Process abstraction	Information availability
Warhammer 40000	Land (mostly)	Variable	Individual-platoons	Set pieces placed before the game	Conglomerated damage	Full information
Dungeons & Dragons	Land	Local	Individual	Predefined	Individual	Incomplete
Axis and Allies	Land, air, sea	Global	Army-sized	Predefined	Army damage	Full information
Kingdom Death	Mystical land	Home base/local	Individual	Predefined	Individual	Incomplete
Little Wars	Land	Local	Squad	Set pieces placed before the game	Unit damage	Full information

Table 2. Taxonomy analysis of war-games **construction** for the games *Dungeons & Dragons*, *Axis and Allies*, *Kingdom Death*, and *Little Wars*

Game	Time Processing	Stochastic/deterministic world	Sidedness
Warhammer 40000	Turn-based with automatic actions	Stochastic	1 vs many
Dungeons & Dragons	Turn-based with initiative	Stochastic	1 player side vs the game
Axis and Allies	Turn-based	Stochastic	1 vs many
Kingdom Death	Turn-based with interrupts	Stochastic	1 player side vs the game
Little Wars	Turn-based	Deterministic	1 vs 1

Table 3. Areas in which **PCG techniques** are used by the games: *Warhammer 40000*, *Dungeons & Dragons*, *Axis and Allies*, *Kingdom Death*, and *Little Wars*

Game	Lookup tables		Tiles		Control structures
	Integrated	External	Geometric	Terrain	
Warhammer 40000	Limited	*		*	
Dungeons & Dragons		*	*	Limited	
Axis and Allies		*			
Kingdom Death	*	*	*		Card based AI
Little Wars		*		*	

4   Conclusions

This paper describes a taxonomy of wargames and a survey of analog procedural content generation methods used by such games/simulations. By providing a taxonomy to break down current games, planners and developers of new games can better transfer both methods and tools to future applications in new domains. Common complaints are applying old methods which work well in one situation into another situation - from both Military and games developers, by better understanding the current sets of games, its highlights were a game is (un)applicable to a situation.

The taxonomy highlights not only the domain of application but the tools available to designers. In future work, an interesting direction would be to apply the taxonomy and look for correlations between the battle field properties and the mechanisms used in play.

References

1. Allen TB (1994) War games: inside the secret world of the men who play at world war III
2. Anderson LB, Cushman JH, Gropman AL, Roske VP (1987) SIMTAX: a taxonomy for warfare simulation. Phalanx 20(3):26–28
3. Anderson LB, Cushman JH, Gropman AL, Roske Jr VP (1989) SIMTAX: a taxonomy for warfare simulation (workshop report). Technical report, Military Operations Research Society Alexandria, VA
4. Ashlock D, Joenks M, Koza JR, Banzhaf W (1998) Isac lists, a different representation for program induction. In: Koza JR, Banzhaf W, Chellapilla K, Deb K, Dorigo M, Fogel DB, Garzon MH, Goldberg DE, Iba H, Riolo R, (eds) Genetic programming 1998: proceedings of the third annual conference. University of Wisconsin, Madison. Morgan Kaufmann, pp 3–10
5. Ashlock D, Lee C, McGuinness C (2011) Search-based procedural generation of maze-like levels. IEEE Trans Comput Intell AI Games 3(3):260–273
6. Cathrine II (1845) Mémoires de l'impératrice catherine ii, écrits par elle-même, et précédés d'une préface par a. herzen, londres
7. Dales GF (1968) Of dice and men. J Am Orient Soc 88(1):14–23
8. Nakamura T (2016) The fundamental gap between tabletop simulation games and the "truth". In: Harrigan P, Kirschenbaum MG, (eds) Zones of control: perspectives on wargaming. MIT Press
9. Sabin P (2012) Simulating war: studying conflict through simulation games. A&C Black, London
10. Smith G (2015) An analog history of procedural content generation. In: Proceedings of the 2015 conference on the foundations of digital games (FDG)
11. Togelius J, Champandard AJ, Lanzi PL, Mateas M, Paiva A, Preuss M, Stanley KO (2013) Procedural content generation: goals, challenges and actionable steps. In: Lucas SM, Mateas M, Preuss M, Spronck P, Togelius J (eds) Artificial and computational intelligence in games, vol 6. Dagstuhl Follow-Ups. Schloss Dagstuhl-Leibniz-Zentrum fuer Informatik, Dagstuhl, pp 61–75
12. Tzu S (2008) The art of war. In: Strategic studies. Routledge, pp 63–91

13. Valtchanov V, Brown JA (2012) Evolving dungeon crawler levels with relative placement. In: Desai BC, Mudur S, Vassev E, (eds) C3S2E 2012 Fifth International C* Conference on Computer Science & Software Engineering. ACM, Montreal, pp 27–35
14. Van Creveld M (2013) Wargames: from gladiators to gigabytes. Cambridge University Press, Cambridge
15. Von Clausewitz C (1940) On war. Jazzybee Verlag, Schwab
16. Wells HG (2013) Little wars. Read Books Ltd., Redditch
17. Wintjes J (2016) Not an ordinary game, but a school of war. Vulcan 4(1):52–75
18. Wintjes J (2017) When a spiel is not a game: the prussian kriegsspiel from 1824 to 1871. Vulcan 5(1):5–28
19. Yannakakis GN, Togelius J (2014) A panorama of artificial and computational intelligence in games. IEEE Trans Comput Intell AI Games PP(99):1

When the Price Is Your Privacy:
A Security Analysis of Two Cheap
IoT Devices

Margherita Favaretto[1], Tu Tran Anh[1], Juxhino Kavaja[1], Michele De Donno[1],
and Nicola Dragoni[1,2(✉)]

[1] DTU Compute, Technical University of Denmark, Kongens Lyngby, Denmark
{s170065,s156015}@student.dtu.dk, {juxk,mido,ndra}@dtu.dk
[2] Centre for Applied Autonomous Sensor Systems (AASS),
Örebro University, Örebro, Sweden

Abstract. The Internet of Things (IoT) is shaping a world where
devices are increasingly interconnected, cheaper, and ubiquitous. The
more we move toward this world, the more cybersecurity becomes
paramount. Nevertheless, we argue that there exists a category of IoT
devices which commonly overlooks security, despite dealing with sensitive information. In order to demonstrate this, in this work, we present
the results of the security assessments we performed on two IoT devices
that we consider emblematic of such category: the Rohs K88h smartwatch and the Sricam SP009 IP camera. The results demonstrate the
existence of critical vulnerabilities that could be easily exploited, even
by non-expert attackers, for extracting sensitive information and severely
impacting on user's privacy.

1 Introduction

The Internet of Things (IoT) is increasingly branching out in our daily lives. It is
changing our cities [1], homes [2,3], industries [4,5], transports [6], and healthcare
[7,8], just to name a few. This trend is not going to stop, as it is estimated that
by 2020 the number of IoT devices is expected to reach 21 billion [9].

This relentless growth of IoT demonstrates that the advantages resulting
from such paradigm are well known. Nevertheless, since the IoT is far from
being perfect, it is paramount to cope with such enthusiasm and increase the
awareness about the issues associated with its usage. Among these issues, one of
the most critical is cybersecurity.

The security concerns associated with the IoT have been extensively discussed in the literature [10–12]. However, it is our strong belief that there still
exists a lack of information about the importance of appropriately securing IoT
devices. This is especially the case for cheap IoT devices, where both users and
manufacturers do not realize that in most cases, even if such devices have a small
economic value, the quantity of sensitive information they store turns them into
invaluable devices that need to be protected at any cost.

© Springer Nature Switzerland AG 2020
P. Ciancarini et al. (Eds.): SEDA 2018, AISC 925, pp. 55–75, 2020.
https://doi.org/10.1007/978-3-030-14687-0_6

1.1   Contribution

The main motivation behind this work is to raise awareness in relation to the severe privacy issues arising when IoT devices are not properly secured. We argue that there exists a category of IoT devices that can be easily exploited by attackers since lacking adequate security mechanisms. This category is composed of those IoT devices which are used by consumers in their daily lives and that are perceived as accessories, and not as primary goods. We claim that there are two properties that make such devices particularly important from a security perspective and at the same time vulnerable:

– Ubiquitousness: although such devices are considered futile, since not vital for consumers, from a security perspective they can not be considered as such. Indeed, by accompanying users in every step of their daily lives, they can acquire large amounts of sensitive information which turns them into critical devices that need to be protected.
– Cheapness: the only criteria employed by consumers for choosing which device to buy, is essentially based on the highest number of functionalities they can get at the lowest price. This implies that producers strive to reduce costs as much as possible, which usually implies sacrificing security.

On the same line of some previous work [13,14], in order to demonstrate the great potential of being attacked that this category of IoT devices has, in this paper we report the results of the security analysis we performed on two different IoT devices: the Rohs K88h smartwatch, the Sricam SP009 IP camera.

Both devices have the property of being among the cheapest in their market segment and of being ubiquitous (i.e. supporting and accompanying users in their daily activities). We emphasize that these were the only criteria we adopted in choosing such devices for performing our security assessments.

1.2   Paper Structure

The rest of this paper is organized as follows. Section 2 presents the security assessment that has been performed on the Rohs K88h smartwatch. Section 3 describes the vulnerabilities we identified on the Sricam SP009 IP camera. Internally, these sections, are both structured in a similar way: initially, the most relevant works dealing with the privacy issues of such devices are briefly presented; then, an overview of the devices and of the tools used for performing their security assessments is given; finally, the identified vulnerabilities are presented together with possible solutions. Section 4 concludes the paper and outlines the topics on which we are currently working for extending this work.

2   Security of Cheap and Ubiquitous IoT Devices: A Smartwatch Case Study

Smartwatches are rapidly becoming widely adopted [15]. These devices are increasingly being used for disparate purposes: remote health monitoring [16,17], activity recognition [18–20], biometric authentication [21,22], supporting outdoor activities [23] and many others. All these applications are enabled by few characteristics that these devices have: the high number of sensors and functionalities they offer, their proximity to end-users and, most importantly, their reduced cost. Nevertheless, if from the one hand these characteristics allow the widespread adoption of smartwatches, on the other hand, the same properties make such devices insecure. Indeed, the fact that they are equipped with many sensors which are constantly located in the proximity of end-users allows such devices to acquire a lot of sensitive data that need to be protected. However, the need to protect such data collides with the need for reducing production costs. In order to demonstrate this, we performed the security assessment of a device that we consider emblematic of all other devices satisfying the aforementioned characteristics.

The goal of this section is to describe the results of the security assessment that we performed on the Rohs K88h smartwatch together with the vulnerability we identified. Before doing this, in the following subsection, we present the main literature works that dealt with the security issues of smartwatches.

2.1   Related Works

In this section, we report some of the literature works discussing the security issues of smartwatches. We mostly focus on manuscripts that take into account how such devices can be exploited for affecting the privacy of users.

Several works in the literature, such as [24–28], demonstrate the possibility to exploit smartwatches sensors for inferring inputs typed on external keyboards. These works show that sensitive information (such as PINs) can be inferred through data sensed from smartwatches. These attacks are considered side-channels attacks and are commonly referred to as keystroke inference attacks.

Differently from the aforementioned works, Lu et al. [29] show that it is possible to use sensors embedded on smartwatches, not only for inferring inputs typed on external keyboards, but even for discovering PINs typed on the same smartwatch where the sensors are installed.

In [30], Do et al. describe the types of sensitive data that are stored in smartwatches. They also present an attack for extracting these data from smartwatches based on the Android Wear platform. For this category of devices, the authors have been able to exfiltrate data by exploiting a security flaw of the devices bootloader.

Liu and Sun [31] present a survey of attack methodologies that can affect wearable devices. They classify attacks in relation to the security property they affect, focusing on three properties: integrity, authenticity, and privacy.

Analyzing the security of smartwatches is a complex task due to the different existing models, vendors, and functionalities. For this reason, Siboni et al. [32] propose a testbed framework for testing the security properties of wearable devices. The goal of such framework is to test the wearable IoT devices against a given set of security requirements. As we are going to show in the next sections, most of the security flaws identified in the devices we considered in our security assessment are basic and common errors that could have been easily discovered by such framework.

Lee et al. [33] perform experimental studies for evaluating the amount of data transferred from host devices (e.g. smartphones) to wearable devices (e.g. smartwatches) after they have been successfully paired with each other. Based on the results of this analysis, a large amount of sensitive information is transferred from host devices to the wearable ones. On top of that, it resulted that such transfers of sensible data are not sufficiently notified to users. Such a situation rises severe privacy concerns as it impedes users from clearly knowing where their data are stored.

In [34], ten different smartwatches are studied from a security perspective. As expected, the results of such study demonstrate a lack of proper security mechanisms and the existence of naive vulnerabilities (such as, insufficient authentication and authorization or lack of mechanisms to protect network services) that can easily be exploited also by non-expert attackers.

In the rest of this section, we are going to describe the results of the security assessment we performed on the Rohs K88h smartwatch. To this aim, in the next subsection, we describe the common scenario in which every smartwatch operates. Hence, we explain how these devices communicate with smartphones and their role in preserving users' privacy. Therefore, we recall some basic concepts of the Bluetooth technology that are going to be fundamental for describing the security vulnerability of the Rohs K88h smartwatch.

2.2   The Typical Setting in Which Smartwatches Operate

Smartwatches are typically connected to a smartphone by means of a Bluetooth connection. Such a connection is used by the smartwatch to receive user's data from the other device (e.g. messages, contacts, notifications and others). However, it is also employed for sending data in the opposite direction. Indeed, by means of their sensors, smartwatches collect several types of data (e.g. sleeping habits, physical activity records and other personal information) that need to be shared with a smartphone for being elaborated and allowing the extraction of useful information. In many cases, the smartphone may also use an Internet connection for outsourcing such data and performing remote elaboration on the Cloud. Figure 1 depicts the relevant entities and communications that have to be considered when working with smartwatches.

In the following, we are only going to consider the Bluetooth communication between the smartwatch and the smartphone. In relation to this, we are going to describe the vulnerability we identified in the smartwatch implementation of the Bluetooth technology.

Bluetooth Background. In this subsection, we present some basic concepts of the Bluetooth technology that are going to be used in the subsequent sections. For a thorough treatment of Bluetooth and its security implications, we suggest the reader refers to the NIST guideline on Bluetooth security [35].

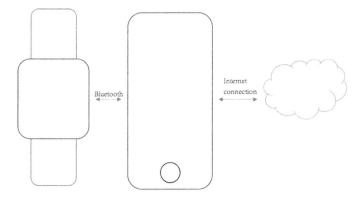

Fig. 1. Relevant entities and communications to consider when dealing with common smartwatches.

Bluetooth is an open standard for short-range radio frequency communication mainly used for establishing wireless personal area networks (WPANs) [35]. It is a low-cost and low-power technology operating in the frequency range between 2.4000 gigahertz (GHz) and 2.4835 gigahertz (GHz) [35].

Discoverable and *connectable* modes are defined by the Bluetooth standard for allowing devices to communicate and connect with each other. In simple, when a device operates in both modes, it replies to inquiries from other devices in such a way to be detectable and to permit the establishment of connections [35]. In a Bluetooth network, it is always possible to distinguish a master device from all other devices which are typically referred to as slave nodes. A master device controls and establishes the network and defines the network frequency hopping scheme [35]. The network frequency hopping scheme is a transmission technique that can be seen as a first minimal security mechanism that avoids inference and makes it a bit more difficult (if compared to fixed-frequency transmissions such as IEEE 802.11 g/n) to capture transmissions [35].

From a security perspective, the Bluetooth standard defines only three security services: Authentication, Confidentiality and Authorization. Each Bluetooth device has to select and operate in one Security Mode. There exists 4 Security Modes which defines when the aforementioned security services (authentication, confidentiality and authorization) are initiated [35]. We briefly summarize the properties of each of them [35]:

1. Security Mode 1: security functionalities are never started. By employing such Security Mode, Bluetooth devices cannot be distinguished, moreover, no mechanisms for preventing the establishment of connections can be applied.
2. Security Mode 2: in this case, security functionalities can be initiated after link establishment but before logical channel establishment. This is a service level-enforced security mode where authentication, encryption and authorization are used.

3. Security Mode 3: in this mode, security functionalities have to be started before the physical link is established.
4. Security Mode 4: this is another case of service-level enforced security mode where security functionalities are initiated after physical and logical link setup.

In order to initiate security services (especially authentication and confidentiality), a secret symmetric key, also known as link key, has to be generated. The process of generation of such key is known as *pairing process*. For the sake of completeness it should be mentioned that Security Modes 2 and 3 perform the pairing process by means of a method called Personal Identification Number (PIN) Pairing (also known as Legacy Pairing); while Security Mode 4 uses the Secure Simple Pairing (SSP) which is meant to simplify the pairing process [35].

We are not going to detail more about the security of the Bluetooth technology since, as we are going to describe in the next sections, the vulnerability of the smartwatch we analyzed is related to the use of the basic and totally insecure Secure Mode 1. For this reason, given the focus of this paper, it makes no sense to dive into more advanced topics when there still exists devices not considering security at all. We strongly believe that this total lack of security awareness is common in these categories of IoT devices. For this reason, we find paramount to document these issues and contribute to make aware of how much security is neglected together with the devastating impact such problems can have on privacy.

Although the issue we identified for the Rohs K88h smartwatch is not strictly related to any vulnerability of the Bluetooth technology itself (but rather to the lack of knowledge in using it), we underline that there exist several others sophisticated attacks and vulnerabilities. An introduction to these issues is presented in the NIST guide to Bluetooth Security [35].

2.3   Rohs K88h Smartwatch Overview

The smartwatch we used in performing our security assessment is the Rohs K88h smartwatch. At the time in which these experiments have been performed (October 2017), this is one of the top 10 cheap smartwatch alternatives in 2017 [36], it is placed 62th in the list of Amazon's bestseller "heart rate monitors", 127th in the category of "Fitness technology", and number 1414th in the "smartwatches" category [37].

The Rohs K88h smartwatch uses a MTK2502C processor which is widely adopted by different smartwatch manufacturers. It runs the Nucleus Operating System (OS),[1] which is commonly used as alternative to Wear OS[2] or watchOS.[3] It is equipped with a touch screen, microphone, speaker, Bluetooth 4.0 adapter, and accelerometer. Among the most interesting functionalities it offers we find: heart rate monitoring, tracking of burned calories, pedometer, sleep monitoring

---

[1] https://www.mentor.com/embedded-software/nucleus/.

[2] https://wearos.google.com.

[3] https://www.apple.com/lae/watchos/.

and remote controlling. In Table 1 a summary of the hardware specifications of this smartwatch are reported, while in Table 2 the main functionalities it offers are summarized.

We chose to evaluate this smartwatch for two main reasons: it is one of the cheapest devices in its market segment and offers numerous features. In other words, we considered it as emblematic of all those devices having the two aforementioned proprieties. It is our strong belief that these characteristics coexist in detriment of security. In order to demonstrate this, we decided to select an IoT device satisfying such criteria and performing a security assessment for verifying what is the real cost associated with the use of cheap and ubiquitous IoT devices. As expected, during our assessment, we found a severe security issue that is not rooted or caused by the used technologies itself, but, instead, caused by misinformation and inattention in implementing such technology in building the smartwatch. Hence, we argue that the potential impact of such naive issues is huge: they can easily be exploited also by non-expert attackers for acquiring a large amount of sensible information.

In the next section we describe the analysis we performed on the Bluetooth connection enabled by the smartwatch and present the vulnerability we identified at this level together with how it could be exploited for affecting the users' privacy.

Table 1. Rohs K88h Hardware specifications.

CPU	MTK2502C
Screen	IPS Capacitive touch screen
Battery	300 mAh Li-battery
Bluetooth	BT4.0LE
Speaker	1511 ACC Speaker
Microphone	YES
Memory	128 MB + 64 MB
Accelerometer	YES

2.4   The Vulnerability

The security assessment we performed on the Rohs K88h smartwatch showed a lack of protection of the Bluetooth communication. In this section, we present the tools, commands and steps we followed for identifying this issue.

In order to analyze the Bluetooth connection of the smartwatch, we used common commands available on Kali Linux 2017.2. By using the `hcitool scan` command, we were able to locate the smartwatch Bluetooth device address (BD_ADDR) and corresponding alias even if the device was not set in discoverable mode. Then, we used the `sdptool browse` command to inspect the services

Table 2. List of Rohs K88h offered features.

Features	Description
Heart rate monitor	Measurement of heart rate
Pedometer	Real time step count
Calories tracking	Counting of burned calories
Synchronization	Possibility to pair the smartwatch with iOS & Android based smartphones
Remote control	Access to personal assistants (such as Siri and Cortana)
Sleep monitor	Track sleep quality
Other	Calculator, Alarm, Remote camera, Remote music, Stopwatch, IP54, Call log, Dial, Phonebook etc.

offered by the device. At this point, we tried to establish a radio frequency communication (RFCOMM) [38] connection in order to verify whether a pairing process would start and notify the end-users of an attempt to connect with the smartwatch. For this purpose, we used the `rfcomm connect` command without the specification of any channel on which to establish the connection. In this way, not only we were able to establish a connection with the smartwatch, but we also did it in a completely silent way, i.e. no messages were generated from the smartwatch to, at least, notify the establishment of a new connection to the end-user. Since no action was required by the end-user (neither to accept the connection nor to enter a PIN code), we infer that no paring process occurs, which implies that the lowest level of security (Security Mode 1) is used by the Bluetooth connection used by the smartwatch [35].

Application Reverse Engineering. In order to take advantage of this vulnerability and extract sensible information from the smartwatch, we decided to perform the reverse engineering of the client application running on the smartphone that enables the exchange of data between the smartwatch and smartphone. Our goal is to use the reverse engineering process for building a new client application that can allow to silently exfiltrate sensible data from the device. Therefore, in this subsection, we detail how we performed the reverse engineering and the code we were able to recover.

The application that allows the smartphone to interact and control the smartwatch is called *Fundo Wear*. By means of this application, end-users can create profiles which permit to synchronize the information acquired by the IoT device with the smartphone. Examples of such information are: heart rate, burned calories or step counts measurements.

We performed the reverse engineering of the *Andorid* version of *Fundo Wear v. 1.6.0* using Apktool.[4] Even if part of the source code resulting from the reverse engineering process is obfuscated, we were able to identify the Java classes

---

[4] https://ibotpeaches.github.io/Apktool/.

that allow to establish a connection with the device and read the heart rate measurements. These classes can allow us to replicate the application so that it can be employed as an exploitation tool against the smartwatch.

Although these classes are a core part required for the implementation of the exploitation tool, we have not developed the complete exploitation tool yet. In relation to this, we emphasize that, even if the identified vulnerability can be exploited in several ways (such as by exploiting the services we identified by means of the command `sdptool browse`), we are currently working on building a complete exploitation tool that takes advantage of the reverse engineering process.

In the next subsection, we present how the vulnerability we identified could be solved. Moreover, we describe simple precautions that can render Bluetooth communications more secure.

2.5   Possible Solution

In the previous subsection, we presented the vulnerability we identified in the implementation of the Bluetooth technology used by the Rohs K88h smartwatch. We identified the existence of an extremely dangerous security flaw which allows external users to easily connect to the smartwatch. We believe that this vulnerability could be easily solved by setting the Security Mode of the device at any level higher than the one that is currently used. Nevertheless, we underline that, even by using higher-level Security Mode, there are still possible attacks that can be performed against a secure Bluetooth connection. In relation to this, a list of Bluetooth-specific attacks can be found in [35].

In the rest of this section, we present some basic precautions that can reduce the risks associated with attacks enabled by the Bluetooth connections of IoT devices. We divide them into two categories: the first considers what can be done from the consumer side, while the other presents what manufacturers can do for enhancing the Bluetooth connection security of IoT devices they produce.

Consumers' Side Precautions

- change default manufacturing settings (such as default authentication pin code);
- disable discovery of device if not required;
- turn off Bluetooth interface if not used;
- do not accept pairing requests, unless they are trusted;
- install firmware updates.

Manufacturers' Side Precautions

- do not implement Security Mode 1;
- notify users of every new connections that is attempted;
- encrypt data that are transmitted.

This completes the description of the security assessment we performed on the Rohs K88h smartwatch. In the next section, we are going to present another security assessment that we performed on a different device: the Sricam SP009 IP camera.

3   Security of Cheap and Ubiquitous IoT Devices: An IP Camera Case Study

With the spread of smart homes, our houses are increasingly populated of intelligent devices. As result, IP cameras, motion sensors, and connected door locks are more and more being used for home security [39].

An Internet Protocol camera, or simply IP camera, is a digital video camera commonly used for surveillance purposes, which may send and receive data via a computer network or the Internet [40]. It usually uses a Wi-Fi connection and implements a two-way audio communication which permits the user to listen and communicate with the environment in which the device is located. Moreover, it also makes possible to create a remote connection so that the customer can watch on a smartphone, or another device, the video being recorded.

There exist many low-cost IP cameras offering several functionalities and whose price is between 30\$ and 80\$. Most of these products typically store the images they acquire on a cloud server that can be accessed from everywhere.

IP cameras are composed by a lens, a sensor, a processor for analyzing images, and an internal memory where the firmware and software are installed. A typical functionality they offer is the possibility to act as *alarm system*. Indeed, the device can detect movements and react to them by sending a message to the registered e-mail addresses. Most of the time, an image is attached to the message so that the user can identify what caused the alarm and decide whether it is necessary to intervene or not. Further, most IP cameras are equipped with infrared sensors (IR sensors) which allow monitoring environments with low brightness. This set of features makes such devices suitable for being deployed in everyday environments.

Despite being used in everyday scenarios, the security of IP cameras is emerging as an important issue. An example of attack on surveillance cameras which occurred in a real scenario has been reported in [41].

In the next subsections, we analyze the security vulnerabilities of a low-cost IP camera. The device we considered is the Sricam SP009 H.264 Wifi Megapixel Wireless ONVIF CCTV IP camera, which was released by Shenzhen Sricctv Technology CO.[5] Before detailing our analysis, in the next subsection, we report some of the most relevant works related to the security of IP cameras.

3.1   Related Works

Although the literature on the topic is not very extended, for the sake of completeness, in this section, we review some works worth to be mentioned.

---

[5] http://www.sricam.com/.

Albrecht and Mcintyre [42] provide examples of two IP cameras that have been hacked. The authors conclude the work proposing some palliative solutions, such as: to change default passwords, to set unique and strong passwords for all home devices, to update the firmware, and to set the devices not to broadcast their SSID.

Tekeoglu and Tosun [40] investigate the security of cloud-based wireless IP cameras. They analyze the traffic generated by a low-end, easy-to-setup, off-the-shelf wireless IP camera for the average home user. The authors identify many security and privacy issues in these devices with different levels of severity. Specifically, they state that if a malicious person can sniff the IP camera network traffic anywhere in between the mobile device-cloud servers-IP camera path, he would be able to reconstruct the transmitted JPEG images, which is surely a serious privacy issue.

At the conference in Amsterdam on Positive Hacking Day, Sergey Shekyan and Artem Harutyunyan presented their analysis called "To Watch or Be Watched: Turning Your Surveillance camera Against You" [43]. The presentation explains how to gain control over a camera in the wild, analyzes security malpractices and how to make the camera less insecure.

Alshalawi and Alghamdi [44] propose a solution against illegal access on wireless surveillance camera. The solution is designed into two stages. First, a new monitoring scheme is built to keep the privacy of data. Second, an investigation process is facilitated, playing an important role in saving users' privacy and highly secure places that use surveillance cameras.

The rest of this section focuses on security issues and vulnerabilities of the SP009 IP camera, concluding with possible solutions to mitigate the issues.

3.2 System Overview

IP cameras are key devices for video surveillance systems. Security and privacy standings of these devices are important and need to be studied in detail.

In our use case, we used the IP camera released by Sricam: the SP009 IP camera. Figure 2 depicts this device and its main components: an IR Sensor, a Lens, a Microphone, a Speaker, a Reset, a MicroSD card slot, and a DC 5V Power Slot. Figure 3 provides an overview of its main functionalities.

The camera presents the following technical requirements. It has to be connected to a router supporting the 2.4 GHz Wi-Fi frequency band (802.11 b/g/n) with DHCP protocol enabled. Moreover, also the smartphone used to access the camera has to be connected to the Internet with a WLAN/Wi-Fi connection.

The device applies the advanced network transmission technology "Cloudlink P2P" which permits information exchange between the IP camera and a P2P Cloud. This P2P architecture makes possible to watch a live feed from the camera. Indeed, using a smartphone, it is sufficient to download the appropriate application, scan the device QR code, and begin to watch a live feed from the camera. An overview of the systems is presented in Fig. 4.

Fig. 2. The Sricam SP009 IP camera.

Fig. 3. Overview of functionalities offered by the Sricam SP009 IP camera.

3.3 Tools and Set-Up

In this section, the tools used to perform our security investigation are described.

The smartphone used for performing the experiments is a Huwawei P8 lite running Android 5.0.1. The mobile application necessary to control the IP camera is called "*Sricam*" and it is available both on Google Play[6] and Apple Store.[7] Moreover, for performing our experiments, we used the Sricam application having version *00.06.00.41*. Sricam is a free mobile application customized for this brand of IP cameras. It allows to remotely monitor the camera, stream video, and organize settings of the camera. The application let also modify administration credentials of the camera and manage a password for visitor users. Furthermore, customers can use the application to change network settings of the camera, such as: IP address, subnet, gateway, and DNS.

---

[6] https://play.google.com/store/apps/details?id=com.xapcamera&hl=it.
[7] https://itunes.apple.com/it/app/sricam/id1040907995?mt=8.

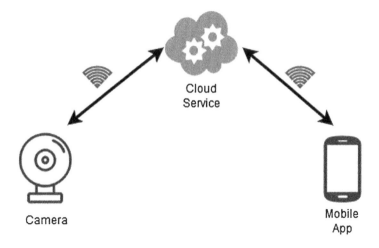

Fig. 4. System architecture of the Sricam SP009 IP camera.

3.4 The Vulnerabilities

The purpose of this section is to describe the vulnerabilities we identified in the remote control system of the Sricam SP009 IP camera. Furthermore, we also present how such vulnerabilities can be exploited for compromising user's privacy.

Before presenting the vulnerabilities we identified, it is necessary to describe how the smartphone application used for controlling the camera works. Hence, in the next paragraphs, we are first of all going to describe how the application running on the smartphone works and which are the steps that users are required to follow for acquiring the rights to control a new camera. Subsequently, we describe the vulnerabilities we identified in this procedure and how a user can gain control of devices that he does not own.

After installing the application on the smartphone, it is required to register a new user account and to add the camera to the list of devices owned by the new user of the application. In order to perform this operation, it is required to specify the device ID and password, which are information reported on a label located beneath the camera support. The ID is a sequence number (in our case its value was: 995811), while, the password of each new camera has a default value of 888888. After registering the device, the final step required for remotely controlling it is to connect the IP camera to a Wi-Fi network. At this point, it is possible to remotely access the camera, change its settings, set alarms and perform a lot of other operations enabled by the device.

After this brief description of the procedure that allows gaining control of a new camera, we present the vulnerabilities we identified in this process.

Lack of Adequate Access Control. In the procedure for gaining control of a new camera, two main issues have been identified:

1. Although the application requires the user to register by means of a valid e-mail, we noticed that no confirmation e-mail is sent to the address used for registration. This implies the possibility to register a new account with a fake e-mail address or an address that is not owned by the user.

   We strongly believe that sending a confirmation e-mail that forces the user to prove its identity is a minimal level of security that must be provided no matter what.

2. More important, once the new camera is added to the list of devices owned by the user, the application does not force him to change the default password of the device.

   In relation to this, we have tested that, by entering a device ID similar to the ID of the device we own and by using the default password, it is possible to connect and control in a completely silent way other cameras that we do not possess. In Fig. 5, we depict the three simple steps required for gaining control of the image stream produced by a camera we do not own. Moreover, in the last image of Fig. 5 it is also possible to see a preview of the image stream we acquired.

If the first issue can be considered as having minor consequences, this is not true for the second one. We claim that if users are not compulsory forced to change the default password, it leads to the existence of a vast number of cameras that are used with a default password. This allows non-expert attackers to connect and gain control of such devices.

In order to have an insight on the number of IP cameras affected by this issue (i.e. the number of sold cameras that are still using the default password), we performed a simple experiment with the goal of estimating this number. Using the smartphone application, we randomly tested the devices with an ID number in the range from 995810 to 995840. We discovered that, on this range of 30 cameras, 11 of them use the default password. In other words, in this small range, more than 30% of devices are affected by the vulnerability. If we assume that this percentage is representative for a larger set of samples, by knowing the number of sold cameras, we can try to empirically evaluate which is the number of cameras affected by the issue. According to the information reported on Amazon and eBay, we can conclude that at least 400 models of this specific camera have been sold. So, we can empirically derive that it could possible to control 146 cameras.

Even if these statistics might not be accurate, we believe that they demonstrate the potential impact this issue can have on the privacy of several families and businesses which use such devices.

We argue that this vulnerability is critical for the privacy of users. Indeed, in order to understand this, it is sufficient to consider the multitude of actions attackers can perform without being real owners of the device:

– see and control the images being streamed by the camera: turn on/off the camera, use the local record, capture videos, and change the resolution of images;

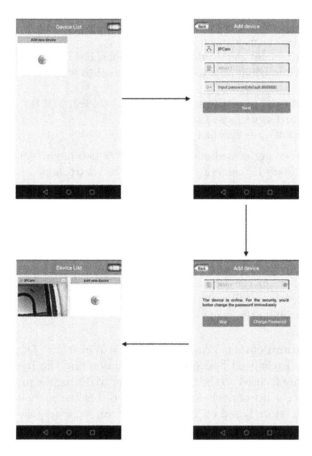

Fig. 5. Steps required to add a device that is not owned by the user of the application from which screenshots are taken. In the last image (located at the bottom, on the left), a preview of the scene captured by the camera we attacked is shown.

- listen to what is happening in the environment in which the camera is placed by using the built-in microphone;
- affect the environment in which the camera is placed by using the built-in speaker (huge impact for baby-monitoring cameras);
- see and modify device information: device version, uBoot version, CPU version, and system version;
- see and modify time zone settings;
- see and modify media settings: video standard (PAL/NTSC), volume, image reverse;
- modify security settings: add a visitor password;
- see and modify network settings: network mode (Wired or Wi-Fi), IP, subnet, gateway, DNS;
- see the name of Wi-Fi list and change Wi-Fi network;

- modify alarm settings: receive notifications (yes/no), e-mail alerts, motion detection alarm, buzzer alarm, alarm switch;
- see and modify email alerts information (sender, SMTP server, port, encryption mode) while being able to add a new email to which alarms have to be sent;
- see and modify record settings: modify record mode, see if the device uses an SD Card, record switch (yes/no);
- learn the version of device firmware.

The analysis we performed allowed to discover two important issues related to the Sricam SP009 IP camera. First of all, the smartphone application used for controlling the camera does not force users to change the default password after the camera has been registered in the system for the first time. Secondly, there is no proper access control, i.e. by only knowing the ID and the password of the camera, it is possible to have complete control of the device. Hence, by exploiting these two vulnerabilities, an external attacker can completely take control of the camera in an easy and silent fashion.

On top of that, in the next subsection, we describe another vulnerability that we identified by sniffing the network traffic generated by the smartphone application controlling the camera.

Insecure Communication. We decided to use Wireshark[8] for inspecting the network packets exchanged between our application and the remote server. In doing this, we only focused on the packets exchanged during the process of adding a new camera to the list of devices. We discovered that the communication is not properly secured, since based on HTTP. Therefore, we were able to extract the ID of the device and its password by simply eavesdropping the communication channel. Hence, even if the password used by the IP camera is not the default one, it can still be possible to get such password by simply sniffing packets that are exchanged during the remote communication of the application.

The presence of all these vulnerabilities, triggered us to perform other experiments for verifying the existence of other issues. For this purpose, we performed the reverse engineering of the smartphone application controlling the IP camera.

Analysis and Reverse Engineering of the Control Application. With the goal of verifying whether other vulnerabilities exist, we decided to analyze the smartphone application controlling the camera for better understanding how the overall control system is built and works.

In order to perform the reverse engineering of the smartphone application, the following tools have been used:

- the application apk extractor[9] has been employed for obtaining the installed apk from the smartphone;
- dex2jar[10] has been used for converting the machine code to Java class files;

---

[8] https://www.wireshark.org/download.html.
[9] https://play.google.com/store/apps/details?id=com.ext.ui&hl=it.
[10] https://sourceforge.net/projects/dex2jar/.

– Java compiler JD-GUI[11] has been used for inspect the Java source code.

Using Android Studio,[12] we have partially analyzed the code of the Sricam application. We have been able to analyze which controls are performed when a user requests to add a new camera on his profile. We have discovered that the controls being performed for this purpose only involve the ID and password of the camera: the system checks if the IP camera details are correct, and successively, if the device is online, it is added in the device database of the user. Therefore, as suspected, we verified that no authentication or authorization of the user is involved in the process of adding a new device.

We are still working on the application source code for trying to infer more information on the way in which the overall system works in order to understand whether other vulnerabilities exist. In relation to this, we have already identified some possible issues that need to be analyzed in more detail.

3.5 Possible Solutions

Several possible solutions can be applied for patching the identified vulnerabilities. In the following, for each vulnerability we discovered, possible solutions are presented:

– *Lack of adequate access control*: this issue has to be addressed in three steps. First of all, the application has to be patched in order to force users to modify the device password after the IP camera has been successfully associated with a user for the first time.
  Secondly, the access control mechanism must also take into account the user requesting to gain access to the device, i.e. it has not to rely solely on information of the device.
  Finally, in order to implement proper access control mechanisms, it is paramount to have adequate authentication mechanisms. Hence, it is essential to solve the vulnerability associated with the improper verification of e-mail address at registration time.
– *Insecure communication*: using secure communication protocols would be sufficient to reduce the risk of eavesdropping sensitive information. In the specific case, we suggest using HTTPS instead of HTTP.

4 Conclusions and Future Works

The objective of this work is to raise the awareness in relation to the poor security state of cheap and ubiquitous IoT devices, with the aim of encouraging and supporting the development of new security solutions for IoT, such as the one presented in [45].

It is our strong belief that this category of devices is a major threat to users' privacy. In order to demonstrate this, we selected two different devices and

---

[11] http://jd.benow.ca/.
[12] https://developer.android.com/studio/index.html.

performed their security assessments. Only two criteria were used in selecting them: (1) they had to be among the cheapest in their market segments while offering several functionalities (2) they had to be ubiquitous (i.e. accompanying and supporting users in their daily activities). As a result, we inspected the security state of the *Rohs K88h smartwatch* and that of the *Sricam SP009 IP camera*. The outcomes of our analysis are clear: the security of such devices is neglected and, as a consequence, users' privacy is seriously threatened. Indeed, the results of both case studies show that even non-expert attackers can exploit such vulnerabilities for negatively affecting users' privacy.

We showed that the Rohs K88h smartwatch implements an insecure Bluetooth communication that allows the establishment of connections without requiring the approval of the end-user. In order to demonstrate how such vulnerability could be exploited by an adversary, we proposed to exploit the client application that interacts with the smartwatch for building an exploitation tool that silently exfiltrates sensitive information.

As regards the Sricam SP009 IP camera, we identified several vulnerabilities. The most important among these issues is the use of an inadequate access control mechanism. We showed that such simple issues can allow to completely take control of the IP camera together with the image stream it produces.

Our goal was to show the security vulnerabilities we identified in such devices. Nevertheless, we are currently working for building a complete data exfiltration tool for the Rohs K88h smartwatch that takes advantage of the reverse engineering of the existing application. Furthermore, in relation to the Sricam SP009 IP camera, we are studying the source code resulting from the reverse engineering of the application controlling the camera, for identifying the exact number of IP cameras that is possible to control. The overall objective of these future works is to demonstrate what is the concrete impact that these vulnerabilities have on privacy.

Based on our experience of assessing the security of IoT devices, future work will focus on how to secure the devices in context-aware systems. To this aim, we are currently working on a security framework based on two key concepts: Security-by-Contract [46,47] and Fog Computing [48].

References

1. Zanella A, Bui N, Castellani A, Vangelista L, Zorzi M (2014) Internet of Things for smart cities. IEEE Internet Things J 1(1):22–32
2. Soliman M, Abiodun T, Hamouda T, Zhou J, Lung C-H (2013) Smart home: integrating Internet of Things with web services and cloud computing. In: 2013 IEEE 5th international conference on cloud computing technology and science (CloudCom), vol 2. IEEE, pp 317–320
3. Kelly SDT, Suryadevara NK, Mukhopadhyay SC (2013) Towards the implementation of IoT for environmental condition monitoring in homes. IEEE Sens J 13(10):3846–3853
4. Da Xu L, He W, Li S (2014) Internet of Things in industries: a survey. IEEE Trans Ind Inform 10(4):2233–2243

5. Sadeghi A-R, Wachsmann C, Waidner M (2015) Security and privacy challenges in industrial Internet of Things. In: Proceedings of the 52nd annual design automation conference. ACM, p 54

6. Kirk R (2015) Cars of the future: the Internet of Things in the automotive industry. Netw Secur 2015(9):16–18

7. Laplante PA, Laplante N (2016) The Internet of Things in healthcare: potential applications and challenges. IT Prof 18:2–4

8. Amendola S, Lodato R, Manzari S, Occhiuzzi C, Marrocco G (2014) RFID technology for IoT-based personal healthcare in smart spaces. IEEE Internet Things J 1(2):144–152

9. Eddy N (2015) Gartner: 21 billion IoT devices to invade by 2020. InformationWeek, vol 10

10. Sicari S, Rizzardi A, Grieco LA, Coen-Porisini A (2015) Security, privacy and trust in Internet of Things: the road ahead. Comput Netw 76:146–164

11. Jing Q, Vasilakos AV, Wan J, Lu J, Qiu D (2014) Security of the Internet of Things: perspectives and challenges. Wirel Netw 20(8):2481–2501

12. Dragoni N, Giaretta A, Mazzara M (2016) The internet of hackable things. In: International conference in software engineering for defence applications. Springer, pp 129–140

13. Goyal R, Dragoni N, Spognardi A (2016) Mind the tracker you wear: a security analysis of wearable health trackers. In: Proceedings of the 31st annual ACM symposium on applied computing, SAC 2016. ACM, New York, pp 131–136

14. Giaretta A, De Donno M, Dragoni N (2018) Adding salt to pepper: a structured security assessment over a humanoid robot. In: Proceedings of the 13th international conference on availability, reliability and security, ARES 2018. ACM, New York, pp 22:1–22:8

15. Rawassizadeh R, Price BA, Petre M (2015) Wearables: has the age of smartwatches finally arrived? Commun ACM 58(1):45–47

16. Mortazavi B, Nemati E, VanderWall K, Flores-Rodriguez HG, Cai JYJ, Lucier J, Naeim A, Sarrafzadeh M (2015) Can smartwatches replace smartphones for posture tracking? Sensors 15(10):26783–26800

17. Wu R, De Lara E, Liaqat D, Thukral I, Gershon AS (2016) Feasibility of using smartwatches and smartphones to monitor patients with COPD. In: A49. COPD: CARE DELIVERY. American Thoracic Society, pp A1695–A1695

18. Weiss GM, Timko JL, Gallagher CM, Yoneda K, Schreiber AJ (2016) Smartwatch-based activity recognition: a machine learning approach. In: 2016 IEEE-EMBS international conference on biomedical and health informatics (BHI), pp 426–429. IEEE

19. Ramos FBA, Lorayne A, Costa AAM, de Sousa RR, Almeida HO, Perkusich A (2016) Combining smartphone and smartwatch sensor data in activity recognition approaches: an experimental evaluation. In: SEKE, pp 267–272 (2016)

20. Casilari E, Oviedo-Jiménez MA (2015) Automatic fall detection system based on the combined use of a smartphone and a smartwatch. PLoS One 10(11):e0140929

21. Johnston, AH, Weiss GM (2015) Smartwatch-based biometric gait recognition. In: 2015 IEEE 7th international conference on biometrics theory, applications and systems (BTAS). IEEE, pp 1–6

22. Lee W-H, Lee R (2016) Implicit sensor-based authentication of smartphone users with smartwatch. In: Proceedings of the hardware and architectural support for security and privacy 2016. ACM, p 9

23. Guerra-Rodríguez D, Granollers A (2016) User experience experiments with mobile devices in outdoors activities: use case: cycling and mountain biking. In: Proceedings of the XVII International Conference on Human Computer Interaction, Interaccion 2016. ACM, New York, pp 26:1–26:4

24. Liu X, Zhou Z, Diao W, Li Z, Zhang K (2015) When good becomes evil: keystroke inference with smartwatch. In: Proceedings of the 22nd ACM SIGSAC conference on computer and communications security, CCS 2015. ACM, New York, pp 1273–1285

25. Sarkisyan A, Debbiny R, Nahapetian A (2015) WristSnoop: smartphone PINs prediction using smartwatch motion sensors. In: 2015 IEEE international workshop on information forensics and security (WIFS), pp 1–6

26. Wang H, Lai TTT, Roy Choudhury R (2015) MoLe: motion leaks through smartwatch sensors. In: Proceedings of the 21st annual international conference on mobile computing and networking, MobiCom 2015. ACM, New York, pp 155–166

27. Maiti A, Jadliwala M, He J, Bilogrevic I (2015) (Smart)watch your taps: side-channel keystroke inference attacks using smartwatches. In: Proceedings of the 2015 ACM international symposium on wearable computers, ISWC 2015. ACM, New York, pp 27–30

28. Maiti A, Jadliwala M, He J, Bilogrevic I (2018) Side-channel inference attacks on mobile keypads using smartwatches. IEEE Trans Mob Comput PP(99):1

29. Lu CX, Du B, Wen H, Wang S, Markham A, Martinovic I, Shen Y, Trigoni N (2018) Snoopy: sniffing your smartwatch passwords via deep sequence learning. In: Proceedings of ACM on interactive, mobile, wearable and ubiquitous technologies, vol 1, pp 152:1–152:29

30. Do Q, Martini B, Choo KR (2017) Is the data on your wearable device secure? An android wear smartwatch case study. Softw: Pract Exp 47(3):391–403

31. Liu J, Sun W (2016) Smart attacks against intelligent wearables in people-centric Internet of Things. IEEE Commun Mag 54:44–49

32. Siboni S, Shabtai A, Tippenhauer NO, Lee J, Elovici Y (2016) Advanced security testbed framework for wearable IoT devices. ACM Trans. Internet Technol. 16:26:1–26:25

33. Lee Y, Yang W, Kwon T (2017) Poster: watch out your smart watch when paired. In: Proceedings of the 2017 ACM SIGSAC conference on computer and communications security, CCS 2017. ACM, New York, pp 2527–2529

34. Internet of Things security study: Smartwatches. https://www.ftc.gov/system/files/documents/public_comments/2015/10/00050-98093.pdf. Accessed 26 Mar 2018

35. Padgette J (2017) Guide to bluetooth security. NIST Special Publication, vol 800, p 121

36. Martínez, E (2017) Best chinese smartwatch of 2017. http://epinium.com/blog/best-chinese-smartwatch/. Accessed Oct 2017

37. Amazon.co.uk (2017) K88h product description. https://www.amazon.co.uk/Waterproof-Dustproof-Notification-Smartphone-Silver-Black-Steel/dp/B01MSYUOFM/ref=sr_1_1?s=electronics&ie=UTF8&qid=1507642684&sr=1-1&keywords=k88h. Accessed Oct 2017

38. Bluetooth Specification (2003) RFCOMM with TS 07.10. Bluetooth SIG

39. Fernandes E, Jung J, Prakash A (2016) Security analysis of emerging smart home applications. In: 2016 IEEE symposium on security and privacy (SP), pp 636–654

40. Tekeoglu, A, Tosun AS (2015) Investigating security and privacy of a cloud-based wireless IP camera: NetCam. In: 2015 24th international conference on computer communication and networks (ICCCN). IEEE, pp 1–6

41. CCTV hack takes casino for $33 million in poker losses. https://www.theregister.co.uk/2013/03/15/cctv_hack_casino_poker/. Accessed 30 Mar 2018
42. Albrecht K, Mcintyre L (2015) Privacy nightmare: when baby monitors go bad [opinion]. IEEE Technol Soc Mag 34(3):14–19
43. Shekyan S, Harutyunyan A (2013) To watch or be watched: turning your surveillance camera against you. HITB Amsterdam
44. Alshalawi R, Alghamdi T (2017) Forensic tool for wireless surveillance camera. In: 2017 19th international conference on advanced communication technology (ICACT), pp 536–540
45. De Donno M, Dragoni N, Giaretta A, Mazzara M (2016) AntibIoTic: protecting IoT devices against DDoS attacks. In: International conference in software engineering for defence applications. Springer, pp 59–72
46. Dragoni N, Massacci F, Walter T, Schaefer C (2009) What the heck is this application doing? A security-by-contract architecture for pervasive services. Comput Secur 28(7):566–577
47. Bielova N, Dragoni N, Massacci F, Naliuka K, Siahaan I (2009) Matching in security-by-contract for mobile code. J. Logic Algebraic Program 78:340–358
48. Bonomi F, Milito R, Zhu J, Addepalli S (2012) Fog computing and its role in the Internet of Things. In: Proceedings of the first edition of the MCC workshop on mobile cloud computing, MCC 2012. ACM, New York, pp 13–16

An Open-Source Software Metric Tool for Defect Prediction, Its Case Study and Lessons We Learned

Bulat Gabdrakhmanov, Aleksey Tolkachev, Giancarlo Succi,
and Jooyong Yi[(✉)]

Innopolis University, Innopolis, Russia
{b.gabdrakhmanov,a.tolkachev,g.succi}@innopolis.ru, jooyongyi@acm.org

Abstract. The number of research papers on defect prediction has sharply increased for the last decade or so. One of the main driving forces behind it has been the publicly available datasets for defect prediction such as the PROMISE repository. These publicly available datasets make it possible for numerous researchers to conduct various experiments on defect prediction without having to collect data themselves. However, there are potential problems that have been ignored. First, there is a potential risk that the knowledge accumulated in the research community is, over time, likely to overfit to the datasets that are repeatedly used in numerous studies. Second, as software development practices commonly employed in the field evolve over time, these changes may potentially affect the relation between defect-proneness and software metrics, which would not be reflected in the existing datasets. In fact, these potential risks can be addressed to a significant degree, if new datasets can be prepared easily. As a step toward that goal, we introduce an open-source software metric tool, SMD (Software Metric tool for Defect prediction) that can generate code metrics and process metrics for a given Java software project in a Git repository. In our case study where we compare existing datasets with the datasets re-generated from the same software projects using our tool, we found that the two datasets are not identical with each other, despite the fact that the metric values we obtained conform to the definitions of their corresponding metrics. We learned that there are subtle factors to consider when generating and using metrics for defect prediction.

1 Introduction

Defect prediction has been researched actively due to its promise that it can predict which part of software is likely to be defective and as a result, improve software quality in a time-pressured business/development environment. Defect prediction is typically performed by learning a prediction model from training data and applying the learned model to the software under investigation.

B. Gabdrakhmanov and A. Tolkachev—These authors contributed equally to the work.

© Springer Nature Switzerland AG 2020
P. Ciancarini et al. (Eds.): SEDA 2018, AISC 925, pp. 76–85, 2020.
https://doi.org/10.1007/978-3-030-14687-0_7

The number of research papers on defect prediction has sharply increased for the last decade or so. One of the main driving forces behind it has been the publicly available datasets for defect prediction such as the PROMISE repository. These datasets make it possible for researchers to conduct various experiments on defect prediction without having to collect data themselves.

Indeed, according to the systematic literature review by Malhotra [7], datasets available through the PROMISE repository are used in 75% of the 64 investigated papers.[1] Another 12% of the papers also use another publicly available datasets extracted from Eclipse [17], and yet another 12% of the papers use publicly available datasets extracted from various open source projects such as Apache Ant and Lucene [5].

However, the practice of reusing datasets also poses new risks. First, there is a risk that the research efforts can be unwittingly customized to commonly used datasets. In other words, there is a potential risk that the knowledge accumulated in the research community is, over time, likely to overfit to the datasets that are repeatedly used in numerous studies. Second, when old datasets are reused, researchers often ignore the possibility that software development practices used in the field evolve over time, and these changes may potentially affect the relation between defect-proneness and software metrics.

Need for Open Source Software Metric Tools. The aforementioned potential risks can be addressd to a significant degree, if new datasets can be prepared easily. As a step toward that goal, we introduce an open-source software metric tool, SMD (Software Metric tool for Defect prediction) that can generate code metrics and process metrics for a given software project; SMD is available in the following Git repository: https://github.com/jyi/SMD. Apart from addressing the potential risks, an easy access to an open-source software metric tool can also facilitate new research and applications. For example, an open-source software metric tool can facilitate the "in vivo" experiments of defect prediction for the latest revisions of software projects. Furthermore, an open-source software metric tool can help researchers maintain the high integrity of their datasets, which is essential to guarantee the quality of empirical studies. Note that a breach of integrity is not unprecedented in the literature of defect prediction. For instance, Shepperd *et al.* [13] cleaned up the noisy NASA dataset that was used in numerous previous studies. Open-source software metric tools can help the research community verify and purify datasets.

The Status Quo of Software Metric Tools. Table 1 shows the list of software metric tools used in the literature, extracted from a systematic literature review on 226 papers about defect prediction [16]. The majority of the tools in Table 1 are not open source. In fact, none of these tools are not originally designed for defect prediction, and lack features for defect prediction, such as a capability to generate process metrics and identify defective past modules. To our knowledge,

---

[1] Note that the NASA datasets mentioned in [7] are available in the PROMISE repository.

Table 1. Software metric tools used in the literature of defect prediction (paraphrased from [16]). The last three columns indicate, respectively, whether the tool is open-source, whether the tool generates process metrics, and whether past buggy modules are labeled by the tool.

Tool	Open-source	Process metrics	Labeling
Understand	N	N	N
CKJM	Y	N	N
Metrics 1.3.6	N	N	N
JHawk	N	N	N
AOPMetrics	Y	N	N
Moose	Y	N	N
CCCC	Y	N	N
Borland together	N	N	N
src2srcml	Y	N	N
Analizo	Y	N	N
Rational software analyzer	N	N	N
SDMetrics	N	N	N
Featureous	N	N	N
CIDE	N	N	N

there is no currently available open-source software metric tool that generates both code metrics and process metrics and identify defective past modules.

Our Contributions (An Open-Source Tool). In this paper, we introduce an open-source software metric tool, SMD (Software Metric tool for Defect prediction), with the following features.

- SMD generates code metrics. Currently, it extracts the popular CK metrics [1] from Java programs. Unlike the existing tool CKJM [15] that requires compilation of Java source code to bytecode, SMD extracts metrics directly from source code.
- SMD generates process metrics. Unlike in many previous studies where process metrics are extracted from old-fashioned version control systems such as CVS and SVN, SMD extracts process metrics from Git repositories.

We note that the labeling feature (a variance of the SZZ algorithm [14] customized to Git-oriented development environments) is under development at the time of writing, and not considered in this paper.

Our Contributions (A Case Study). In an attempt to assess SMD, we conduct a case study where we compare existing datasets with those re-generated from the same software projects using SMD. We found that the two datasets are not identical with each other, despite the fact that our manual inspection shows that the metric values we obtained conform to the definitions of their corresponding metrics. We learned that there are subtle factors that can affect the values of metrics, as detailed in Sect. 4. This suggests that recent efforts in the research community to encourage reproducible research should also include data generation tools.

2 Description of SMD

Our open-source tool, SMD, generates both code metrics and process metrics. This section provides more detailed descriptions about SMD.

2.1 Extracting Code Metrics

SMD currently supports the well-known CK metrics suite [1], which has been used most frequently in the literature of defect class [16].

- **WMC (weighted methods per class)**. As a weight, we use value 1 for each method.
- **DIT (depth of inheritance tree)**. Since Java has an Object class as the parent to all other classes, the minimum value is 1.
- **NOC (number of immediate descendants of the class)**.
- **CBO (coupling between object classes)**. We consider coupling that occurs through method calls, field accesses, inheritance, arguments and return types.
- **RFC (response for a class)**. We consider only method calls from class's body, as done in CKJM [15].
- **LCOM (lack of cohesion in methods)**.

Currently, SMD extracts code metrics from Java programs. Internally, SMD exploits ANTLR4 [9] to extract raw data necessary to compute the metrics. Since SMD does not require compilation of Java files to bytecode—unlike CKJM [15] which has been often used in the literature—it can be more easily applied to previous versions of source code, without suffering from build dependencies. Also, the fact that ANTLR4 supports not only Java but also other languages such as Python and Javascript makes it manageable to extend SMD into other languages.

To extract code metrics, SMD requires the following input: the path of the source code, and an optional information about the output directory (by default, an output csv file is written in the current directory).

2.2 Extracting Process Metrics

SMD currently supports the following process metrics proposed by Moser [10] (except for the last one), which have been used in a number of studies (e.g., [2,6,8,11]). Currently, SMD extracts process metrics from Git repositories.

- **COMM (number of commits)**. Also known as REVISIONS or NR. We count it as number of lines with "commit SHA" in git log message.

- **NFIX (number of bugfixes).** Also known as BUGFIXES or NDPV. We increase the value each time we find "fix" in a commit message, while excluding "prefix" or "postfix".
- **NREF (number of refactorings).** We increase the value each time we find "refactor" in a commit message.
- **AUTH (number of distinct authors).** Also known as AUTHORS or NDC.
- **ALOC/RLOC (numbers of added/removed lines of code).** Also known as LOC_ADDED/LOC_DELETED. We parse a line from git log output, which contains data about number of added and removed lines of code.
- **AGE (age of a file in weeks).** We extract the age counted between two releases if the starting and the ending releases are provided. If only the ending release is provided, we extract the age between the first commit and the ending release.
- **NML (number of modified lines of code)** [4]. We parse a line from git log output, which contains data about number of changed lines of code and file name.

To extract process metrics, SMD requires the following input: the path of the Git repository, the path of the directory in which the project is cloned, source code file extensions, and the release or tag ID. If two release/tag IDs are given, the process metrics are extracted between these two releases/tags. If one release/tag ID is given, the process metrics are extracted between the first commit and the provided release/tag.

The process metric part of SMD is implemented by weaving Git commands using Python. While SMD currently supports only Git, other version control systems can also be supported by replacing Git commands with the corresponding commands of the other version control systems. We note that SMD is a part of an ongoing umbrella project, and we plan to extend SMD to support other version control systems and other programming languages.

2.3 Comparison Datasets

In an attempt to assess our tool SMD, we compare the datasets generated from SMD with the following two existing datasets: (1) the Jureczko dataset [4], and (2) the D'Ambros dataset [2]. The software projects in these datasets are shown in Table 2. Both datasets satisfy the following our dataset selection criteria.

1. These datasets are publicly available. The Jureczko dataset is available at http://snow.iiar.pwr.wroc.pl:8080/MetricsRepo/, and the D'Ambros dataset is available at http://bug.inf.usi.ch/download.php. The Jureczko dataset is also available in the PROMISE repository.
2. They have been used in the literature of defect prediction.
3. The metrics available in these comparison datasets overlap with those generated from SMD.
4. Source code in this dataset is written in Java. Note that SMD currently supports Java programming language.

Table 2. Software projects from which we extract metrics. All projects appear in the existing datasets used in the literature; their origins are shown in the "Origin" column. The last two columns show the release period used to extract process metrics, and the source code versions to extract code metrics.

Origin	Project	Website
Jureczko	Apache Lucene	lucene.apache.org
Jureczko	Apache Ant	ant.apache.org/
Jureczko	Apache Camel	camel.apache.org/
Jureczko	Ckjm	github.com/dspinellis/ckjm/
Jureczko	Apache Tomcat	tomcat.apache.org
Jureczko	POI	poi.apache.org
D'Ambros	Eclipse JDT Core	www.eclipse.org/jdt/core/
D'Ambros	Eclipse PDE UI	www.eclipse.org/pde/pde-ui/
D'Ambros	Equinox	www.eclipse.org/equinox/
D'Ambros	Apache Lucene	lucene.apache.org

Algorithm 1. Calculating similarity score

1: **function** CALCULATE-SIMSCORE(m_1, m_2) // $m_1 \geq 0$, $m_2 \geq 0$
2:     $score \leftarrow 0$
3:     **if** $m_1 = m_2$ **then** $score \leftarrow 100$
4:     **else** $score \leftarrow 100 * (1 - |m_1 - m_2|/Max(m_1, m_2))$
5:     **end if**
6:     **return** $score$
7: **end function**

---

5. The software projects in these datasets are being maintained in Git repositories. Note that SMD currently supports the Git version control system.
6. Information to reproduce these datasets (such as version/release numbers) are available in their accompanying papers.

3   A Case Study

3.1   Mean Similarity Score

In our case study, we measure the similarity between two datasets (i.e., the existing dataset and the re-generated dataset using SMD), by measuring the similarity between metric values using Algorithm 1. Given a metric value m_1 in the existing dataset and another value m_2 of the same metric in the re-generated dataset, Algorithm 1 computes the distance between them, and returns the similarity score between m_1 and m_2, 100 being no distance (thus, equal to each other) and a lower value being further distance. In this paper, we report the mean similarity scores averaged over all files in each software project under investigation.

Table 3. The mean similarity scores of the code metrics for projects in Jureczko dataset [4]. The values in parentheses represent the source code versions from which the code metrics are extracted.

Metric	Lucene (2.0, 2.2, 2.4)	Ant (1.6, 1.7)	Camel (1.0, 1.2, 1.4, 1.6)	Ckjm (1.8)	Tomcat (6.0.3)	POI (2.5.1, 3.0)
WMC	87.66	85.96	87.18	92.26	81.19	94.50
DIT	97.81	97.00	86.79	93.75	92.93	96.00
NOC	89.59	96.07	95.05	100.0	95.36	99.16
CBO	68.84	59.50	55.86	59.41	51.15	69.02
RFC	68.84	73.91	76.59	77.46	68.17	71.31
LCOM	64.87	57.65	59.18	81.12	53.53	62.51

Table 4. The mean similarity scores of the code metrics for projects in D'Ambros dataset [2].

Metric	JDT Core (3.4)	PDE UI (3.4.1)	Equinox (3.4)	Lucene (2.4.0)
WMC	49.73	62.70	65.86	57.21
DIT	89.48	82.41	92.20	95.93
NOC	93.09	98.35	99.03	98.72
CBO	49.31	58.27	61.41	50.62
RFC	81.32	79.59	82.23	81.80
LCOM	39.39	60.40	64.89	42.00

We measure the similarity using the mean similarity score mainly because of its conceptual simplicity. We mention that as an alternative similarity score, we could have used another distance calculating method with clearly defined boundary values, which could give us a clear representation of how good the resulting measurements are. For example, the cosine distance could take as input two vectors (first - recalculated metrics, second - benchmark metrics) and give us the result between 0 and 1. The closer result is to 0, the closer recalculated metrics is to the existing one. Note that using such scores is appropriate only in case of calculating distances between vectors.

3.2   Results

Tables 3, 4 and 5 show the mean similarity scores for code metrics and process metrics, respectively. We compute code metrics for the versions shown in Tables 3 and 4 (shown inside the parentheses), and report their mean similarity scores when compared with the metrics of the matching versions of the compared datasets. Meanwhile, we compute process metrics between the release periods shown in Table 5, and similarly report mean similarity scores. Our results show

Table 5. The mean similarity scores of the process metrics of the projects in the Jureczko dataset (the upper table) and the D'Ambros dataset (the lower table). The values in parentheses represent the release periods from which the process metrics are extracted. In the D'Ambros dataset, the starting date is the initial commit date. The NML metric of POI is not available in the Jureczko dataset.

Metric	Lucene (2.2.0 − 2.4.0)	POI (2.5.1 − 3.0)	Ant (1.3 − 1.4)	Camel (1.4.0 − 1.6.0)
COMM (NR)	15.83	82.46	98.98	54.63
AUTH (NDC)	15.83	73.05	90.02	54.54
NML	17.99	N.A	99.64	54.34
NFIX (NDPV)	50.36	80.12	93.31	74.78

Metric	JDT Core (− 3.4)	PDE UI (− 3.4.1)	Equinox (− 3.4)	Lucene (− 2.4.0)
COMM	76.42	80.23	23.86	2.31
NFIX	94.02	97.32	69.28	66.28
NREF	100	95.58	92.17	93.95
AUTH	88.40	86.21	29.10	2.31
ALOC	66.76	42.31	34.22	2.31
RLOC	82.70	83.41	37.73	2.31
AGE	98.83	86.44	29.82	2.31

that in general the re-generated datasets are not identical with the existing datasets, both in the code metrics and the process metrics. The degree of difference varies across metrics and projects. Note that similarity score can be low, even for small differences—e.g., the similarity score between metric values 1 and 2 is 50%. We manually checked our results for randomly sampled files, and our results conform to the definitions of the metrics we used. We discuss the sources of the differences in the next section.

4   Discussion

Sources of the Differences. One source of the differences is the different representation of the raw information from which metrics are extracted. More specifically, while SMD extracts code metrics from source code directly, the compared code metrics were extracted from Java bytecode. Meanwhile, SMD extracts process metrics from Git repositories, whereas the compared process metrics were extracted from SVN/CVS. The projects in the compared datasets were originally maintained in SVN/CVS repositories, and later migrated into Git. Note that at the time of migration, there is no guarantee that the original information is preserved in the new repository. Indeed, in Lucene and Equinox, significant information changes (both in files and logs) are observed after migration. In these projects, for example, many of the files are recreated and their original history information is accordingly lost. We suspect that low similarity scores for the process metrics in these projects are largely due to these information changes and losses.

Lessons We Learned. Overall, we learned the following lessons from our case study:

1. Reproducing the existing dataset is not as straightforward as we initially expected.
2. The efforts to reproduce the existing dataset are rewarding. We could find and fix problems in our tool while trying to reproduce the existing dataset. The results shown in this paper are our best-effort results.
3. There are subtle factors that can affect metric values such as the representation of code (whether it is source code or bytecode) and the kind of version control system from which metrics are extracted.
4. When version-control-system migration happens, process information is not necessarily preserved. In such cases, process metrics are likely to be severely distorted, losing information that can be useful for defect prediction.

Despite the utmost importance of the integrity of metric data in the research on defect prediction, there is little effort to check the reproducibility of the existing metric data. Nor are there substantial efforts to share metric generation tools and have them investigated publicly. Considering the fact that data necessary for defect prediction can be automatically extracted from a plethora of open source software projects, the current common practice of simply reusing the existing datasets without checking their integrity should be reconsidered. Encouraging researchers to share not only the dataset they generate but also tools they use to generate the dataset will help with improving the validity of defect prediction research results.

5   Related and Future Work

Software Metric Tools. To our knowledge, there is no currently available open-source tool that supports both code metrics and process metrics, as mentioned in Sect. 1. Though a process metrics generation tool called BugInfo is mentioned in the literature [4], its URL (http://kenai.com/projects/buginfo) is currently not active. Our tool is hosted in GitHub for permanent archival.

Integrity of Datasets. Shepperd *et al.* [12] conducted a meta-analysis to make sense of many conflicting research results on defect prediction. They found that research results are strongly affected by who conduct research and which datasets are used in the experiments. Their results suggest the importance of the reproducibility of results and the integrity of datasets. The importance of dataset integrity is also shown in the study of Ghotra *et al.* [3] where they observed that after using the cleaned NASA datasets, they obtained results different from the previous results obtained from the original NASA datasets.

In this work, we introduce an open-source software metric tool, SMD, that supports both code metrics and process metrics. We plan to further investigate the integrity of the existing datasets and our tool, in addition to planned extension of our tool.

References

1. Chidamber SR, Kemerer CF (1994) A metrics suite for object oriented design. IEEE Trans Softw Eng 20(6):476–493
2. D'Ambros M, Lanza M, Robbes R (2010) An extensive comparison of bug prediction approaches. In: Proceedings of MSR 2010, 7th IEEE working conference on mining software repositories. IEEE CS Press, pp 31–41
3. Ghotra B, McIntosh S, Hassan AE (2015) Revisiting the impact of classification techniques on the performance of defect prediction models. In: 37th IEEE/ACM international conference on software engineering, ICSE 2015, Florence, Italy, 16–24 May 2015, vol 1, pp 789–800
4. Jureczko M (2011) Significance of different software metrics in defect prediction. Softw Eng Int J 1(1):86–95
5. Jureczko M, Madeyski L (2010) Towards identifying software project clusters with regard to defect prediction. In: Proceedings of the 6th international conference on predictive models in software engineering, PROMISE, p 9
6. Madeyski L, Kawalerowicz M (2017) Continuous defect prediction: the idea and a related dataset. In: Proceedings of the 14th international conference on mining software repositories. IEEE Press, pp 515–518
7. Malhotra R (2015) A systematic review of machine learning techniques for software fault prediction. Appl Soft Comput 27:504–518
8. Osman H (2017) An extensive analysis of efficient bug prediction configurations. In: Proceedings of the 13th international conference on predictive models and data analytics in software engineering. ACM, pp 107–116
9. Parr T (2013) The definitive ANTLR 4 reference, 2nd edn. Pragmatic Bookshelf
10. Moser R, Pedrycz W, Giancarlo S (2008) A comparative analysis of the efficiency of change metrics and static code attributes for defect prediction. In: Proceedings of ICSE, the ACM/IEEE international conference on software engineering, pp 181–90
11. Rahman F, Devanbu PT (2013) How, and why, process metrics are better. In: 35th International conference on software engineering (ICSE), pp 432–441
12. Shepperd MJ, Bowes D, Hall T (2014) Researcher bias: the use of machine learning in software defect prediction. IEEE Trans Softw Eng 40(6):603–616
13. Shepperd MJ, Song Q, Sun Z, Mair C (2013) Data quality: some comments on the NASA software defect datasets. IEEE Trans Softw Eng 39(9):1208–1215
14. Śliwerski J, Zimmermann T, Zeller A (2005) When do changes induce fixes? In: ACM SIGSOFT software engineering notes, vol 30. ACM, pp 1–5
15. Spinellis D (2005) Tool writing: a forgotten art? (software tools). IEEE Softw 22(4):9–11
16. Varela ASN, Pérez-González HG, Martínez-Perez FE, Soubervielle-Montalvo C (2017) Source code metrics: a systematic mapping study. J Syst Softw 128:164–197
17. Zimmermann T, Premraj R, Zeller A (May 2007) Predicting defects for eclipse. In: Proceedings of the third international workshop on predictor models in software engineering

Community Targeted Phishing

A Middle Ground Between Massive and Spear Phishing Through Natural Language Generation

Alberto Giaretta[1(✉)] and Nicola Dragoni[1,2]

[1] Centre for Applied Autonomous Sensor Systems (AASS), Örebro University, Örebro, Sweden
alberto.giaretta@oru.se
[2] DTU Compute, Technical University of Denmark, Kongens Lyngby, Denmark
ndra@dtu.dk

Abstract. Looking at today phishing panorama, we are able to identify two diametrically opposed approaches. On the one hand, massive phishing targets as many people as possible with generic and preformed texts. On the other hand, spear phishing targets high-value victims with hand-crafted emails. While nowadays these two worlds partially intersect, we envision a future where Natural Language Generation (NLG) techniques will enable attackers to target populous communities with machine-tailored emails. In this paper, we introduce what we call Community Targeted Phishing (CTP), alongside with some workflows that exhibit how NLG techniques can craft such emails. Furthermore, we show how Advanced NLG techniques could provide phishers new powerful tools to bring up to the surface new information from complex datasets, and use such information to threaten victims' private data.

Keywords: Phishing · Security · NLG · Natural Language Generation

1 Introduction

Even though many years have passed since the Internet has come into our lives, some of its atavistic problems still have to be addressed. Acquisition of trustable information is increasingly important in both professional and private contexts. However, establishing what information is trustable and what is not, is still an open and challenging question [4]. The advent and success of phishing techniques is making the trust problem even more challenging, and we foresee that its severity could even worsen in the coming years.

On the one hand, we have a massive phishing approach where attackers fill simple templates with some basic information about the target, aiming to create simple and sound emails with the minimum effort possible.

On the other hand, we have *spear phishing* emails where attackers experienced in social-engineering techniques write articulate and specific emails. Even though these emails are carefully forged and hard to detect, the creation process

© Springer Nature Switzerland AG 2020
P. Ciancarini et al. (Eds.): SEDA 2018, AISC 925, pp. 86–93, 2020.
https://doi.org/10.1007/978-3-030-14687-0_8

requires such an effort that they are reserved only for high-level targets. Spear phishing techniques are more and more popular into hackers' toolboxes, as an example they are essential to strike *Advanced Persistence Threats (APTs)* [7].

Although they have the same goal, until today these two worlds have been somehow distinct from each other: massive production on one side, craftsmanship on the other one. We envision a future where a middle ground will be prominent, thanks to sophisticated production processes that go under the name of *Natural Language Generation* techniques, which will enable to target specific categories with machine-tailored messages. We call this new approach to phishing *Community Targeted Phishing (CTP)*, and we show in Fig. 1 an example of this evolution, as we foresee it.

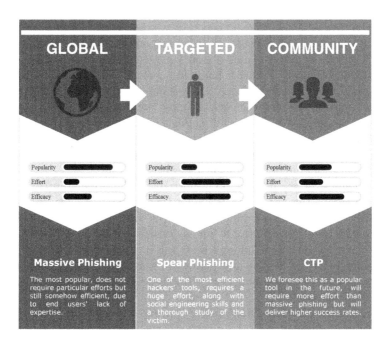

Fig. 1. Phishing naturally evolved from a massive approach to a manually hand-crafted tool, used by hackers to target high value victims. We foresee a midway approach in the future, which combines effectiveness and cheapness. Value bars present an estimate of each characteristic, not actual statistic data.

To the best of our knowledge, at the time of writing only one recent work about NLG applied to emails generation has been published [2]. In that work, authors show that NLG-based emails can be extremely effective to deceive people. Apart from such work, no other studies dig into how NLG fits into the big picture, nor how smart partitioning the targets could dangerously combine with NLG techniques. Strictly related, Natural Language Processing (NLP) goes in the opposite direction of NLG by extracting information from written texts, and it has been proposed as a solution to develop spam filters [9].

This paper is organized as follows: Sect. 2 gives a basic overview about NLG and its core concepts, useful to read through this paper. Section 3 digs into our predictions introducing our CTP concept with related workflows, in order to clarify how Template-Filling NLG could enable attackers to target the victims in the future. Then, Sect. 4 goes even further, making some hypothesis about how Advanced NLG techniques could enable attackers to gain victims' trust and increase the success rate of their phishing attempts. Finally, Sect. 5 wraps up our conclusive thoughts and lists some directions that we would like to take in the future.

2   Natural Language Generation (NLG)

First of all, what do we talk about when we talk about Natural Language Generation? A simple definition is: "Natural language generation (NLG) systems generate texts in English and other human languages" [6], yet this task is easier said than done. As a matter of fact, NLG systems have to make quite a number of different choices, in order to achieve the most suitable text for the specific purpose. As an example, psycho-linguistic models of language comprehension show that reflective pronouns usually improve a text readability and thus should be normally used, but such pronouns should be usually avoided in safety-critical texts, like operation manuals for nuclear power plants [6].

From a high-level perspective, NLG is traditionally divided in three macro-categories, namely *Canned Text*, *Template-Driven (or Template Filling)*, and *Advanced (or Proper) NLG*. Canned Text is the most basic approach, where a text is pre-generated (usually by a human being) and is used when the right moment comes. Even though different texts can be merged with some *glue texts*, this technique is hardly flexible and unable to adapt even to little domain variations. As previously pointed out [8], "Canned text is the borderline case of a template without gaps".

Talking about Template-Driven, this approach consists into defining some templates that exhibit *gaps*, to be filled with the correct information gathered from an external source, such as a database. The third approach, the Advanced NLG, is sometimes called proper NLG to highlight the fact that a NLG system implies a large variety of choices, and the two aforementioned approaches oversimplify the matter by using preformed texts. In particular, some authors argue that Template-Driven approaches are not proper NLG and tend to dismiss them, while others affirm that Template-Driven systems have been so much refined through the years that the distinction between them and pure NLG is too blurred to hold, nowadays [8].

No matter what kind of approach is chosen, even though NLG systems are far from being widespread, they have been used in many applications. One of the classic examples are weather forecasts, which are automatically generated from meaningful database entries; other fields implemented NLG in their routines, such as soccer, financial, technical, and scientific automatic reports.

3   Aiming at Groups: Community Targeted Phishing (CTP)

What massive phishing lacks is a deeper comprehension about recipients' relationships, characteristic which is totally neglected nowadays. Yet, dividing users in meaningful subsets could help phishers to impersonate some group members, and attack the others, in a much more effective way. This is not done yet, mostly because traditional phishing is still effective, but spam filters are proving themselves more and more reliable [1, 3]. One popular approach to develop such filters, is to apply the Bayesian filtering. In order to discern a legitimate email from a malicious one, this process checks every single word against a table, which contains the most commonly used words in spam emails. Even though advanced adversarial machine learning techniques can be used to poison the whole process [5], this approach is effective.

Such filters require training and leverage the strong points of phishing, which are massiveness and repetitiveness, to get statistically meaningful data. If phishers could smartly divide this huge user blob in smart subsets, they could forge more complex emails and deprive Bayesian spam filters of the aforementioned vital statistical data. We call this approach Community Targeted Phishing (CTP), and in the following subsection we provide some examples on how it could be applied.

3.1   Case Study: Scientific Community

One community which we are familiar with is the scientific one. As a matter of fact, researchers working in the same research field mostly share many interests and know quite a number of colleagues. Furthermore, the email exchange for collaborations, sharing of ideas and call for papers are the routine. On the average, if computer science researchers are enough knowledgeable about security and spam, members of other scientific fields usually exhibit very basic computer skills, which makes them easy targets for malicious attackers.

Scientists already receive a lot of scam emails from shady conferences and journals, which name resembles a legitimate venue and invite the recipients to submit their work. Sometimes people willingly submit to pay-for-publish venues, some other times they are fooled into believing that they are actually submitting to an honest venue. But these phishing emails are really simple in their form, as attackers just fill the first line with something along the line of "Dear Prof. Smith" without any kind of variation in the preformed text and without trying to impersonate an actual researcher. Furthermore, in many cases the attackers even fail to parse correctly the title, as it often happens for PhD Students to be addressed as Dr., Prof., or similarly wrong titles. Succeeding into impersonate a real person through an automated process, on the other hand, is a far more serious (and hard to achieve) threat and this is what we address in our paper.

As a case study, we propose a mechanism that fills two proposed templates by utilizing Template-Driven NLG techniques, and the related workflows that show how the whole email forging process could take place. As a matter of fact, we believe that even such basic templates could successfully fool a good number of people. The main problem that we see with this approach is that it still requires some human effort, to understand how a community usually interacts and what subsets should be excluded or included into the attack. As an example, in Subsect. 3.1 we will see that targeting very close colleagues with that particular template should be avoided.

Template 1. The first proposal takes advantage of the fact that researchers know many names in their field, but the relationships tend to be very weak. Yet, networking is an important part in scientists' lives, therefore a proposal from a respected name in the field could be an easy lure for a young researcher. The template is: "Dear [Colleague Name], I've read your recent work entitled [Manuscript Title] and I found it quite interesting. I've come up with some ideas about that same topic and I would be enthusiastic to work with you on this. At [Fake Google Drive Link] you will find a brief recap about my insights. Hope to hear from you soon, best regards [Scholar Name]". As before, the link will redirect the user to a fake login page which resembles Google login page. Figure 2 shows the workflow to produce the proposed text.

Template 2. The second approach is riskier because the attacker has to impersonate someone close to the victim. A forged email could sound way too formal (e.g., nicknames are normally used between acquaintances), but it could pay since colleagues often read privately each others' works, as a form of preliminary peer-review. The template is: "Hi [Colleague Name], as you know I'm working on [Topic Name] and I've written the attached paper. I would like to have some suggestions from you about it, since I'm a little uncertain about the solidity of the whole work. At [Fake Google Drive Link] you will find my draft. Thanks so much in advance, see you soon [Scholar Name]". As before, the link will redirect the user to a fake login page which resembles Google login page. Figure 3 shows the workflow to produce the proposed text.

4   Advanced NLG

In this section, we take one step more in our predictions and depict a scenario where Advanced NLG is used to create more convincing phishing emails.

As an example, an attacker could parse Google Scholar and look for researchers' h-index trend during the past 5 years and their performance with respect to their colleagues in the same field. With such data, the attacker could use NLG techniques to summarize the trends in a natural language and incorporate this new information in a phishing email. The result would be something along the line of "Dear [Target Name], I am Prof. [Reputable Name] from [Reputable University], we are currently expanding our lab and we are evaluating some possible candidates. I have personally checked your Google Scholar account and noticed that you h-index is high with respect of the average in your field, and considerably grew over the last 5-years span. Therefore, we would like to

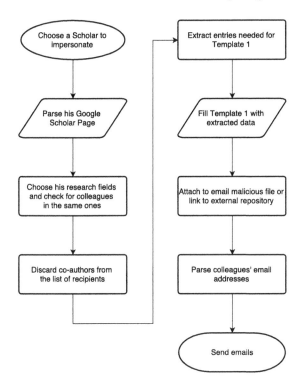

Fig. 2. Workflow of template-driven NLG applied to Template 1.

propose you a position. You can enrol at the following link: [Phishing Link]". If the victim is successfully tricked, they might fill the recruitment form and give up valuable information to the attacker.

Besides the research community, similar techniques could be applied to social network users. For example, an attacker could use NLG techniques to identify and summarize a specific interest of a social network account and forward a private message along the line of: "Hello [Victim], I've seen that you're really into [Music Genre]. I'm an event organizer and I would like to inform you that the next weekend, at the [Big Square] in [Victim Hometown], [Good Music Player] will give a short exhibition. All the details in the following link, feel free to attend! [Phishing Link]". The phishing link would be a fake login page which mimics the social network one, created by the attacker. If the victim believes that a disconnection due to technical issues happened, they might try to login again, which would result in stolen credentials.

Despite the very simplistic nature of the aforementioned examples, the capability of translating heterogeneous information into natural language text can play a huge part in the next future of automatic phishing generation. Indeed, through Advanced NLG techniques attackers do much more than creating automatic text: they pinpoint submerged information which would take a huge effort to manually discover.

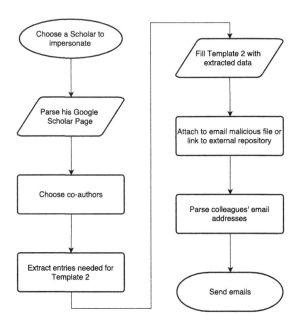

Fig. 3. Workflow of template-driven NLG applied to Template 2.

5   Conclusion

In this paper, firstly we have given a brief overview about the concept of Natural Language Generation (NLG) techniques and how these could be used to forge phishing emails. Moreover, we have expressed our concerns about how NLG could help hackers to easily target a community, such as the scientific one, proposing the concept of Community Targeted Phishing (CTP). We have provided two workflows that show how a Template-driven approach could be used to create phishing messages, and then we analysed how Advanced NLG techniques could enable phishers to achieve complex (thus, harder to spot) emails.

Future works will investigate different NLG techniques, in order to understand which ones provide the best results in terms of credibility and effectiveness. Moreover, we would like to investigate the variety that Advanced NLG techniques can ensure in text generation, as suggested in Sect. 4, and evaluate how traditional Bayesian spam filters perform against such phishing texts.

This work is just a first step, useful to introduce the very concept of CTP and raise awareness about a problem that might soon show up. We believe that using a NLG approach to target people with similar interests could be worthwhile, since it would allow to create emails more effective than the general ones, and cheaper than the spear phishing ones.

References

1. Almomani A, Gupta BB, Atawneh S, Meulenberg A, Almomani E (2013) A survey of phishing email filtering techniques. IEEE Commun Surv Tutor 15(4):2070–2090
2. Baki S, Verma R, Mukherjee A, Gnawali O (2017) Scaling and effectiveness of email masquerade attacks: exploiting natural language generation. In: Proceedings of the 2017 ACM on Asia conference on computer and communications security, ASIA CCS 2017. ACM, New York, pp 469–482. https://doi.org/10.1145/3052973.3053037
3. Blanzieri E, Bryl A (2008) A survey of learning-based techniques of email spam filtering. Artif Intell Rev 29(1):63–92. https://doi.org/10.1007/s10462-009-9109-6
4. Mazzara M, Biselli L, Greco PP, Dragoni N, Marraffa A, Qamar N, De Nicola S (2013) Social networks and collective intelligence: a return to the agora (chap. 5). In: Fagerberg J, Mowery DC, Nelson RR, Caviglione L, et al (eds) Social network engineering for secure web data and services. Oxford University Press, IGI Global, pp 88–113
5. Nelson B, Barreno M, Chi FJ, Joseph AD, Rubinstein BIP, Saini U, Sutton C, Tygar JD, Xia K (2008) Exploiting machine learning to subvert your spam filter. In: Proceedings of the 1st USENIX workshop on large-scale exploits and emergent threats, LEET 2008. USENIX Association, Berkeley, pp 7:1–7:9. http://dl.acm.org/citation.cfm?id=1387709.1387716
6. Reiter E (2010) Natural language generation. In: The handbook of computational linguistics and natural language processing. Wiley-Blackwell, pp 574–598. https://doi.org/10.1002/9781444324044.ch20
7. Sood AK, Enbody RJ (2013) Targeted cyberattacks: a superset of advanced persistent threats. IEEE Secur Priv 11(1):54–61
8. Van Deemter K, Krahmer E, Theune M (2005) Real versus template-based natural language generation: a false opposition? Comput Linguist 31(1):15–24. https://doi.org/10.1162/0891201053630291
9. Verma R, Shashidhar N, Hossain N (2012) Detecting phishing emails the natural language way. In: Foresti S, Yung M, Martinelli F (eds) Proceedings of computer security – ESORICS 2012: 17th European symposium on research in computer security, 10–12 September 2012, Pisa, Italy. Springer, Heidelberg, pp 824–841. https://doi.org/10.1007/978-3-642-33167-1_47

Semantic Query Language for Temporal Genealogical Trees

Evgeniy Gryaznov[✉] and Manuel Mazzara[✉]

Innopolis University, Innopolis, Russian Federation
e.gryaznov@icloud.com, m.mazzara@innopolis.ru

Abstract. Computers play a crucial role in modern ancestry management, they are used to collect, store, analyse, sort and display genealogical data. However, current applications do not take into account the kinship structure of a natural language. In this paper we propose a new domain-specific language KISP, based on a formalisation of the English's kinship system, for accessing and querying traditional genealogical trees. KISP is a dynamically typed LISP-like programming language with a rich set of features, such as kinship term reduction and temporal information expression. Our solution provides a user with a coherent genealogical framework that allows for a natural navigation over any traditional family tree.

1 Introduction

With the advent of computers, we are able to manage genealogies of incredible size. Now, due to the progress in storage engineering, it became possible to maintain and expand existing ancestries of considerable size. But merely keeping data on physical disks is not enough. To sufficiently realize the full potential of computers in genealogy management, one should also provide means of inquiry in ancestral data.

There is another important concept to consider when working with family trees, namely, the concept of time. A genealogy can exist only in specific temporal framework that is imposed on it by the very nature of history itself. As a result, any computer representation of an ancestry that lacks this framework is exorbitantly inadequate. Therefore, its preservation is a crucial feature for any software that is aimed for effective genealogy management.

If we want to teach computers understand lineage, we need to construct some type of artificial language that will allow us to effectively navigate and query any possible family tree. But observe, that an ancestry already has its own idiosyncratic terminology and grammar, which can be successfully utilized as a natural basis for such a language. Our research is an attempt to do exactly that.

© Springer Nature Switzerland AG 2020
P. Ciancarini et al. (Eds.): SEDA 2018, AISC 925, pp. 94–109, 2020.
https://doi.org/10.1007/978-3-030-14687-0_9

2   Related Work

Ontologies have been used in several application areas (including by the authors of this paper [1]) and find their natural application in the context of our work. Since the original formulation of the concept, software has been developed to manage ontologies, including such systems as Protege, Inge and others. These systems have already been heavily used in the variety of different fields. For instance, Tan [2] designed and implemented a genealogical ontology using Protege and evaluated its consistency with Pellet, HermiT and FACT++ reasoners. He showed that it is possible to construct a family ontology using *Semantic Web* [3] technologies with full capability of exchanging family history among all interested parties. However, he did not address the issue of navigating the family tree using kinship terms.

Ontologies can be used to model any kind of family tree, but the problem arises when a user wants to query his relatives using kinship terms. No standard out-of-the-box ontological query language is able to articulate statements such as in our example above. Although an ontology can be tailored to do so, it is not in any way a trivial task. Marx [4] addressed this issue, but in the different area. He designed an extension for XPath, the first order node-selecting language for XML.

Lai and Bird [5] described the domain of linguistic trees and discussed the expressive requirements for a query language. They presented a language that can express a wide range of queries over these trees, and showed that the language is first order complete. This language is also an extension of XPath.

Artale et al. [6] did a comprehensive survey of various temporal knowledge representation formalisms. In particular, they analysed ontological and query languages based on the linear temporal logic LTL, the multi-dimensional Halpern-Shoham interval temporal logic, as well as metric temporal logic (MTL). They noted that the W3C standard ontological languages, OWL 2 QL and OWL 2 EL, are designed to represent knowledge over a static domain, and are not well-suited for temporal data.

Modelling kinship with mathematics and programming languages, such as LISP, has been an extensive area of research. Many people committed a lot of work into the field, including Bartlema and Winkelbauer, who investigated [7] structures of a traditional family and wrote a simple program that assigns fathers to children. Their main purpose was to understand how this structure affects fertility, mortality and nuptiality rates. Although promising, small steps has been made towards designing a language to reason about kinship. Also, their program cannot express temporal information.

Another prominent attempt in modelling kinship with LISP was made [8] by Findler, who examined various kinship structures and combined them together to create a LISP program that can perform arbitrary complex kinship queries. Although his solution is culture-independent, he did not take the full advantage of LISP as a programming language, and because of that it is impossible to express queries which are not about family interrelations. In contrast, our system does not suffer from that restriction.

More abstract, algebraic approach was taken [9] by Read, who analysed the terminology of American Kinship in terms of its mathematical properties. His algebra clearly demonstrates that a system of kin terms obeys strict rules which can be successfully ascertained by formal methods. In another article [10] he discusses how software, in the broader sense of intelligent systems, can help anthropologists understand foreign cultures.

Periclev et al. developed [11] a LISP program called KINSHIP that produces the guaranteed-simplest analyses, employing a minimum number of features and components in kin term definitions, as well as two further preference constraints that they propose in their paper, which reduce the number of multiple component models arising from alternative simplest kin term definitions conforming to one feature set. The program is used to study the morphological and phonological properties of kin terms in English and Bolgarian languages.

According to authors [9] and [11] there have been many attempts to create adequate models of kinship based on mathematics and computational systems. For instance, a prominent french mathematician Weil analysed [12] the Murngin system of kinship and marriage using *group theory*. In particular, he showed how one can embed the nuptial rules of a particular society into the framework of permutation groups S_n. His work was later extended by Bush [13], who proposed the concept of *permutation matrices* as a more effective tool for analysis. Kemeny, Snell and Thompson were [14] first to systematize the properties of prescriptive marriage systems as an integrated set of axioms. All distinct kinship structures which satisfy these axioms were systematically derived and described by White [15], whose more practical *generators* set the structural analysis on a more concrete basis.

Similar attempt was made by Boyd, who also used [16] the apparatus of group theory to give a mathematical characterization of the conditions under which groups become relevant for the study of kinship. He argued that the concept of group extension and its specialization to direct and semi-direct products determine the evolutionary sequences and the coding of these kinship systems.

Gellner discusses [17], from the pure philosophical point of view, the possibility of constructing an *ideal language* for an arbitrary kinship structure.

However, despite remarkable progress, there is a certain doubt in the mathematical community about the applicability of such abstract approaches to the study of kinship. For example, White [15] recognized the failure of his structural analysis of societies like Purum or Murngin, which practice matrilateral cross-cousin marriage. Liu addresses [18] this problem with establishing a new mathematical method for the analysis of prescriptive marriage systems.

We also note that none of the works include the temporal element into their formalisms, which provides novelty for our research.

3   Formal Language of Kinship

The study of kin structures has its roots in the field of anthropology. Among the first foundational works was Henry Morgan's *magnum opus* "Systems of

Consanguinity and Affinity of the Human Family" [19], in which he argues that all human societies share a basic set of principles for social organization along kinship[1] lines, based on the principles of **consanguinity** (kinship by blood) and **affinity** (kinship by marriage).

Following Henry Morgan, we recognize two primary types of family bonds: marital (affinity) and parental (consanguinity). These bonds define nine basic kin terms: *father, mother, son, daughter, husband, wife, parent, child and spouse.* Observe that combining them in different ways will yield all possible kinship terms that can and do exist.

For instance, *cousin* is *a child of a child of a parent of a parent* of a particular person. Another example: *mother-in-law* is just *a mother of a spouse.*

Now let us represent a traditional Christian family tree as a special type of *ontology* with its' own concepts, attributes, relations and constraints. Concepts are people in a family, their attributes are: *name, birth date, birthplace, sex* and relations are parental and marital bonds with a wedding date.

Together with the everything stated above, we have the following cultural constraints imposed on our genealogy:

1. Each person can have any finite number of children.
2. Each person can have at most two parents of different sex.
3. Each person can have at most one spouse of different sex.
4. A spouse cannot be a *direct relative*, i.e. a sibling or a parent. In other words, direct incest is prohibited.

When considering those prerequisites one should bear in mind that we deliberately focused only on rules, taboos and customs of one particular culture, namely American culture in the sense of Read [9]. Under different assumptions and in further studies, these conditions can be relaxed and revisited.

Apart from these four, here are two additional temporal constraints that express the interrelation between birth and wedding dates:

1. No one can marry a person before he or she was born, i.e. a wedding date can only be strictly after a birth date of each spouse.
2. A parent is born strictly before all of his (her) children.

Due to the general nature of these two constraints, they are always true in every culture and therefore can be safely assumed in our work.

Every genealogy that meets these six requirements we shall call a **traditional family tree**. As the name "tree" suggests, we can indeed view this structure as a graph with its vertices as people and edges as bonds. Observe that every kinship term corresponds exactly to a *path* between ego and specified relative. Under such view, kin term becomes a set of instructions, telling how to get from the starting vertex A to the end vertex B. For example, consider the term *mother-in-law*. What is it if not precisely a *directive*: "firstly, go to my spouse, then proceed to her mother". The wonderful thing is that, due to the nature of kinship terminology, we can *compose* old terms together to create new, even

---

[1] Recall that in this paper, the word "kinship" includes relatives as well as in-laws.

those which do not have their own name. This simple observation shows that we can see kinship terms as paths in a family tree underpins our entire research.

Now, if we want to efficiently query a traditional family tree, we need to further investigate the mathematical features of the language of kinship terms.

Here we present our attempt to model the language of traditional American, in the sense of Read [9], kinship terminology. There are three main characteristics that define every formal language: its syntax (spelling, how words are formed), semantics (what does particular word mean) and pragmatics (how a language is used).

3.1   Syntax

We use Backus-Naur Form to designate the syntax for our formal language. Let Σ be the set of six basic kinship terms: *father, mother, son, daughter, husband, wife.* Then we can express the grammar as follows:

$$term ::= \Sigma | (term \cdot term) | (term \vee term) | (term)^{-1} | (term)^{\dagger}$$

The first operation is called *concatenation*, second – *fork*, third – *inverse* and the last – *dual.* We denote this language by $\mathcal{L}$.

Here are some examples of ordinary kinship terms expressed in our new language. Note that we deliberately omit superfluous parentheses and the composition sign for the sake of simplicity:

- Parent is $father \vee mother$.
- Child is $son \vee daughter$.
- Brother is $son(father \vee mother)$.
- Sibling is $(son \vee daughter)(father \vee mother)$.
- Uncle is $son(father \vee mother)(father \vee mother)$.
- Daughter-in-law is $daughter \cdot husband$
- Co-mother-in-law is $mother(husband \vee wife)(son \vee daughter)$

From these examples you can see the real power of this language – the power to express all possible kinship terms. Now the important step towards solving our main goal, developing a language for managing temporal genealogies, is to assign meaning to these words. From now on we distinguish between *artificial* kinship terms, i.e. well-formed terms of our formalization, and *natural* kin terms used in ordinary English. By referring to just terms, we mean the former, if nothing else is stated.

3.2   Semantics

Let Σ^* stand for the set of all possible kin terms generated from the basis Σ using the previously defined syntax. Let $\mathcal{G} = (V, E)$ be a traditional family tree with V as a set of its vertices (people) and E as a set of its edges (bonds). Moreover, because $\mathcal{G}$ is traditional, every person from the set V have the following attributes:

- A father. We will denote him as $father(p)$, a function that returns a *set* containing at most one element.
- A mother. We will denote her by $mother(p)$.
- A set of his or her children: $children(p)$.
- A set of his or her sons:

$$son(p) = \{c|c \in children(p) \land Male(p)\}$$

- A set of his or her daughters:

$$daughter(p) = \{c|c \in children(p) \land Female(p)\}$$

- A spouse: $spouse(p)$.
- A husband:

$$husband(p) = \{s|spouse(s) \land Male(p)\}$$

- A wife:

$$wife(p) = \{s|spouse(s) \land Female(p)\}$$

Due to the cultural constraints stated above, result-set of $father$, $mother$, $spouse$, $husband$ and $wife$ can contain at most one element.

Now we are ready to introduce **Denotational Semantics** for Σ^*. This name was chosen because it highly resembles its namesake semantics of programming languages. Note that we regard kinship terms as *functions* on subsets of V. Each function takes and returns a specific subset of all relatives, so its type is $f : \mathcal{P}(V) \to \mathcal{P}(V)$.

We proceed by induction on the syntactic structure of $\mathcal{L}$. Let t be an element of Σ^*, then:

1. If $t \in \Sigma$, then $[\![t]\!] = F(t)$, where $F(t)$ assigns to each basic kin term its corresponding function from the list 3.2.
2. Term concatenation is a composition of two functions:

$$[\![(t_1 \cdot t_2)]\!] = [\![t_1]\!] \circ [\![t_2]\!]$$

3. Fork is a set-theoretic union of results of its sub-functions:

$$[\![(t_1 \lor t_2)]\!] = p \mapsto [\![t_1]\!](p) \cup [\![t_2]\!](p)$$

4. Term inverse is exactly the inverse of its function:

$$[\![t_1^{-1}]\!] = [\![t_1]\!]^{-1}$$

The *dual* operator (†) is more difficult to define. We want it to mean exactly the same as the term, where the gender of each its basic sub-term is reversed, e.g. dual of "uncle" is "aunt", dual of "brother" is "sister" and so on. Here we can use induction once again:

1. If $t \in \Sigma$, then $[\![t]\!] = D(t)$, where $D(t)$ is a basic term of opposite sex.
2. Dual is distributive over concatenation, i.e. dual of concatenation is a concatenation of duals:
$$[\![(t_1 \cdot t_2)^\dagger]\!] = [\![(t_1^\dagger \cdot t_1^\dagger)]\!]$$
3. Dual is distributive over forking:
$$[\![(t_1 \vee t_2)^\dagger]\!] = [\![(t_1^\dagger \vee t_2^\dagger)]\!]$$
4. Inverse commutes with dual:
$$[\![(t^{-1})^\dagger]\!] = [\![(t^\dagger)^{-1}]\!]$$

Observe that we also have the distributivity of concatenation over forking. This semantics allows us to efficiently navigate any family tree.

3.3   Pragmatics

Now, when syntax and semantics has been defined, let's discuss the applications of our new formalism.

One of the main goals in constructing our language was achieving *cultural independence*. That is, the language should assign a unique term to every relative in genealogy without relying on labels and kinship words from a particular society. Although we use terms such as `father`, `daughter` and `husband` from English for the basis of the language, we do that only for readability, since they can be easily replaced with abstract placeholders like x or y.

The proposed formalism finds its natural application in the field of *machine translation*. Languages drastically, and often even incompatibly, differ in the way they express kinship information. The correct translation of kinship terms still poses a challenge for linguists and anthropologists. For example, in Russian there is a word for a son of a brother of ones' wife, but no corresponding term in English. With our formalism we can encode the meaning of such words and use it to provide a more adequate translation between any possible pair of natural languages.

We can also apply the formalism to the problem of *cross-cousin marriage*: given the description of a particular society, a genealogy of a family from that society and two individuals from it, ascertain whether the rules of their community allow them to marry. This problem is extensively studied in the field of computational anthropology. What makes it especially difficult is that each culture has its own peculiar set of regulations and laws regarding this subject. In every case, one of the important steps towards solving the problem is to research and establish a correct model of that society. With our new formalism we can facilitate this process.

4   Term Reduction

Our artificial language has a problem: its too verbose. Indeed, to encode such ubiquitous kin terms as "uncle" or "great-nephew" one must use quite lengthy phrases that are hard to write and read. It is therefore important to have some sort of reduction mechanism for our language that will shorten long terms into a small set of common kinship relations to help understand them.

Firstly, let us analyse the problem. We have the following mapping ω between Σ^* and the set of English kinship terms $\mathcal{W}$

$$son(father \lor mother) \mapsto \text{brother}$$
$$daughter(father \lor mother) \mapsto \text{sister}$$
$$father(father \lor mother) \mapsto \text{grandfather}$$
$$mother(father \lor mother) \mapsto \text{grandmother}$$
$$son(son \lor daughter) \mapsto \text{grandson}$$

$$\vdots$$

$$father(son \lor daughter)(wife \lor husband)(son \lor daughter) \mapsto \text{co-father-in-law}$$

This dictionary allows us to effectively translate between kin terms of our artificial language $\mathcal{L}$ and their usual English equivalents. We can also view this mapping as a *regular grammar* in the sense of Chomsky hierarchy [20]. However, note that we strictly prohibit mixing these two collections and therefore we deliberately avoid using words from the RHS in the LHS, because otherwise the grammar will lose its regularity and become *at least* context-free, making the problem even more challenging. Let us define another function on top of ω that will replace the first sub-term $u \subset t$ in a term $t \in \Sigma^*$:

$$\Omega_u(t) = t[u/\omega(u)]$$

Here the only change in meaning of $t[u/\omega(u)]$ is that the substitution takes place only once.

Now the task can be stated thusly: given a term $t \in \Sigma^*$ find its shortest (in terms of the number of concatenations) translation under Ω, i.e. which sub-terms need to be replaced and in what order.

This problem can be easily reduced to that of finding the desired point in the tree of all possible substitutions. Moreover, this vertex is actually a *leaf*, because otherwise it is not the shortest one. But the latter can be solved by just searching for this leaf in-depth. Unfortunately, the search space grows exponentially with the number of entries in the dictionary ω, thus making the naive brute-force approach unfeasible.

Here we propose a heuristic greedy algorithm shwon in Fig. 1 that, although does not work for all cases, provides an expedient solution to the reduction problem in $O(n^2)$ time. Firstly, it finds the longest sub-term u that exists in the dictionary ω, then divides the term into two parts: left and right from u, after that it applies itself recursively to them, and finally it concatenates all three sub-terms together.

Now let's analyze the time complexity of this algorithm:

Theorem 1. *The execution time of the algorithm listed in Fig. 1 belongs to* $\Theta(n^2)$.

```
1   input: kinship term t.
2   note: ω is a dictionary of kinship terms.
3   note: "leftPart(t, u)" and "rightPart(t, u)" return the sub−term of t
4   note: from the left of sub−term u or from the right respectively.
5   note: Function "subterm(t, i, j)" returns the
6   note: sub−term of the kinship term t between indices i and j.
7   output: reduced kin term.
8   function shorten(t)
9   begin
10          maxShortenableSubterm ← empty
11          currentSubterm ← empty
12          for i ← 0 to length(t) do
13          begin
14              for j gets length(t) − i to 0 do
15              begin
16                  currentSubterm ← subterm(t, i, j)
17                  if length(currentSubterm) > length(maxShortenableSubterm)
18                      and ω(currentSubterm) is not empty
19                  then
20                      maxShortenableSubterm = currentSubterm
21              end
22          end
23          return shorten(leftPart(t, maxShortenableSubterm))
24              · ω(maxShortenableSubterm)
25              · shorten(rightPart(t, maxShortenableSubterm))
26   end
```

Fig. 1. Kinship term reduction

Proof. Let $T(n)$ be the execution time of the algorithm, where n stands for the number of concatenations in a kin term. First of all, observe that $T(n)$ obeys the following recurrence:

$$T(n) = 2T(n/2) + O(n^2) \tag{1}$$

Indeed, we make a recursive call exactly *two* times and each call receives roughly the half of the specified term. During execution the function passes through two nested cycles, so one call costs us $O(n^2)$.

Secondly, to solve a recurrence, we use the **Master Method** from the famous book *Introduction to Algorithms* [21] by Cormen et al. In our case $a = 2$, $b = 2$, and $f(n) = O(n^2)$. Observe, that if we take ϵ to be any positive real number below one: $0 < \epsilon < 1$, then $f(n) = \Omega(n^{\log_b a + \epsilon}) = \Omega(n^{1+\epsilon})$.

Let us show that $f(n)$ satisfies the *regularity* criterion: $af(n/b) \leqslant cf(n)$ for some constant $c < 1$. Indeed, just pick $c = 1/2$:

$$2f(n/2) \leqslant cf(n),$$

$$2\frac{n^2}{4} \leqslant cn^2,$$

$$\frac{1}{2}n^2 \leqslant cn^2,$$

$$\frac{1}{2}n^2 \leqslant \frac{1}{2}n^2$$

Thus, we can use the third case from the Master Method, which tells us that $T(n) = \Theta(n^2)$.

4.1   Pursuing Confluence

The current approach has one major disadvantage: like any other greedy algorithm it can fail to choose a correct reduction path between two terms with equal amount of concatenations. We can alleviate this by augmenting our rewriting system, based on ω dictionary, with a feature called *confluence*, also known as *Church-Rosser* property:

Definition 1. *An abstract term rewriting system is said to possess* **confluence**, *if, when two terms N and P can be yielded from M, then they can be reduced to the single term Q. Figure 2 depicts this scenario.*

Not only we can fix our reduction algorithm by introducing this property, but also we can improve the time complexity, making it linear.

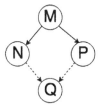

Fig. 2. Confluence in a term rewriting system.

One way to achieve confluence is to attach a single kinship term to any possible path in a family tree. Observe, that English kinship terms have a specific pattern that we can exploit. All relatives who are distant enough from ego have the following structure of their kin term:

$$n^{th} \text{ cousin } m^{th} \text{ times removed}$$

In-laws also have their own pattern, where the ending "-in-law" is appended to a valid consanguine kinship term. However, this applies only to people, who are linked together by only one nuptial bond. For instance, there is no single term for a husband of ego's wife's sister. These relations can be accounted for by *prefixing* "-in-law" with an ordinal, which shows the number of marital bonds that one should pass in order to go to such person. Under this representation, last example will receive the term "brother *twice*-in-law". After generalizing that scheme we get a pattern that looks like this:

$$\langle \text{Consanguine kinship term} \rangle k^{th} \text{ times-in-law}$$

We can also view this as an attribution of a distinct *natural number* to every vertex with ego as an *offset*, thus imposing a natural ordering on the set of all vertices. This assignment can be made in such a way that reducing a kinship term n will correspond *exactly* to the calculation of n from some arithmetic expression like $5 \cdot (2+3)+4$, thus providing a **translation** between the language of all valid arithmetic expressions and our formal language of kinship $\mathcal{L}$.

However, it is not the topic of this paper, so we are leaving it to the considerations of future researches.

5   Incorporating Time

Now the only matter that is left to address is an adequate representation of time. Historically, there are two main approaches for modelling time: point-based and interval-based. The former treats time as a single continuous line with distinguished points as specific *events*, and the latter uses *segments* of that line to represent time entries. The latter method was used in Allen's interval algebra [22]. For the sake of simplicity we chose the former approach, because it can easily imitate intervals by treating them as endpoints of a line segment.

Not only we want to be able to express different events by modelling them as points on a line, but also we want to orient ourselves on that line, i.e. to know where we are, which events took place in the past and which will happen in the future. Thus, we need to select exactly one point that will stand for the present moment and we call it "now". Then all points to the left will be in the past, and all point to the right will be in the future. Also, notice that any set with total ordering on it will suffice, because the continuous nature of a line is redundant in point-based model. Collecting everything together, we have the following formalisation of time:

$$\mathcal{M} = \langle T, now, \leqslant \rangle$$

Where T is a non-empty set with arbitrary elements, $now \in T$, and $\leqslant$ is a total ordering relation on T.

Within this model we can reason about which event comes *before* or *after*, what events took place in the past or in the future, and so on.

When considering family trees it is necessary to define only five predicates:

1. $Before(x, y)$ is true iff $x < y$.
2. $After(x, y)$ is true iff $x > y$.
3. $During(x, s, f)$ is true iff $s \leqslant x \leqslant f$.
4. $Past(x)$ is true iff $Before(x, now)$.
5. $Future(x)$ is true iff $After(x, now)$.

Those relations are the basis from which all other operations on $\mathcal{M}$ can be defined. It is also interesting to note that, since any ordering relation generates a *topology* over its structure, we can speak about time in terms of its topological properties.

6   KISP Language Specification

6.1   Grammar and Lexical Structure

Since KISP is a dialect of LISP, it inherits some syntax from the predecessor, but generally it is a new programming language. The grammar is presented with the help of Bacus-Naur notational technique. For the sake of simplicity, we omit angle brackets and embolden non-terminal words. A plus, a star sign in a superscript and a question mark have the same meaning as in regular expressions.

$$\textbf{term} ::= \textbf{literal}|\textbf{lambda}|\textbf{define}|\textbf{atom}|(\textbf{term}^{+})$$

$$\textbf{lambda} ::= (\text{lambda } (\textbf{reference}^{*}) \textbf{ term})$$

$$\textbf{define} ::= (\text{define } \textbf{reference term})$$

$$\textbf{reference} ::= \textbf{word}\{\text{-}\textbf{word}\}\text{'?'?}$$

$$\textbf{word} ::= \textbf{letter}^{+}$$

$$\textbf{letter} ::= a|b|...|z|A|B|...|Z$$

$$\textbf{atom} ::= *|+|\text{concat}|\text{list}|\text{append}|...$$

$$\textbf{literal} ::= \text{void}|\text{true}|\text{false}|\text{people}|\text{vacant}|\text{now}|\textbf{numeral}|\textbf{string}$$

$$\textbf{numeral} ::= \text{-?}\textbf{digit}^{*}$$

$$\textbf{digit} ::= 0|1|...|9$$

$$\textbf{string} ::= \text{'}\textbf{symbol}^{*}\text{'}$$

$$\textbf{symbol} ::= \textit{any non-blank ASCII symbol}$$

As we can see from the definition, there are three kinds of terminals in the grammar: literals, references and atom functions, which are called simply **atoms**. Literals are instances of primitive types, such as *Numeral*, *String* or *Boolean*, or special keywords. They stands for the following: "void" represents NULL type, "people" – a list of all persons in a family tree, "now" – the current time entry and "vacant" – an empty list. References are used as definientia in "define" terms and as names for parameters in lambda terms. It is possible for a reference to end in a question mark, which means that it denotes an instance of *Boolean* type. References can we written in so-called *dash case*, so "long-name" and "very-long-name" are both legal. The only exception are names which start with the dash like "-illegal", they are invalid.

Note that we allow niladic lambdas, so, for instance, this is a valid expression: (lambda () 'Hello, World!'). But at the same time () is not a well-formed term. We also prohibit "define" terms inside other terms, so this would not work: (+ 2 (define three 3)). Strings are nested in single quotes. Integers, in KISP we call them "numerals", can start with a zero and be prefixed by a negative sign.

Here is the complete list of all keywords in KISP: **true, false, define, lambda, people, now, void, if, vacant**. The rule is that you can use as a reference everything you want as long as it is not a keyword, so you cannot redefine their standard behaviour, thus a programmer is unable to tamper with inner workings of the interpreter.

As in all other dialects of LISP, a term (f a b c ...) means the *execution* of a function f with the specified arguments $f(a, b, c, ...)$. Of course, we can construct and call the higher-order functions as usual: ((twice square) 2) will yield 16, or ((compose inc inc) 0) which just prints 2.

6.2   Query Examples

In this section we will demonstrate how one can use KISP to perform various queries in a genealogical tree. Particularly, we focus our attention on statements that express kinship terms.

Let's start with a simple task of selecting people based on a certain boolean condition. Suppose we want to query only those, who have at least one child. This can be accomplished as follows:

```
(filter (lambda (p) (< 0 (count (children p)))) people)
```

Here we iterate through the list of all people in a tree and take only those, on who defined lambda predicate evaluated to **true**. The number of children for a particular person is calculated by counting elements of the list (children p).

The next task is to select all husbands, that is, all men who are married. This can be done in two ways: either select only males and then discard all bachelors, or combine the two operations together in a single boolean predicate using **and** clause:

```
(filter (lambda (p) (not (= void (spouse p))))
        (filter (lambda (p) (= 'MALE' (attr p 'sex')))
        people))
(filter (lambda (p) (and (= 'MALE' (attr p 'sex'))
                         (not (= vacant (spouse p)))))
people)
```

The advantage of the second approach is that the list **people** will be iterated only once.

Now to the more advanced queries; suppose that the term **ego** stands for the user's node in an ancestry, and he wants to know how many cousins he has:

```
(define parents (lambda (p) (join (mother p) (father p))))
(define cousins
(lambda (p) (children (children (parents (parents p))))))
(- (count (cousins ego)) 1)
```

This is where the expressive power of KISP truly comes into play. Although cousins is not a standard KISP function, we can easily implement it using kinship framework of KISP, which successfully utilizes the structure of natural kinship terms. Moreover, notice how the function **parents** is expressed. Since a parent is either a mother or a father, it corresponds to the formal kinship term $(mother \lor father)$, which is implemented as a *join* of two or more lists. And because every cousin is a grand-child of one's grandparents, it corresponds to:

$$(son \lor daughter)(son \lor daughter)(mother \lor father)(mother \lor father)$$

The last decrement was made because in this scheme the **ego** itself will be included to the resulting list.

Finally, temporal queries can be expressed with the help of the type *Date*. For instance, if we need to know, who, among our relatives, was born during the WWII, we just need to evaluate:

```
(define WWII-start (date '01.09.1939'))
(define WWII-end (date '02.09.1945'))
(filter (lambda (p) (during (attr p 'birthdate') WWII-start WWII-end))
people)
```

The type *Date* provides all the necessary functions for working with temporal information.

Because KISP is Turing-complete and inherits LISP's capabilities of meta-programming, one can easily extend it with any functionality that one wants.

7   Conclusion

In this work we designed a new programming language KISP for effectively navigating and querying temporal family trees. We described a formal mathematical model of traditional kinship, on which KISP is based. Additionally, we tackled the problem of term reduction and discussed the possibilities for achieving confluence.

There are three requirements that we want our language to satisfy:

1. **Expressiveness.** The language should allow for any possible consanguine as well as affinal relations to be described.
2. **Speed.** The response time must not exceed the standard for an interpreted language.
3. **Simplicity.** Language should be able to express natural kinship and temporal terms as straightforward as possible.

Here the phrase "response time" stands for the time passed between the start and finish of a programs evaluation. The last quality is what truly distinguishes our approach from the rest, allowing for the most obvious representation of genealogical and temporal information. We are certain that our presented solution fully covers every one of them.

However, there are some topics yet left to tackle in the area of kinship and genealogy management. On the theoretical side, there is a problem of total term reduction and formal language enrichment. It is also interesting to shift attention to other languages and cultures with different kinship structures, such as Russian or Hawaiian. The constructed formalism can be considered from the algebraical side, focusing on its many mathematical properties as a special type of an algebraic system.

On the practical side, one can consider to improve the virtual assistant component. Besides already mentioned Voice Generation & Recognition technology, it can be made context-aware, which will increase its intelligence. Additionally, the family ontology can be enhanced to incorporate information about divorces and deaths. The performance of the interpreter can be significantly improved by introduction of *Just-in-Time* compilation.

Another important step towards improvement of existing system is addressing its current limitations, such as static nature of our genealogical database and execution time of KISP queries. It's no doubt that the schema-less approach is much more versatile, because it does not depend on traditions and customs of a particular culture.

Further advancement may also include new data types and standard functions for KISP language. Specifically, it is beneficial to add a **char** type that represents individual characters in a string. Another useful feature is support for *variadic lambdas* and *closures*, which will significantly increase the versatility of KISP.

Moreover, one can also consider including capabilities for a logical reasoning into KISP. They will be applicable for inferring implicit time constraints for events, whose exact date is unknown. For instance, if we are uninformed about a birthday of a person, but we do know his parents and his children birthdays, we can justifiably bound this missing date to a specific time interval.

Applications in the medical field could be studied, in particular in the monitoring and assistance of the elderlies and in domotics [23]. The ability to add health-related information to the ancestry management may enable to formulate queries providing useful answers. From the point of view of programming paradigms, features related to the (micro)service paradigm could be considered in order to extend the applicability of the language to service-based contexts [24].

References

1. Yan Z, Cimpian E, Zaremba M, Mazzara M (2007) BPMO: semantic business process modeling and WSMO extension. In: 2007 IEEE international conference on web services (ICWS 2007), 9–13 July 2007, Salt Lake City, Utah, USA, pp 1185–1186
2. Tan MT (2015) Building a family ontology to meet consistency criteria. Master's thesis, University of Tun Hussien
3. Bailey J, Bry F, Furche T, Schaffert S (2005) Web and semantic web query languages a survey
4. Marx M (2004) XPath the first order complete XPath dialect
5. Lai C, Bird S (2009) Querying linguistic trees
6. Artale A, Kontchakov R, Kovtunova A, Ryzhikov V, Wolter F, Zakharyaschev M (2017) Ontology-mediated query answering over temporal data: a survey. In: 24th international symposium on temporal representation and reasoning, vol 1, no 1, pp 1–37
7. Bartlema J, Winkelbauer L (1961) Modelling kinship with lisp; a two-sex model of kin-counts. In: IIASA working papers, vol WP-96-069, no 1, p 48
8. Findler NV (1992) Automatic rule discovery for field work in antropology. Comput Hum 26(1):285–292
9. Read DW (1984) An algebraic account of the american kinship terminology. Curr Antropol 25(49):417–429
10. Read D (1990) Computer representation of cultural constructs: new research tools for the study of kinship systems. Comput Apl Anthropol 1(1):228–250
11. Periclev V, Valdes-Perez RE (1998) Automatic componental analysis of kinship semantics with a proposed structural solution to the problem of multiple models. Anthropol Linguist 40(2):272–317
12. Weil A (1969) On the algebraic study of certain types of marriage laws, pp 221–229
13. Bush RR (1963) An algebraic treatment of rules of marriage and descent. In: White HC (ed) An anatomy of kinship, Appendix 2. Prentice-Hall, Englewood Cliffs

14. Kemeny J, Snell L (1972) Mathematical models in the social sciences. MIT Press, Cambridge
15. White HC (1963) An anatomy of kinship. Prentice-Hall, Englewood Cliffs
16. Boyd J (1972) The algebra of group kinship. J Math Psychol 6(1):139–167
17. Gellner E (1975) Ideal language and kinship structure. Philos Sci 24(3):235–242
18. Liu P (1973) Murngin: a mathematical review. Curr Anthropol 14(1–2):2–9
19. Morgan LH (1870) Systems of consanguity and affinity of the human family. Smithsonian Institution, Washington, D.C
20. Chomsky N (1956) Three models for the description of language. IRE Trans Inf Theory 2(1):113–124
21. Cormen TH, Leiserson CE, Rivest RL, Stein C (2009) Introduction to algorithms, 3rd edn. The MIT Press, Cambridge
22. Allen JF (1983) Maintaining knowledge about temporal intervals. Commun ACM 26(11):832–843
23. Nalin M, Baroni I, Mazzara M (2016) A holistic infrastructure to support elderlies' independent living In: Encyclopedia of E-health and telemedicine. IGI Global
24. Guidi C, Lanese I, Mazzara M, Montesi F (2017) Microservices: a language-based approach. In: Present and ulterior software engineering. Springer

Recruiting Software Developers a Survey of Current Russian Practices

Vladimir Ivanov$^{(\boxtimes)}$, Sergey Masyagin, Marat Mingazov, and Giancarlo Succi

Innopolis University, Universitetskaya St, 1,
420500 Innopolis, Respublika Tatarstan, Russia
{v.ivanov,s.masyagin,m.mingazov,g.succi}@innopolis.ru

Abstract. Software is mostly a "people" business: the single and most important asset of a software company is its developers. Finding an appropriate software developer is a problem that has created the whole area (and business) of IT recruiting, which is mostly an "art" involving a set of practical techniques and approaches. This paper discusses the typical practices of IT recruiting in Russia. Our aim is to describe its baseline from the point of view of its practitioners and to study ways to improve it. To achieve this aim we conducted a survey among more than 70 professional recruiters, we analysed the results, and we have identified mechanisms to improve it, also partially automating it. The results of our research appear to be applicable and beneficial for professionals and researchers living in several other parts of the world, given the evident similarity of the market structure.

Keywords: Software industry · Recruiting · Human resources · Survey · GQM

1 Introduction

It is well known that the single, most important component of a successful software endeavor is the presence in the development team of skilled and motivated software developers and engineers; such people are in very high demand all across the world. According to the survey conducted by PageGroup[1] in 2016, twenty four countries had severe needs for software engineers. On one side, IT jobs are expanding world-wide much faster than the number of college students graduating in IT sector [17] and on the other, the process of searching and recruiting suitable candidates is becoming more difficult.[2] Thus, building a strong IT team is very hard.

Perhaps, the best solution to the issue is a significant changes in IT education. However, worldwide improvement of IT education is a long lasting activity, while

---

[1] http://www.michaelpage.co.uk/minisite/most-in-demand-professions/.

[2] Jobvite, http://www.jobvite.com; LinkedinTalent, https://business.linkedin.com/talent-solutions.

© Springer Nature Switzerland AG 2020
P. Ciancarini et al. (Eds.): SEDA 2018, AISC 925, pp. 110–127, 2020.
https://doi.org/10.1007/978-3-030-14687-0_10

immediate solutions for business are provided by recruiting. Therefore, improvement of IT recruiting appears the only viable approach in the short term. Still, very few systematic approaches to recruiting software developers and engineers have been proposed and investigated so far [13,17,24,31]; research in this area is scarce [18,23,29].

Such apparent lack of studies could be explained by an ad-hoc nature of the hiring process. Indeed, large IT companies have their own practices, that sometimes become legendary, quite like religion. Google, for instance, puts a special committee in charge of making hiring decisions; and it does not delegate this task to HR managers [26]. For medium IT companies and startups, the recruitment process is also ad-hoc; it involves certain amount of tacit expert knowledge from both domain experts and HR-experts. The knowledge stays with the individuals having it and does not become a common asset of the recruitment agency or the IT company. Thus, systematic studying the area is hindered, but it is very important to understand the current HR practices.

In this paper we study what are the advantages of such practices, their limitations, and whether there is a possibility for improving current hiring processes using automated systems and taking advantages of information stored in public source code repositories. We study Russian IT market, which is also representative of a large number of cases throughout the world: as will be detailed later, it is very similar to the one in Ireland, China, India, and so on, and consequently our considerations can be applied to a large number of situations.

The investigation is based on a survey we have ran in collaboration with Russian recruiting agencies and IT companies. The survey involved 56 recruiters. Some of these recruiters do not operate only in Russia, but also in Ukraine (16%), Belarus (23%), European Union (12.5%), and the United States (7%). Our empirical investigation has evidenced that most recruiters follow a very similar (and simplistic) approach:

1. advertising the vacancies on job sites,
2. collecting and reviewing resumes, and
3. interviewing the most promising candidates in order to make a final decision.

We found, however, that all such tasks are performed manually and, thus, very time consuming. Some recruiters use specialized platforms for developers such as Stackoverflow, Github, and Bitbucket. Still, the main obstacle to the diffusion of such approach is technical structure of such platforms and the skewed data present on them. For people without technical background it is not easy to use the platforms and to interpret content. The most promising approaches to recruiting appear in the line of extensive collecting and mining large datasets about developers as well as development of various applications, e.g. recommendation services, on top of these datasets.

This paper is organized as follows. Section 2 introduces the results of recent studies of IT-recruiting. Then, the design and the administration of the survey are described (Sect. 3) and the collected results are presented (Sect. 4). Finally, the validity of the findings is discussed (Sect. 6) and the lines for future research and development are drawn along with conclusions in Sect. 7.

2   The State of the Art

In general, there is a vast body of research on recruiting. Most works cover general issues of the recruiting process, discuss ideas and best practices in the whole area, but they do not touch IT-specific issues. There is also a substantial number of studies covering modern trends, such as e-recruiting, social media recruiting [2,6,7] and mobile recruiting [5,10]. However, there is very little fraction of research papers on recruiting in the IT labor market, including the methods of hiring software developers and other IT specialists.

Most issues in IT recruiting are similar across countries, where a gap between supply and demand in IT professionals constantly grows. Searching for candidates becomes more and more difficult task. This, in turn, forces recruiters to apply new approaches into recruitment process. While improvement of IT education gives long term results, studying and facilitating IT recruiting gives an immediate effect.

In 2011 Lang et al. presented a comprehensive survey of the last 20 years of research in this area [20]. They reviewed 80 journals and proceedings articles and found 23 research papers discussing drivers, challenges and consequences of e-recruiting, on the base of which they proposed a suitable model aiming at[3]:

– reduction of recruitment costs,
– increasing of the number of applicants, and
– reduction the processing time for each application.

A fundamental contribution to IT recruiting research made in the work of Agarwal and Ferratt [1]. They focused on collecting best practices from 56 senior IT and HR executives from 32 successful companies. The practices were grouped in 4 categories:

– sourcing (source from which a need for IT professionals is met),
– skills sought (mix of skills sought to identify prospective hires for the IT organization),
– competitive differentiation elements (basis upon which the IT organization positions itself as an attractive place to work during its recruiting efforts), and
– one-time inducements (any inducements offered on a limited basis at the time an offer is made to attract a potential hire to the company).

An interesting pattern was revealed in the study: successful organizations tend to exploit partnerships and help their partners locate potential hires. These partnerships frequently include relationships with academic institutions from which students could be hired. It was reported that up to 60% of new hires came from internal referrals. Now, this practice is evolving into usage of social media and locating potential hires through the contacts from social networks as well as professional networks.

---

[3] These three point are clearly represent the most significant goals in IT recruiting.

Social recruiting is one of the fastest growing trends mentioned by Amadoru et al. [2]. The global spread of social networks made a significant change in methods of searching and analysis of candidates. Billions of posts, messages and photos posted by users in social networks, helping recruiters. In software engineering, there are professional social media resources that can be used by recruiters: (open) source code repositories (such as Github), issue trackers, bug databases, Q&A services (such as Stackoverflow), etc [12, 19, 28]. These repositories have the potential of supporting early screenings of candidates before an actual interview takes place.

Thus, in the context of social recruiting, IT recruiting typically turns into a specific form, due to existence of open source code repositories and collaborative development. Indeed, in recruiting IT professionals the social media services and software repositories play an important role. Popularity of the collaborative software development platforms (such as Github) allows recruiters to overcome main obstacle they face during pre-selection (or screening) of candidates: assessment of professional skills of developers.

However, browsing software repositories to find developers with good profiles requires additional technical skills; a recruiter must understand technical terms such as 'version control systems', 'commit', 'pull request'. This gap could be covered by special tools for skills-aware analysis of source code repositories. Unfortunately, such tools have limitations due to the lack of data [4].

Development activity on Github has been studied by Marlow and Dabbish [22]. They considered Github profile as a set of signals, each of which has a certain level of 'reliability'. The more 'reliable' signals the developer has, the more interest of recruiters he should attract. Another feature of a signal is the 'ease of verifiability'. Authors identified the following typical groups of signals:

- active open source involvement,
- contributions accepted to high status project,
- project ownership,
- side projects, and
- number of watchers or forks of project.

These signals imply a certain set of competences and skills of a person involved in a Github project (e.g., 'project ownership' signal imply 'initiative, project management' skills). This inference could be not purely reliable, but still it may address some of recruiter's questions. A limitation of the study is the sample size. There were 65 volunteers who agreed to discuss the topic of GitHub and recruiting. Out of these 65 people only 7 were employers who actively use Github to search and assess candidates.

It is clear that issues of IT recruiting are insufficiently investigated. In Russian IT recruitment scientific studies are also rare. Most works represent case studies for a particular hiring practice.

3   Design of the Survey

To the best of our knowledge, there is no holistic study of Russian IT recruiting processes. Thus, the main goal of our work is twofold: (a) to describe a baseline of IT recruiting to raise awareness of it in scientific community and to study ways to (partially) automate it; (b) to identify areas for any improvement with respect to the points stated in the work of Lang et al. [20].

The exploration of IT recruiting in Russia is a challenge due to a lack of explicit expert knowledge. This essentially distributed knowledge mostly exists as recruiters' experience. Therefore, the first step of our investigation was to understand a recruiting process in general.

Our study of the IT recruiting process begins with building a model of the recruiting process as a sequence of steps that are applied by recruiters at certain time. The whole process is guided by a customer's need which is usually represented as a necessity to close a position. Given the goal, IT recruiter carries out activities required to fulfill the customer's need. Hence, successful recruiters first apply hiring techniques aimed at revealing what customer needs and which skills a candidate should have to get a position.

We have identified approaches and techniques that IT recruiters use during searching for software developers. After that, we have used the Goal, Question and Metric (GQM) approach proposed by Basili et al. [3]. By means of GQM we have designed a survey of Russian IT recruiters and collected data about particular steps of the process.

3.1   Analysis of Typical IT Recruiting Process

Here we aim at a reconstruction of typical process of recruiting software engineers and developers. We assume that hiring procedure consists of several steps preformed by most of recruiters. To build such a common view we interviewed HR experts to reveal the steps of a typical recruiting process in IT. Based on the preliminary interviews we have identified five general steps of IT recruitment process:

1. identification of customer's (or an employer's) needs,
2. selection of a recruitment strategy,
3. formation of a "long list" of candidates,
4. dissemination of the call,
5. final check of a candidate and closing a position.

3.2   Design of a Questionnaire: Goal Question Metric

We consider the five steps as a general view shared by most IT recruiters. The actual sequence of the steps depends on environment and context. We will use these steps as a skeletal framework to guide a questionnaire design. The questionnaire focuses on the most important challenges in IT recruiting and analysis of time consumption. The goal is to facilitate understanding of the IT recruiting in order to find the areas for possible improvements.

The questionnaire design is guided by the Goal Question Metric (GQM) approach. This approach was successfully applied in recent works [15,25] aimed at software process and product measurement. According to the GQM, we first define goals, then we design questions and metrics that help measurable evaluation of the selected goals. Definition of the goal is a crucial to the GQM approach. The goal is usually formulated using a template in which all important parts of the investigation are present. The template includes three main parts: Purpose, Perspective an Environment. We formulate the goal as follows.

Goal: Analyze the IT recruiting process for the purpose of improvement with respect to cost and time consumption from the point of view of the HR company in the following context: quickly growing economy of a developing country with high demand and low supply of IT personnel.

The purpose part is quite clear, as time factor is the key to successful hiring. In the perspective part of the goal we focus on HR company, but it could be a perspective of a single IT recruiter as well. The environment part of the goal formulation is justified by the previous discussion. Given the goal we propose the following two questions.

Question 1: How much investment is required for companies to proceed with recruitment of new employees?

Question 2: What challenges IT-recruiter is faced to when looking for developers?

These two questions should be addressed to the primary source of the information, i.e. IT recruiters. The first question aims at the measurement of steps of the process, with respect to the time consumption. Here we will use the framework with five general steps and measure each of them using a questionnaire. The second question aims at reveling the hardest issues IT recruiters are faced to, which could be used for improvement of the process. Following the GQM approach, using the questions, we define a set of metrics that should produce measurable results. The first set of metrics tailored to the Question 1. These metrics facilitate quantitative analysis of the recruiting process, especially in terms of time consumption and performance. The following list of metrics correspond to the first question.

Metric 1: Time spent developing a portrait of the required candidate.

Metric 2: Time required (on average) to generate a "long-list" of candidates.

Metric 3: Number of resumes (on average) considered per month.

Metric 4: Tools for verification of the information provided in candidates' CVs.

Metric 5: Number of vacancies closed per month?

The second set of metrics helps to reveal challenges experienced by IT-recruiters. The following three metrics tailored to the Question 2 and are related to challenges of recruitment process.

Metric 6: The most difficult, most time-consuming parts of the process.
Metric 7: Kinds of services which are most popular when searching for candidates.
Metric 8: The necessity in a web-service enabling search for developers.

Obviously, the metrics as well as the outcomes of the metrics are subjective, but we doubt that it is possible to run a purely objective measurements in recruiting domain. In order to decrease the extent of subjectivity, for each metric we defined a fixed and exhausted list of possible values.

This enables a certain amount of consistency in the survey. Finally, the metrics can be naturally transformed into a questionnaire that consists questions to recruiters. As we prefer to ask closed questions, the list of questions and possible answers (options) is presented in Table 1.

Table 1. Questions for the survey with options

A question from the survey	Possible answers
How much time do you spend on making a portrait of the candidate you are looking for?	One day One week One month
How much time do you spend on average on the formation of "long-list" candidates?	1–2 days One week One month More than one month
How many resumes (on average) do you consider for a month?	50 100 500 More than 1000
How do you verify the credibility of the candidates' CVs?	Traditional interview Phone call Recommendations Online tests Online services Other
How many vacancies do you close per month?	1–10 10–100 More that 100

The questionnaire we use to address the Question 2 is the following.

- What are the most difficult problems in your work that takes most of time?
- What services do you use most often when searching for candidates?
- Which tasks do you consider most appropriate for automation in a web-service?

For the sake of space we do not provide all options for these three questions here; we list them in the Sect. 4.6 along with the distribution of answers. We intentionally made the list of questions short. First of all, it is more likely to get answers on a short survey than on a longer one. Second, we focus our study on the time/effort aspect and main challenges of the recruiting process; this short questionnaire represents serves this goal well.

3.3 Collecting Data for the Survey

First, we have collected contacts of potential participants of the survey form social networks. In Russia the following social networks are most popular: LinkedIn,[4] VKontakte, MoiKrug and Facebook. We searched each network with keywords "IT-recruiter" and "IT-recruitment" and have retrieved 300 personal e-mails. Finally, we have sent the survey and received responses in a time frame of one month: from February 15, 2016 till March 15, 2016.

We received answers to the survey from 56 HR managers. More than a half of them (32) did not share their contacts and left anonymous. We have analyzed open social profiles of the remaining 24 recruiters and collected the demographics information here. Out of 24 recruiters 23 are from Russia. Interesting that these recruiters work on both Russian (82%) and on CIS IT markets: Ukrainian (16%) and Belorussian (23%). We analyzed their social accounts in order to collect information about years of experience, and company names. Years of experience ranges from 3 to 9, but most values are around 6 years, plus or minus 1 year.

The following list contains companies names of participants. Mail.Ru Group, Crossover, DigitalHR, EPAM Systems, RNG & Associates, M.Video, MGCom, Spice IT, Ok-konsalt, Odnoklassniki, Appodeal, Deluxe Interactive Moscow, NVIDIA, atsearch, Yandex, Acronis, Astrosoft, Innopolis University, Harman, Kelly Services, Prologics, Wargaming.net, and few others.

4 Results of the Survey

In this section we represent results of the survey. We go step by step through all the major steps of recruiting process and analyze answers we received from the respondents.

4.1 Identification of the Need of the IT Company (An Employer's Need)

At this step, a recruiter receives from a client exhaustive description of a required specialist, including competences and skills, personal qualities and experience. This initial step is very important because it can dramatically affect the following steps and final result of the process. A common tool to gather details about required candidates is an "application card". It contains a set of questions, the

---

[4] At the time of the survey web-site of LinkedIn was available in Russia.

answers to which should give an precise view of a required specialist's profile. In the simplest case, application cards exist in the form of simple paper documents. In other cases, application cards can be represented as a record in an automated system for applications collection or another software for HR management.

According to results of the survey (Fig. 1), 66% of respondents spend 1 day to generate a portrait of the desired candidate, 28,5% of respondents said that they spend 1 week (here we also include four recruiters who spend more than 1 day[5]). Less then 2% of respondents need 1 month to generate candidate portrait and 3,5% did not answer the question.

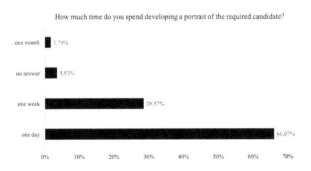

Fig. 1. Most IT recruiters spend less than a week to create a portrait of a required candidate

4.2   Selection of Recruitment Strategy

Given a vacancy description a recruiter needs either to post a job vacancy online, or to look for relevant candidates in online services. In general, the following web resources to post a job vacancy are of interest: bulletin boards, mass media, internal services for candidate search, web services, and social networks. Meanwhile, a more efficient way is to look for a relevant candidate at professional web services and special job-sites for software developers like job4it, it-jobs, it-stars, github, stackoverflow, etc. In the questionnaire we provided respondents with 6 options to select to answer the question about their preferences. These options correspond to categories of online services (see Table 2).

The question about the services permits multiple answers, therefore here we expected to investigate the popularity of HR-services and other tools among the IT-recruiters (see Fig. 2).

According to results of our survey, the majority of recruiters from Russia (96.4%) use "traditional" HR-services to search for candidates (such as LinkedIn, HeadHunter and MoiKrug), which are not specialized on IT recruiting. Most of

---

[5] For instance, one of respondents wrote "In general, it takes 3–4 days, but it does not happen in a moment: first, forming an application, then discuss it, then generate job description, then correct it, then after first interviews correct job description again".

Table 2. Categories of online services used by IT recruiters in Russia

Category	Online services
Business-oriented services	LinkedIn, Viadeo, XING, professionali.ru, MoiKrug, etc.
Job-sites	headhunter.ru, job.ru, superjob.ru, rabota.ru, jobsmarket.ru, etc.
Web-based source code repositories	github, google code, bitbucket, etc.
General purpose social networks	BranchOut (Facebook), TalentME (Facebook), SmartStart (VKontakte), etc.
Job-sites for IT professionals	job4it.net, it-jobs.com.ua, it-stars.com, developers.org.ua, webdev.org.ua, etc.
Other	Slack, justLunch, sql.ru, recommendations from colleagues, searching in own database, participating conferences and events, etc.

respondents use job-sites (76.8%). Many respondents (33.9%) use social networks for searching candidates that illustrate the general trend of social recruiting.

Interesting that 37.5% of IT-recruiters use web-based source code repositories for searching relevant candidates, but only 21.4% of respondents actively use professional web services. The group "others" (16.1%) includes answers that do not correspond to initially selected groups, for example, visiting events, using corporate messengers such as Slack, and mobile applications such as justLunch.

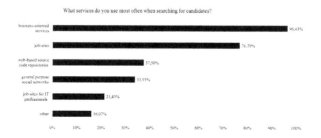

Fig. 2. Most popular online services of IT-recruiters in Russia

4.3 Formation of "a Long List" of Candidates

At this step, the main goal of the recruiter is to create an initial list of candidates who have required skills. In general, there are two approaches can be used to achieve this goal: active search and passive search. One of the most effective ways to attract Russian IT-specialists is headhunting. Luring a specialist from one company to another is often used to find the most competent IT professionals. Indeed, this is a common practice in foreign markets as well. Actually,

Russian developers frequently look for jobs abroad. Thus, headhunting considered as a worldwide practice. In addition to headhunting, Russian IT-recruiters use methods such as recruiting preliminaring, screening and executive search. The goal of these methods is to allow the recruiter quickly identify those candidates who can satisfy the requirements of a position.

According to the results of survey, generating a long list of candidates takes not more than one week. However, sometimes this process may take up to a month or more (Fig. 3), especially when looking for the most competent specialists.

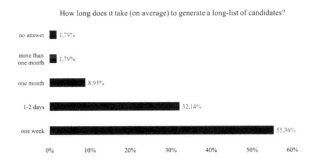

How long does it take (on average) to generate a long-list of candidates?

Fig. 3. Most IT recruiters spend about a week to create a long list of candidate.

We also asked respondents: "How many resumes (on average) do you consider per month?". The answers show that the majority of respondents (69%) are browsing more than 500 resumes per month, or 15–20 resumes per day on average (Fig. 4).

4.4   Dissemination of the Call

After completion of a long list, a recruiter analyzes all resumes again and selects candidates to call for an interview. In 2007 Cole et al. [8] studied the problem of assessing candidates using their resumes. The study revealed that recruiters' perceptions of a job applicant employability typically relate to three sections of a resume: academic qualifications, work experience, and extracurricular activities. During a traditional interview a recruiter has two goals. First, to verify the information provided in the resume. Second, to understand whether a candidate has the skills for the current job (listed in an application card).

In order to explain how recruiters verify the credibility of the candidates' CVs, we provide the following answers (Fig. 5). According to results of our survey, recruiters use the following methods to evaluate the candidate's skills: a traditional interview, a telephone call, recommendation letters and other references, online tests and online services for skills assessment. In the group "other"

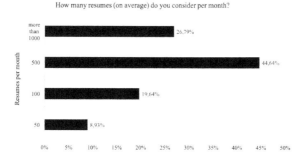

Fig. 4. Average number of resumes reviewed by Russian IR-recruiters monthly

respondents have specified the following methods and tools: evaluation in the agency by a senior-developer, practical assignments and written technical tests, interview via Skype, e-mails, and even delegating the verification process to the customer's company.

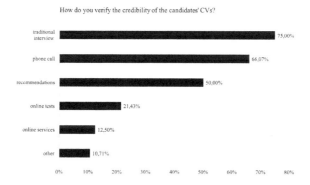

Fig. 5. Methods for verification of candidates' CVs

A disadvantage of a traditional type of interview is its subjectivity: an opinion of a recruiter about competences of the candidate is very significant. Therefore, recruiters also use different types of interviews such as behavior interviews, stress interviews and brainteaser interviews.

4.5 Final Checking of a Candidate and Closing a Position

Looking for suitable candidates is only a part of recruiting. Another main concern of any recruiter is bringing a candidate into the company and providing all information relevant to finalize hiring decision (e.g. represent skills and results of candidate assessment). In-house recruiters are responsible for new employee adaptation to the company, and ensure the employee will not suddenly quit leaving the position open. In other words, the ultimate goal is to close given position with the most suitable candidate.

In our study, we evaluated the number of positions closed per month, which is one of the most important business metrics of recruiter's performance. Almost all IT-recruiters (94.6%) from Russia close from 1 up to 10 positions per month. It is only 2% of the total number of candidates found and present in long lists of the recruiters. This fact indeed needs an attention. Clearly, there are problems that cause such low performance.

4.6   Analysis of Difficulties of IT Recruitment

In order to find the roots of the low performance, we asked respondents: "What part of your work is most difficult, most time-consuming?". Summarizing responses, we have identified and ranked the following difficulties:

– checking skills of developers (66%),
– hiring those who are not looking for a job at the moment (30%),
– too few developers with the required experience and skills (15%),
– analyzing candidates' activity in GitHub, Bitbucket, etc. (14%), and
– searching information about the candidate in social networks (11%).

Finally, we asked respondents: "Would you like to use a web service enabling search for developers?". Most respondents (77%) gave a positive answer. This leads to a conclusion that the search for talented IT professionals in Russia is an urgent problem and there is a high demand in services gathering information about candidates' skills. The use cases important for automation in such a service are listed below:

– searching for candidates with similar CV (71.4%),
– ranking the developers according to the relevance to a job (57.1%),
– analysing the source code of candidates (35.7%),
– analysing the history of closed positions (32.1%),
– analysing the candidate activities on the web (30.4%), and
– comparing several candidates (23.2%).

In summary, based on the analysis of the IT recruiting domain, we assembled a diagram of activities, resources and challenges (Fig. 6). It is clear from the diagram that later steps of the IT recruiting process typically take more time, because they involve more complex decisions than the steps in the beginning. The amount of input information grows fast; usual long lists of candidates contain more than 500 resumes. However, there is a lack of precise signals and tools to make right hiring decisions in a timely manner.

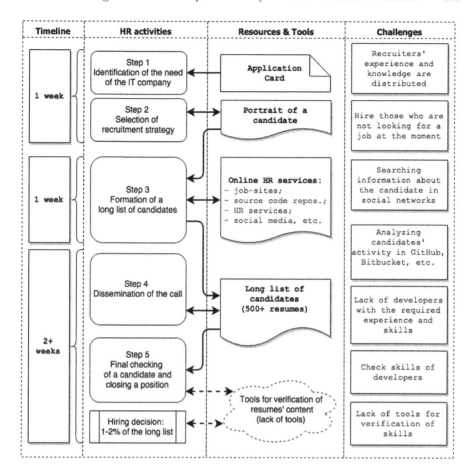

Fig. 6. View of the IT-recruiting process and its challenges

5   Discussion of Results

According to the results of our study, 70% of IT-recruiters analyse more than 500 resumes per month. At the same time, only 8% of the candidates are invited to an interview, and only 2% of them get a job offer. One of the reasons for this is that recruiters cannot get enough information about candidate's skills by inspecting their resume only. Consequently, recruiters conduct more interviews than it is necessary (10+ interviews per position on average). Due to a very specific demand (typically a company requires specific technical skills) most of such interviews are time waste. This is a clear evidence of the gap between a candidate's portrait (or employer's demand) the content of a resume. Thus, lack of information and verification tools leads to low performance of recruiting in general. The results of the survey presented above clearly reveal the lack of tools and methods for semi-automated analysis of data coming from open source code repositories.

There is also a lack of methods that IT recruiters can apply to make better hiring decisions. The primary task of a recruiter is to find and attract the most qualified person available on the market. Talented professionals are rarely looking for a job. Only few of them publish and regularly update their resumes on job cites. Therefore, recruiters often use active search technology, which means that a recruiter collects information about a candidate from all possible sources. However, performance of a recruiter naturally drops, when the number of different tools he uses grows. This requires to aggregate heterogeneous data relevant to recruiting in a single system. A possible solution to this issue could be development of a systems for automatic collection and assessment of programmers' skills. The most important features of such system should implement searching, ranking and comparison of developers with respect to multiple criteria, including analysis of the source code.

During the study we realized that IT recruiters use a composite approach and apply many techniques. To achieve their primary goal IT recruiters combine online and offline tools to find relevant candidates. As IT-specialists in Russia are among the most demanded and fast growing occupations, IT-recruiters prefer tools that allow not only to find a suitable candidate, but also to show detailed information about skills, experience, professional achievements, etc.

Therefore, an additional important tool could be developed upon the methods and algorithms for collection and analysis of professional information about the candidate from professional developer platforms such as Stackoverflow, Github and Bitbucket. The results of these tools can be useful for both recruiters and developers. Recruiters will be able to reduce time of developers searching and even before the interview to gather additional information about professional skills. These techniques can help developers simplify the challenge of looking for a job, and to find jobs best suited to their current skills and abilities.

6   Validity and Generalizability of the Findings

We use as case study for our study the Russian IT market, which is also representative of a large number of cases throughout the world. In 2006, Cusumano claimed that Russia's advantage is the ability to do sophisticated technical work at relatively low cost [11]. The cost of production and the salaries have significantly increased since then, but, in general, statement is true. This makes Russian software development industry very similar to that of several other locations in the world, like Ireland, China, India, and so on, and consequently our considerations can be applied to a large number of situations.

Nowadays, Russian IT sector grows quite fast as the entire economy relies largely on the oil and gas industry, which requires a significant involvement of IT. Even though the number of Russian IT specialists has been growing very fast for the past decade, there is still a huge demand of skilled IT personnel. Russia has a population of 140 million people, but less than one million work in IT, which amount to about 7%; this is already a significant result, since in 2012 there were only 300,000 (around 0,2%), while most developed countries have figures around 1%.[6]

These figures are indirectly confirmed by other sources; for instance, according to Lepinay et al. [21], the Russian GitHub community held the 11th place with respect to the number of developers in 2012, while Russia is the 4th industrialized country in terms of population[7]).

We assume that the results of this investigation could be applicable, at least to a partial extent, to other regions of the world with a similar industry structure, like many areas in continental EU (Paris in France, North East of Italy, the area around Munich in German, etc.), the area around London in UK, the Bay Area or the Washington area in US, the province of Guangdong in China, etc.

7   Conclusions

In the paper we addressed an issue IT recruiting in Russia in two aspects: description and measurement of a typical practices in IT recruiting and looking for possible ways to improve the process using online tools and services. According to the goals of the study, we have answered two questions by running a survey among HR managers and experts. The first question: how much investment is required for companies to proceed with recruitment of new employees? The second question: what challenges IT-recruiter is faced to when looking for developers? We use Russian Federation as a showcase and conclude with a certain quantitative and qualitative results.

Our future work is twofold. First, we are looking forward to build a prototype of a web-service to facilitate IT recruiting process . This system will implement the highly demanded features that were identified during the survey and existing measures [27,30]. Another direction of the work is extending the GQM model and study of IT-recruiting from other perspectives, including pair programming perspective [9,14] and software engineering perspective [16].

References

1. Agarwal R, Ferratt TW (1998) Recruiting, retaining, and developing it professionals: an empirically derived taxonomy of human resource practices. In: Proceedings of the 1998 ACM SIGCPR conference on computer personnel research, SIGCPR 1998. ACM, New York, pp 292–302
2. Amadoru M, Gamage C (2016) Evaluating effective use of social networks for recruitment. In: Proceedings of the 2016 ACM SIGMIS conference on computers and people research. ACM, pp 125–133
3. Basili VR (1992) Software modeling and measurement: the goal/question/metric paradigm. Technical report
4. Begel A, DeLine R, Zimmermann T (2010) Social media for software engineering. In: Proceedings of the FSE/SDP workshop on future of software engineering research. ACM, pp 33–38

---

[6] TAdviser survey of Russian IT market http://tadviser.ru/a/53628.

[7] https://medium.com/@hoffa/github-top-countries-201608-13f642493773.

5. Böhm S, Niklas SJ (2012) Mobile recruiting: insights from a survey among German HR managers. In: Proceedings of the 50th annual conference on computers and people research, SIGMIS-CPR 2012. ACM, New York, pp 117–122

6. Bohnert D, Ross WH (2010) The influence of social networking web sites on the evaluation of job candidates. Cyberpsychol Behav Soc Netw 13(3):341–347

7. Broughton A, Foley B, Ledermaier S, Cox A (2013) The use of social media in the recruitment process Institute for employment studies, Brighton

8. Cole MS, Rubin RS, Feild HS, Giles WF (2007) Recruiters perceptions and use of applicant résumé information: screening the recent graduate. Appl Psychol 56(2):319–343

9. Coman ID, Sillitti A, Succi G (2008) Investigating the usefulness of pair-programming in a mature agile team. In: Agile processes in software engineering and extreme programming: 9th international conference, XP 2008, Limerick, Ireland. proceedings. Springer, Heidelberg, June 2008, pp 127–136

10. Corral L, Sillitti A, Succi G, Garibbo A, Ramella P (2011) Evolution of mobile software development from platform-specific to web-based multiplatform paradigm. In: Proceedings of the 10th SIGPLAN symposium on new ideas, new paradigms, and reflections on programming and software, Onward! 2011. ACM, New York, pp 181–183

11. Cusumano MA (2006) Where does Russia fit into the global software industry? Commun ACM 49(2):31–34

12. di Bella E, Sillitti A, Succi G (2013) A multivariate classification of open source developers. Inf Sci 221:72–83

13. Ferratt TW, Agarwal R, Moore JE, Brown CV (1999) Observations from the front: it executives on practices to recruit and retain information technology professionals. In: Proceedings of the 1999 ACM SIGCPR conference on computer personnel research. ACM, pp 102–112

14. Fronza I, Sillitti A, Succi G (2009) An interpretation of the results of the analysis of pair programming during novices integration in a team. In: Proceedings of the 2009 3rd international symposium on empirical software engineering and measurement, ESEM 2009. IEEE Computer Society, pp 225–235

15. Ivanov V, Pischulin V, Rogers A, Succi G, Yi J, Zorin V (2018) Design and validation of precooked developer dashboards. In: Proceedings of the 2018 26th ACM joint meeting on European software engineering conference and symposium on the foundations of software engineering. ACM, pp 821–826

16. Ivanov V, Rogers A, Succi G, Yi J, Zorin V (2017) What do software engineers care about? Gaps between research and practice. In: Proceedings of the 2017 11th joint meeting on foundations of software engineering. ACM, pp 890–895

17. Janz BD, Nichols EL (2010) Meeting the demand for it employees: can career choice be managed? In: Proceedings of the 2010 special interest group on management information system's 48th annual conference on computer personnel research on computer personnel research. ACM, pp 8–14

18. Jermakovics A, Sillitti A, Succi G (2011) Mining and visualizing developer networks from version control systems. In: Proceedings of the 4th international workshop on cooperative and human aspects of software engineering, CHASE 2011. ACM, pp 24–31

19. Kovács GL, Drozdik S, Zuliani P, Succi G (October 2004) Open source software for the public administration. In: Proceedings of the 6th international workshop on computer science and information technologies

20. Lang S, Laumer S, Maier C, Eckhardt A (2011) Drivers, challenges and consequences of e-recruiting: a literature review. In: Proceedings of the 49th SIGMIS annual conference on computer personnel research, SIGMIS-CPR 2011. ACM, New York, pp 26–35

21. Lepinay V, Mogoutov A, Cointet JP, Villard L (2014) Russian computer scientists, local and abroad: mobility and collaboration. In: Proceedings of the 10th central and eastern European software engineering conference in Russia. ACM, p 18

22. Marlow J, Dabbish L (2013) Activity traces and signals in software developer recruitment and hiring. In: Proceedings of the 2013 conference on computer supported cooperative work. ACM, pp 145–156

23. Maurer F, Succi G, Holz H, Kötting B, Goldmann S, Dellen B (May 1999) Software process support over the internet. In: Proceedings of the 21st international conference on software engineering, ICSE 1999. ACM, pp 642–645

24. Minardi J (2016) How to get hired at a startup as a software developer. XRDS: Crossroads, ACM Mag Students 22(3):14–15

25. De Panfilis S, Sillitti A, Ceschi M, Succi G (2005) Project management in plan-based and agile companies. IEEE Softw 22:21–27

26. Schmidt E, Rosenberg J (2014) How Google works. Hachette UK

27. Sillitti A, Janes A, Succi G, Vernazza T (2004) Measures for mobile users: an architecture. J Syst Arch 50(7):393–405

28. Succi G, Paulson J, Eberlein A (2001) Preliminary results from an empirical study on the growth of open source and commercial software products. In: EDSER-3 workshop, pp 14–15

29. Valerio A, Succi G, Fenaroli M (1997) Domain analysis and framework-based software development. SIGAPP Appl Comput Rev 5(2):4–15

30. Vernazza T, Granatella G, Succi G, Benedicenti L, Mintchev M (July 2000) Defining metrics for software components. In: Proceedings of the world multiconference on systemics, cybernetics and informatics, vol XI, pp 16–23

31. Yager SE, Schambach TP (2002) Newly minted it professionals: a conversation with their perspective employers. In: Proceedings of the 2002 ACM SIGCPR conference on Computer personnel research. ACM, pp 103–105

Comparison of Agile, Quasi-Agile and Traditional Methodologies

Vladimir Ivanov[(⊠)], Sergey Masyagin, Alan Rogers, Giancarlo Succi,
Alexander Tormasov, Jooyong Yi, and Vasily Zorin

Innopolis University, Universitetskaya St, 1,
420500 Innopolis, Respublika Tatarstan, Russia
{v.ivanov,s.masyagin,g.succi,tor,j.yi,v.zorin}@innopolis.ru

Abstract. In this study we were able to gather a substantial quantity
of detailed responses from a group of individuals and companies that are
broadly quite similar to those found in several of the major world centers
of technological innovation. As such, our analysis of the results provides
some tantalizing hints to organizational and methodological challenges
and practices of a broad range of groups.

One intriguing suggestion is that while "traditional" and well defined
Agile groups function according to the standards established to support those approaches, Quasi-Agile groups do not. Instead, Quasi-Agile
groups seem to pursue the goals of Agile using measures and underlying
approaches more similar to traditional methods. One might expect that
such a discordance to affect the effectiveness of a group's efforts.

Keywords: Agile methods · Empirical study · Adoption of technology

1   Introduction

Agile development methodologies continue to gain popularity. In a recent Stack
Overflow survey conducted in 2017, more than 80% of the responding professional developers stated they are using Agile development methodologies. At the
same time, challenges and problems in transforming to Agile are also reported
anecdotally and in the literature. The tension created by these two contradicting
phenomena—high popularity of Agile and difficulties in adopting Agile—causes
a creation of a group of developers who adopt Agile only superficially. Some
people call the Agile practices performed by these developers Quasi-Agile. How
different are organizations practicing Quasi-Agile from those practicing Agile or
traditional development methodologies? This is the question we ask in this study
to better understand an emerging phenomena of Quasi-Agile. To find answer,
we conduct an observational study with more than 40 companies sampled in
the Innopolis Special Economic Zone. In this paper, we report pervasiveness of
Quasi-Agile and the unique patterns of organizations practicing Quasi-Agile differentiated from those practicing Agile or traditional methodologies. We believe

© Springer Nature Switzerland AG 2020
P. Ciancarini et al. (Eds.): SEDA 2018, AISC 925, pp. 128–137, 2020.
https://doi.org/10.1007/978-3-030-14687-0_11

our research results can shed light on better understanding of the current situation of Agile adoption and future research directions.

In this paper, we ask the following research question:

Research Question: How different are companies practicing Quasi-Agile from those practicing Agile or traditional methodologies?

In attempt to answer our research question, we conducted a series of interviews based on a fixed set of questions with representatives 44 companies located in the Innopolis Special Economic Zone. To distinguish Quasi-Agile from Agile, we asked in our questionnaire whether the company uses Agile and whether it follows any specific Agile methodology or a properly crafted methodology also with the help of mentors and/or internal experts or it selects individual practices identified on the web or on books taking advantage of completely self-trained internal expertise. We categorize a company as Agile in the former case and Quasi-Agile in the latter. While this distinction may not be the best way to identify Quasi-Agile, we believe this method is more objective than asking the company to identify whether it uses Agile or Quasi-Agile. In any case, we asked the aforementioned questions only at end not to reveal the intention of the study.

To answer the research question about the difference between companies practicing three different categories of development methodologies (i.e., Agile, Quasi-Agile, and Traditional), we asked the following four categories of questions (the overall questionnaire is designed following the GQM methodology [2]) (details are in Sect. 4):

1. The external aspect category, that is, how to evaluate the overall success of the project from the external point of view. Examples are on-time delivery of the product, satisfying the standard, etc.
2. The internal aspect category, that is, how to assess the process of software development and the quality of software.
3. The planning aspect category. Unlike the prior two categories that focus on the present, the questions in this category ask about future plans.
4. The monitoring aspect category, that is, which concrete information to use in the development process to ensure the success of the project.

This is an observational study, so by itself its generalizability is not of primary importance. Our goal is rather to identify interesting patterns that can then be verified in larger studies. We have therefore analysed the data using visual inspection to gain an intuitive visual understanding of the mutual relationships existing between the different approaches, and with a statistical tool to provide a quantified evaluation of the relationships.

The existence of Quasi-Agile is reported at least as early as 2009, in which year Timothy Korson presented a talk in the Quality Engineered Software and Testing Conference 2009 where he used the term Quasi-Agile to refer to the Agile practiced in traditionally structured organizations. However, to the best of our knowledge, no previous research has been conducted to examine how pervasive Quasi-Agile is and how different Quasi-Agile is from Agile and traditional methodologies.

Our main findings are as follows:

1. Quasi-Agile is quite pervasive. 34% of the 44 organizations that participated in our study are categorized into Quasi-Agile.
2. The answering pattern of the Quasi-Agile companies is closer to the Traditional companies than to the Agile companies in the first and the second categories of questions (i.e., the external aspect category and the internal aspect category).
3. Conversely, the answering pattern of the Quasi-Agile companies is closer to the Agile companies than to the Traditional companies in the third and the fourth categories of questions (i.e., the planning aspect category and the monitoring aspect category).

Our results evidence that the Quasi-Agile group Korson mentioned indeed exists pervasively. While some answer patterns of the Quasi-Agile companies in our sample show similarity to those of the Agile companies (in particular in terms of making an improvement plan), the processes of the Quasi-Agile companies in terms of assessing their products and the success of the projects appear to be more akin to those of the companies practicing traditional methodologies.

The pervasiveness of Quasi-Agile revealed in our study raises critical questions about the current practices and researches conducted in the arena of Agile. The current dichotomous view of development methodologies as either Agile or traditional does not seem up to date in the era when the influence of Agile has already permeated the entire software industry. Instead, paying more attention to Quasi-Agile is called for, considering the pervasiveness of this new phenomenon – note that in our investigation, Quasi-Agile is shown to be more common than the Traditional group. The fact that the overall pattern of the Quasi-Agile group is distinguished from that of the Agile group and the Traditional group prompts the need for further investigations—to understand how the practice of Quasi-Agile affects the organization.

For starters, we would like to ask the following question in future work: should Quasi-Agile be considered harmful? Though further research is required, we briefly discuss this question in Sect. 5.

2   Background and Motivation

Agile is one of several approaches to software development that emerged in the 1980s and 90s to address the challenges of successfully delivering software using the then-dominant Waterfall approach [27]. Agile is not simply the absence of Waterfall; strictly speaking it refers to a methodology guided by the Agile Manifesto [4], a framework synthesizing key elements of several "agile" approaches to software development that emerged in the previous decade. Among the best known Agile methodologies are Scrum, XP, Crystal Clear, as well as several others. The precise definition and boundaries of Agile are open to discussion and have continued to evolve but include far more than a few common rituals such as iterations (or "sprints") and daily stand-up meetings.

For the last decade, the software industry has adopted agile development practices quite extensively [6], often coupled with Open Source tools [7,13,25]. According to a survey conducted in 2009 [30], about 35% of the survey respondents stated that agile development practices most closely reflect their development process, which is about the same percentage (34%) who stated that traditional practices (such as waterfall and iterative) most closely reflect their development process. In a later survey conducted in 2012 [20], 72% of the survey respondents stated their organizations use agile development methodologies, while only 24% of the respondents stated their organizations do not use agile development methodologies. This trend is also observed in the annual survey results from Microsoft developers [18] and is confirmed in many works, like [29].

Regarding the adoption of individual agile methodology, Scrum has shown to be most popular in multiple surveys [21,30]. Similarly, Scrum has been most extensively studied in the literature, according to the literature review of Diebold et al. [9]. While the popularity of other methodologies such as XP and Kanban is not as high as that of Scrum in those aforementioned surveys and literature review, the usage rate of the hybrid of Scrum and XP is as high as XP in [21] and even higher than XP in [9].

3 Methodology

Designing an effective set of questions (or questionnaire) is challenging; the way questions are defined, organized, and laid down may influence significantly the overall result. To address such limits, we followed the rules found in the literature and similar experiences [2,3,5,14–16,26,28].

We selected a closed questionnaire, as recommended for the features of our research in [1,23], using the techniques proposed by Furnham [11] and of Podsakoff et al. [22] to minimize the biases in the responses, including redundancy and replication. In particular, we have devoted significant care to preventing biases in the responses and confirmation bias using the prescriptions detailed in [11,22]. In addition, in this study, to propose our intuitions on Quasi-Agile and Agile, we only analyse distributions of the answers and we refrain from commenting the rankings of the specific questions. Therefore, the risk of a respondent bias is further minimized.

Moreover, to ensure quality responses, the questionnaire was first sent to the participants and then a member of the research team interviewed them face-to-face. The interview results were recorded in a written document, the written document was shared back to the participants to ensure that the right information was captured, and the results were corrected if required. Each interview lasted about one hour and the overall preparation for it and followup check required about two hours.

The industry in the city of Innopolis consists of a few large companies ($\geq$250 employees) and a larger number of smaller companies. About half of employees work in large companies, a quarter in small (10–49), and the rest are split between medium size (50–249) and micro-sized (<10).

The companies in Innopolis typically employ male software engineers who hold a graduate degree in a STEM field and have more than 5 years of industrial experience.

To explore the differences in approaches between agile and not agile teams we decided to identify how the companies achieve the following goals [12]:

G1. to evaluate results of a project from an external perspective,
G2. to assess the internals of a process,
G3. to select improvement strategies,
G4. to monitor the evolution of a project.

To approach the question formulation we decided to use the GQM approach [2]. The model resulted in the following key questions characterizing the development process. The first question refers to the goal of determining how the "external" success of a project is evaluated based on its impact on the surrounding environment, like customers, market penetration, etc. (G1).

Q1. Which criteria do you use to analyze software project results?

The second and the third questions refer to the goal of determining how the "internals" of a process are assessed (G2).

Q2. What are the major obstacles that affect your ability to deliver software?
Q3. What are the most important parameters you use to control the quality of a software product?

The fourth and fifth questions refer to the goal of determining how the improvement strategies are selected (G3).

Q4. What are your planned improvement for the next project?
Q5. What are your planned technological innovations for the next project?

The last two questions to the goal of monitoring how projects evolve (G4).

Q6. Which information do you use to ensure that the project terminates successfully?
Q7. Which numeric parameters do you use real-time monitoring software process quality?

Finally, the questions were grouped into three GQM-based sections: the first included questions aimed to detect obstacles and improvement initiatives (Q4, Q5, Q2), the second was aimed to identify criteria, characterizing basic teams goals (Q1, Q6) and the last group consisted of questions to detect metrics that are used to track the progress and project health (Q7) [10,24,29].

We have analysed 44 companies, 39% large (more than 250 employees), 15% medium (from 50 to 249 employees), 34% small,(from 10 to 49), 12% micro (less than 10 employees); 52% have a well defined agile process in place, 34% of the company have a "Quasi-Agile" approach, and 14% follow a traditional, waterfall process.

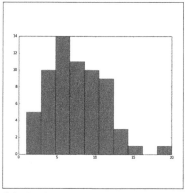

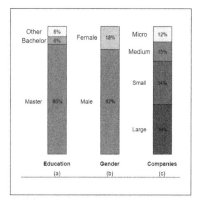

(a) Working experience of the participants

(b) Profiles of the participants

Fig. 1. Information about participants

The subjects we interviewed were developers of these companies, with a working experience ranging from 1 to 20 years, with a median value of 7 (Fig. 1(a)), and mostly with a graduate degree and male (Fig. 1(b)).

The participants were homogeneously distributed across the participating companies and they appear a reasonable representation of the overall population of the developers and software engineers present in the area and around the world.

Altogether, this sample represents well the distribution of the companies in the City of Innopolis.

4   Analysis of the Results

For our research question, we have analysed the answers using a statistical test, to have some more quantified details of the results.

Research Question: How different are companies practicing Quasi-Agile from those practicing Agile or traditional methodologies?

Given the small size of a sample, the answers have been analysed using the Fisher Exact Test to measure of proximity of the distributions instead of the more common (but less precise in this case) Chi-square test [17]. We have used a Python mathematical package adopted for the computation of the results [19], originally developed for Python 2[1] and adapted to Python 3 by the authors. We have considered the three possible cases (Table 1).

- Quasi-Agile vs. Agile (QA-A)
- Quasi-Agile vs. Traditional (QA-T)
- Agile vs. Traditional (A-T)

---

[1] The source code is available online in github: https://github.com/maclandrol/FisherExact.

Table 1. Results of the Fisher test for the proximity of the distributions (0 is the lowest and 1 the highest)

	Questions	QA-A	QA-T	A-T
Q1.	Criteria to analyze software project results	0.91	0.99	1
Q2.	Obstacles affecting the ability to deliver software	0.75	**0.91**	0.88
Q3.	Parameters to control the quality of a product	0.54	1	0.80
Q4.	Planned improvement	**0.72**	0.11	0.39
Q5.	Planned technological innovations	**0.87**	0.32	0.51
Q6.	Information used to ensure that the project terminates successfully	**0.77**	0.49	0.61
Q7.	Numeric parameters used for real-time monitoring software process quality	**0.31**	0.29	*0.08*

The Fisher test confirms the visual intuition. The strongest connection is between Agile and Quasi-Agile, even if in the third case (the one about numerical parameters) the three kinds of processes are quite apart from one another, to the point that the difference between agile and traditional in the seventh question is statistically significant at the 0.1 level.

The significant difference between Agile and Traditional in the numeric parameters used for real time monitoring of software process quality is not surprising, since this is exactly one of the areas where most dissimilarities lie. However, the remarkable even if not significant difference also present between Agile and Quasi-Agile evidences once more that those two have little in common when we deal with the concrete management of a project. It could almost be said that Quasi-Agile is as distinct from agile as from Traditional.

5   Discussion and Conclusion

A summary of the analysis that we have just performed is in Table 2. In our study it appears evident that: **(a) Agile and Traditional companies (the A-T link)** tend to be almost indistinguishable on the concrete criteria used to assess the success of a project, like any company should be, since they are quite independent of whether the process has been carried out in an Agile, Traditional, or any other way; **(b) Quasi-Agile and Agile companies (QA-A)** tend to have similar aims for future improvements of their activities and for what they think they should collect to control the evolution of the process; **(c) Quasi-Agile and Traditional companies (QA-T)** are strongly aligned on the identification of problems during the development process and on the parameters adopted to ensure the quality of a process.

Table 2. Closest links. (*) almost identical.

Goal	Questions	Strongest	Weakest
G1.	Q1.	A-T	QA-A
G2.	Q2.	QA-T	QA-A
	Q3.	QA-T	QA-A
G3.	Q4.	QA-A	QA-T
	Q5.	QA-A	QA-T
G4.	Q6.	QA-A	QA-T
	Q7.	QA-A, QA-T(*)	A-T

Quasi-Agile Considered Harmful? We conjecture so. The aforedescribed analysis reinforces the idea we have previously elaborated, that is, that quasi-agile appears to be an exercise of wishful thinking made by companies with a traditional mindset. While agile is not necessarily better than other development methodologies, the fact that quasi-agile and agile companies tend to have similar aims for future improvements seem to indicate that quasi-agile companies are struggling to get closer to agile companies. We leave further investigation to confirm our conjecture as future work to conduct on the lines of [8].

Acknowledgments. We thank Innopolis University for generously supporting this research.

References

1. Ackroyd S, Hughes JA (1981) Data collection in context. Longman, London
2. Basili VR (1992) Software modeling and measurement: the goal/question/metric paradigm
3. Basili VR, Caldiera G, Rombach HD (1994) The goal question metric approach: in encyclopedia of software engineering. Wiley, Hoboken
4. Beck K, Beedle M, van Bennekum A, Cockburn A, Cunningham W, Fowler M, Grenning J, Highsmith J, Hunt A, Jeffries R, Kern J, Marick B, Martin RC, Mellor S, Schwaber K, Sutherland J, Thomas D (2001) Manifesto for agile software development. http://www.agilemanifesto.org/. Accessed 16 Aug 2017
5. Bond TG, Fox CM (2013) Applying the Rasch model: fundamental measurement in the human sciences. Psychology Press, London
6. Coman ID, Sillitti A, Succi G (2008) Investigating the usefulness of pair-programming in a mature agile team. In: Agile processes in software engineering and extreme programming: proceedings of 9th international conference, XP 2008, Limerick, Ireland. Springer, Heidelberg, pp 127–136
7. Corral L, Sillitti A, Succi G, Garibbo A, Ramella P (2011) Evolution of mobile software development from platform-specific to web-based multiplatform paradigm. In: Proceedings of the 10th SIGPLAN symposium on new ideas, new paradigms, and reflections on programming and software, Onward! 2011. ACM, New York, pp 181–183

8. di Bella E, Sillitti A, Succi G (2013) A multivariate classification of open source developers. Inf Sci 221:72–83

9. Diebold P, Dahlem M (2014) Agile practices in practice: a mapping study. In: 18th international conference on evaluation and assessment in software engineering, EASE 2014, London, England, United Kingdom, 13–14 May 2014, pp 30:1–30:10

10. Fronza I, Sillitti A, Succi G (2009) An interpretation of the results of the analysis of pair programming during novices integration in a team. In: Proceedings of the 2009 3rd international symposium on empirical software engineering and measurement, ESEM 2009. IEEE Computer Society, pp 225–235

11. Furnham A (1986) Response bias, social desirability and dissimulation. Pers Individ Differ 7(3):385–400

12. Jermakovics A, Sillitti A, Succi G (2011) Mining and visualizing developer networks from version control systems. In: Proceedings of the 4th international workshop on cooperative and human aspects of software engineering, CHASE 2011. ACM, pp 24–31

13. Kovács GL, Drozdik S, Zuliani P, Succi G (2004) Open source software for the public administration. In: Proceedings of the 6th international workshop on computer science and information technologies

14. Krosnick JA, Presser S (2010) Question and questionnaire design. Handb Surv Res 2:263–314

15. Lietz P (2008) Questionnaire design in attitude and opinion research: current state of an art. Citeseer, Princeton

16. Maurer F, Succi G, Holz H, Kötting B, Goldmann S, Dellen B (1999) Software process support over the Internet. In: Proceedings of the 21st international conference on software engineering. ACM, pp 642–645

17. McDonald JH (2009) Handbook of biological statistics, 2nd edn. Sparky House Publishing, Baltimore

18. Murphy B, Bird C, Zimmermann T, Williams L, Nagappan N, Begel A (2013) Have agile techniques been the silver bullet for software development at microsoft? In: 2013 ACM/IEEE international symposium on empirical software engineering and measurement, Baltimore, Maryland, USA, 10–11 October 2013, pp 75–84

19. Noutahi E (2016) Fisher's exact test for mxn contingency table. https://mrnoutahi. com/2016/01/03/Fisher-exac-test-for-mxn-table/, Accessed 03 Sept 2017

20. Papatheocharous E, Andreou AS (2013) Evidence of agile adoption in software organizations: an empirical survey. In: Proceedings of systems, software and services process improvement - 20th European conference, EuroSPI 2013, Dundalk, Ireland, 25–27 June 2013, pp 237–246

21. Papatheocharous E, Andreou AS (2014) Empirical evidence and state of practice of software agile teams. J Softw Evol Process 26(9):855–866

22. Podsakoff PM, MacKenzie SB, Lee JY, Podsakoff NP (2003) Common method biases in behavioral research: a critical review of the literature and recommended remedies. J Appl Psychol 88(5):879–903

23. Popper KR (2002) The logic of scientific discovery. Routledge, London

24. Sillitti A, Janes A, Succi G, Vernazza T (2004) Measures for mobile users: an architecture. J Syst Archit 50(7):393–405

25. Succi G, Paulson J, Eberlein A (2001) Preliminary results from an empirical study on the growth of open source and commercial software products. In: EDSER-3 Workshop, pp 14–15

26. Thayer-Hart N, Dykema J, Elver K, Schaeffer NC, Stevenson J (2010) Survey fundamentals: a guide to designing and implementing surveys. Office of quality improvement

27. Valerio A, Succi G, Fenaroli M (1997) Domain analysis and framework-based software development. SIGAPP Appl Comput Rev 5(2):4–15
28. Vannette DL, Krosnick JA (2014) Answering questions: a comparison of survey satisficing and mindlessness. Wiley, Hoboken, pp 312–327
29. Vernazza T, Granatella G, Succi G, Benedicenti L, Mintchev M (2000) Defining metrics for software components. In: Proceedings of the world multiconference on systemics, cybernetics and informatics, vol XI, pp 16–23
30. West D, Agile TG (2010) Development: mainstream adoption has changed agility – trends in real-world adoption of agile methods. Technical report, Forrester Research

Innopolis Going Global
Internationalization of a Young IT University

Sergey Karapetyan, Alexander Dolgoborodov, Sergey Masyagin,
Manuel Mazzara$^{(\boxtimes)}$, Angelo Messina, and Ekaterina Protsko

Innopolis University, Innopolis, Russian Federation
{s.karapetyan,a.dolgoborodov,s.masyagin,m.mazzara,a.messina,
e.protsko}@innopolis.ru

Abstract. Innopolis is a new IT city incorporating a technopark and a university, aiming at prioritizing the development of IT and software engineering in Tatarstan and in the Russian Federation. Innopolis University (IU) is a young university pioneering several research and pedagogical projects and experiments with innovative teaching methods and curricula. This paper describes the first five years of life of the university in terms of internationalization strategies and faculty and students recruitment.

1 Introduction

The phenomenon of globalization leads these days to the creation of an unified educational space and market dominated by international rankings. Universities compete for *internationalization* with associated risks and benefits [12]. Russian education has a recognized reputation around the world, especially in fundamental science, which presents itself as a good selling point on international market. In 2012 Russian government proclaimed *"internationalization"* to be one of the major objectives of higher education development [3]. To support the idea, the government launched a so-called *"5top100"* project [5], which is aimed at "maximizing the competitive position of a group of leading Russian universities in the global research and education market." To put it in a simpler way, a group of leading Russian universities started to receive additional funding in order to facilitate their international attributes, particularly, increase positions held by this group of universities in the world university rankings (QS, Times Higher Education, Shanghai ranking of world universities).

Innopolis University (IU) is a young and ambitious university in Tartarstan in the Russian Federation, which has a strong focus on education, and scientific research in the field of IT and robotics [14]. It is located in the newly created Innopolis City (near the capital city Kazan) which also comprises ICT companies and the Innopolis Special Economic Zone. Innopolis aims to be the major Russian IT hub. In its development the University was trying to follow the main trends of IT education borrowed from the world's leading higher education institutions. One of these trends was and still is internationalization. Since the very

© Springer Nature Switzerland AG 2020
P. Ciancarini et al. (Eds.): SEDA 2018, AISC 925, pp. 138–145, 2020.
https://doi.org/10.1007/978-3-030-14687-0_12

foundation, the university tried to hunt international faculty members as well as attracting international students. Additional, the university has been developing various international initiatives including summer schools, exchange programs, conferences etc. Existing for five years, the university succeeded in creating an international environment on campus.

In this article, we discuss the internationalization process of the university during its five years of existence. After this introduction the paper is structured as follows: Sect. 2 describes the educational model of the university and how new students are recruited; Sect. 3 defines the internationalization strategy; Sect. 4 provides data about enrolment of international students; Sect. 5 offers details on the strategies adopted to hire international faculty while Sect. 6 draws some conclusions on the story of the university and the issue of internationalization.

2   Educational Model of IU and Students Recruitment

Innopolis University follows a model according to which all students are allowed to free education as long as they pass the selection process. IU has a bachelor program in Computer Science and Engineering and four master programs:

- Software Engineering
- System and Network Engineering
- Data Science
- Robotics.

The first two of these programs were transferred respectively from Carnegie Mellon University (CMU) and Amsterdam University (UvA). Both of them are quite unusual for typical Russian Master's programs and designed for students with initial industrial experience. One of the main peculiarities is so-called "Industrial project" integrated in study plans. In case of Master's program in Software Engineering (SE), students for two semesters work on a team project supervised by mentors from IU and industry. This practice was borrowed from CMU too and the project presents a real-industry task. Students have a chance to apply obtained knowledge during courses to the task solving. The Master Program in System and Network Engineering has a similar approach transferred from UvA. In the frames of this program students are also to work on indusial projects, but unlike SE program, here the length of the projects is just 2 months.

3   Internationalization Strategy of IU

From the first step, IU showed its internationalization invitations. In December 2012, just few days after its creation, the University concluded an agreement with the Carnegie Mellon University (CMU). Under the agreement, the sides agreed to transfer a master's degree program on Software Engineering from CMU to IU. In spite of the fact, that IU transferred its first educational program from CMU, IU always took as a role model a slightly different type of a higher education institute, i.e. young technical universities. The best examples worldwide are

KAIST (South Korea), IT University of Copenhagen (Denmark), and the Hong Kong University of Science and Technology (Hong Kong).

Innopolis University believes that internationalization will help with achieving following goals:

- Improvement of educational and research quality;
- Focus on innovative and emerging technologies (i.e.: Improved "agile" software development)
- Inclusion in many core courses of "real life" style production cycle exercises including a final product delivery (i.e. Course projects)
- Competitivenesses of IU's graduates and research results in international academic and IT market;
- Guaranteeing to IU a place as equal participant of international cooperation.

To achieve these aims, IU set its main directions of internationalization that have been developed for last 5 year:

- Recruitment of foreign students and faculty members;
- Employment of foreign faculty members;
- a scheme of visiting professorship from abroad institution
- Mobility programs (student exchange; Erasmus+; research internships; joint PhD supervision)
- International olympiads and competitions;
- Joint research and publications.

Despite the fact that IU is not part of 5top100 project, it does not mean that it has ignored world university rankings. In fact, IU valued them high and believed they potentially might help with following issues:

- proving IU's academic and employment reputation;
- raise awareness about IU and Innopolis projects;
- requirement of international students;
- requirement of international faculty.

For a young university entrance into rankings is a fascinating challenge and IU already makes its first positive steps in rankings. In 2018, it entered Round University Ranking [7], where it reached a high score in internationalization indicators. It received maximum possible points for "Share of international co-authored papers" becoming the best result among Russian universities [4]. Additionally, Round University Ranking has a ranking table evaluating universities according "International Diversity", where IU came the second best among all Russian universities with only Lomonosov Moscow State University ahead. This particular ranking includes following indicators: Share of international academic staff, Share of international students, Share of international co-authored papers, Reputation outside region, International level [9].

4   Internationalization Programs

IU had to work out its recruitment of international students always with a limited budget. Despite of this, increasing interest has been shown by students abroad for the university.

The key elements of the recruiting strategies have been:

– Use of Internet channels
– Word by mouth of current students
– Incentive to current students to invite their talented friends
– Recruiting campaigns in foreign countries (for example Italy in 2017)
– Program of visiting professors from abroad
– Program of international internships
– Students exchange (in and out)
– Erasmus+ with countries such Italy, UK, Ireland, Turkey
– Competitions and olympiads for students and schoolchildren
– International agents.

The growth of numbers and diversity of international students is shown in the following tables and it is a good indicator of the success of the process. Figure 1 shows the diversity of international students in the Bachelor program and its progression and Fig. 2 emphasizes the growth for Master program. Figure 3 summarizes the total.

Table 1. Diversity of international BS students

Year	2016	2017	2018
Europe	9	8	5
America	0	4	0
Asia and Middle Est	19	29	31
Africa	0	4	6
Total	28	45	42

Table 2. Diversity of international MS students

Year	2016	2017	2018
Europe	2	1	4
America	0	4	7
Asia and Middle Est	9	9	18
Africa	1	4	7
Total	12	18	36

Table 3. Diversity of international students (total)

Year	2016	2017	2018
Total per year	40	63	78

As you may see from the Tables, the largest amount of students is coming from Asia and Middle East. Only a small past is actually coming from America and Europe. This is due to the lack of awareness about IU in these continents. Dissemination activities from PR offices have been indeed still limited in these geographical areas. The countries offering the large amount of students to the university are typically CIS countries, Egypt and Pakistan. However, the last year has seen students enrolled also from Italy, Spain and Latin America.

4.1 Recruitment of Foreign Students

The exchange students and the international internships played an important role in placing Innopolis in the world academic map, as shown in the following table (data from [1,2]).

Table 4. Student exchange and internships

Year	2016	2017	2018
Outgoing exchange students	6	13	20
Incoming exchange students	1	3	5
Incoming foreign interns	0	1	10

In the framework of international program students of Innopolis University have a chance to visit world's leading university in Computer Science sphere: National University of Singapore, KAIST, Seoul National University, Hong Kong University of Science and Technology, Polytechnic University of Milan etc. The benefits that exchange brings to the University and participants:

- An opportunity for students to extend academic curriculum thought courses given in the partner universities
- International experience is a significant advantage from employment perspectives
- Experience of working in an international environment and projects
- Significantly increasing language skills
- Broaden horizons and inter-cultural understanding.

On the way of setting a completely new exchange network, IU faced several obstacles. Firstly, difference of study plans and recognition of received courses. To avoid this problem, from the start IU looked for exchange cooperation only with universities that have a strong IT background and relevant degree programs focused on IT, Computer Science, Data Science, Robotics, AI, Software Engineering and Cyber Security. Additionally, IU helps to build up an appropriate individual study plan for each exchange student in order to make recognition as smooth as possible. All exchange partners provide English taught courses at least in Master's level, that is why IU students do not need to learn a new language to participate in the program. Second major problem is that due to the young age IU is not widely recognized by the foreign students. As you can see from the Table 4, the incoming mobility was not developing as fast as outgoing. That is why, IU undertook several methods to change the situation: each participant was used as an ambassador of IU to maximize recognition of IU in partner universities; IU organized webinars for potential exchange students, where they were to learn about the university; exchange coordinators disseminated various materials with partner universities; starting from 2018 IU set a scholarship for incoming students that covers transportation and accommodation. In terms of financial support of mobility, IU was also involved in Erasmus+ scheme with partners from Turkey, UK, Ireland and Italy, which provides funding for supporting not just students exchange but also staff mobility. Outgoing students have not faced serious financial problems, since they receive a monthly scholarship from IU.

The fact that IU is a English speaking university eases things in developing exchange programs: incoming students do not need to know Russian language and outgoing students already have a solid knowledge of English since all their courses at IU are given in this language.

4.2   Olympiads and Competitions

Olympiads and competitions have an important role in IU internationalization and student recruitment. For instance, in 2014 IU became a national provider of the World Robot Olympiad, which allows IU to host Russian stage of the Olympiad (it brings hundreds of schoolchildren ever year [6] and train Russian national team. By training the national team, IU helped it to achieve high results in WRO, thus, in 2018 the national team came first in the competition by winning 1/3 of all medals: 5 gold, 1 silver and 2 bronze medals [8]. Such results have positive influence on potential students, especially in the field of Robotics, and affect their choice of future place of study.

5   Faculty Members

IU started an international recruiting campaign for faculty members as early as 2013, just after its foundation. The initial phase of recruiting was difficult, mainly addressing potential faculty one by one by email after an extensive web search. This is how one of the authors of this paper has been contacted in October 2013. The second round was done using LinkedIn, in particular the contacts of the people already hired. Rising awareness about Innopolis in 2013 and 2014 was a difficult exercise. As soon as a core of specialist known in their areas was hired, the visibility of the university quickly boosted and a regular hiring campaign via traditional channels (journals, conferences, mailing lists...) become possible. The establishment of a program of visiting professors from abroad also helped dramatically to this regard. At the moment the faculty includes more than 20 members, of which the majority are foreigners.

Table 5 shows the dynamics of publications growth. At first glance the number might seem as not substantial, however, if you consider that during last years IU had less than 20 full-time faculty members, it will show a vary productive papers per faculty ratio.

Table 5. Publications in Scopus

Year	2013	2014	2015	2016	2017
Total number of publications	1	5	41	73	78
Publications in computer science	1	2	30	55	62

6 Conclusion

Innopolis University is now in its fifth year and reached a students population of about 700 with 300 employees of which about 20 faculty members and teaching and research stuff up to 80 people. It has an extensively developed network of international institutions collaborating under different forms: students exchange, Erasmus+, visiting professors, joint PhD supervision and joint projects. One of the collaborative projects sees as partners CERN and Newcastle university [10,11] and collaborative PhD supervisions involve several universities including Toulouse, University of Southern Denmark and University of Messina. All these activities dramatically supported the internationalization of the project. In turn, the internationalization of the university also helped the development of the city itself.

With the prerogative of a selective education free of charge and a constant attention on the internationalization of teaching and research, Innopolis is aiming at exploiting the benefit of the global trends without suffering the risks, such as excessive emphasis on the market aspects of education and the death run to global rankings which would put at risk the fundamental academic values [13].

References

1. About IU. https://university.innopolis.ru/en/about/. Accessed 26 Aug 2018
2. Academic Exchange 2017. https://university.innopolis.ru/news/academic-exchange-2017/. Accessed 26 Aug 2018
3. Decree of the president of the Russian Federation. http://www.kremlin.ru/acts/bank/35263. Accessed 23 Aug 2018

4. Innopolis University in round university ranking. http://roundranking. com/universities/innopolis-university.html?sort=O&year=2018&subject=SO. Accessed 26 Aug 2018
5. Project 5top100. https://5top100.ru/en/. Accessed 23 Aug 2018
6. Robot Olympiad. https://university.innopolis.ru/en/news/iu-rro-2018/. Accessed 26 Oct 2018
7. Round university ranking. http://roundranking.com/. Accessed 26 Oct 2018
8. The Russian Federation came in first at the world Robot Olympiad. https:// university.innopolis.ru/en/news/first-place-wro2017/?sphrase_id=11095665. Accessed 26 Oct 2018
9. World university rankings. http://roundranking.com/ranking/world-university-rankings.html#nternational-2018. Accessed 26 Oct 2018
10. Bauer R, Breitwieser L, Di Meglio A, Johard L, Kaiser M, Manca M, Mazzara M, Rademakers F, Talanov M, Tchitchigin AD (2017) The biodynamo project: experience report. In: Advanced research on biologically inspired cognitive architectures. IGI Global, pp 117–125
11. Breitwieser L, Bauer R, Di Meglio A, Johard L, Kaiser M, Manca M, Mazzara M, Rademakers F, Talanov M (2016) The biodynamo project: creating a platform for large-scale reproducible biological simulations. In: 4th workshop on sustainable software for science: practice and experiences (WSSSPE4)
12. Gao Y, Baik C, Arkoudis S (2015) Internationalization of higher education. Palgrave Macmillan UK, London, pp 300–320
13. Jibeen T, Khan MA (2015) Internationalization of higher education: potential benefits and costs. Int J Eval Res Educ (IJERE) 4:196–199
14. Kondratyev D, Tormasov A, Stanko T, Jones RC, Taran G (September 2013) Innopolis University - a new it resource for Russia. In: 2013 international conference on interactive collaborative learning (ICL), pp 841–848

A Conjoint Application of Data Mining Techniques for Analysis of Global Terrorist Attacks

Prevention and Prediction for Combating Terrorism

Vivek Kumar[1]([⊠]) [iD], Manuel Mazzara[2], Angelo Messina[2], and JooYoung Lee[2]

[1] National University of Science and Technology-MiSiS,
Moscow, Russian Federation
vivekkumar0416@gmail.com
[2] Innopolis University, Kazan, Russian Federation
{m.mazzara,a.messina,j.lee}@innopolis.ru

Abstract. Terrorism has become one of the most tedious problems to deal with and a prominent threat to mankind. To enhance counter-terrorism, several research works are developing efficient and precise systems, data mining is not an exception. Immense data is floating in our lives, though the scarce availability of authentic terrorist attack data in the public domain makes it complicated to fight terrorism. This manuscript focuses on data mining classification techniques and discusses the role of United Nations in counter-terrorism. It analyzes the performance of classifiers such as Lazy Tree, Multilayer Perceptron, Multiclass and Naïve Bayes classifiers for observing the trends for terrorist attacks around the world. The database for experiment purpose is created from different public and open access sources for years 1970–2015 comprising of 156,772 reported attacks causing massive losses of lives and property. This work enumerates the losses occurred, trends in attack frequency and places more prone to it, by considering the attack responsibilities taken as evaluation class.

Keywords: Data mining · United Nations · Lazy tree · Multilayer Perceptron · Multiclass Classifier · Naïve Bayes

1 Introduction

"Our responsibility is to unite to build a world of peace and security, dignity and opportunity for all people, everywhere, so we can deprive the violent extremists of the fuel they need to spread their hateful ideologies."

— António Guterres, Secretary-General of the United Nations.

Uncountable losses of lives and assets happen every year all around the globe. The interesting thing to note is sometimes some organizations claims their hand behind or involvement in its execution while sometimes nobody takes the responsibilities of the attack. The objective of this work is getting more concrete idea, by analyzing all the

© Springer Nature Switzerland AG 2020
P. Ciancarini et al. (Eds.): SEDA 2018, AISC 925, pp. 146–158, 2020.
https://doi.org/10.1007/978-3-030-14687-0_13

incidents to build an intelligent model which can be engaged institutive predictions. It has been observed that terrorists usually do not claim responsibility for attacks and from the statistics says that terrorist groups claim credit for only one out of seven attacks [1–5].

Before we go deeper from physiology point of view, the questions that can come across the mind are: - What is the need for claiming an attack? What difference claiming and attack or not will make? The possible answer is claiming an attack publicly in the present time is a way to attract attention, to be taken seriously, to terrorize a whole range of people and different countries at the same time. Aftermath of Paris and Brussels attacks shocked not only France and Belgium, but Europe, America, and the whole world. Similarly, there are several reasons for not claiming an attack, such as to avoid retaliation or to jeopardize the hideout bases. Also, it is evident that they do not need the publicity in the regions of their stronghold. With the increasing reach to cutting edge-technology, and evolution of new ways of advance warfare, the whole planet is endangered. The motivations of terrorist organizations are clear, the question that naturally arises is: what can be done to strengthen counter terrorism? The answer is predicting the attacks beforehand and prevent them in time. However, the path to execution is not as simple as the answer seems. It requires several functional divisions like homeland security, intelligence agency, armed forces working together to connects the dots.

United Nation is also playing a vital role in combating terrorism. Peacekeeping operations have evolved to adapt and adjust to hostile environments, emergence of asymmetric threats and complex operational challenges that require a concerted multidimensional approach and credible response mechanisms to keep the peace process on track. The military component, as a main stay of a United Nations peacekeeping mission plays a vital and pivotal role in protecting, preserving and facilitating a safe, secure and stable environment for all other components and stakeholders to function effectively. The Global Counter-Terrorism Strategy in the form of a resolution and an annexed Plan of Action (A/RES/60/288) that included an overview of the evolving terrorism landscape, recommendations to address challenges and threats, and a compilation of measures taken by Member States and United Nations entities to fight against terrorism. And composed of 4 pillars:

- Addressing the conditions conducive to the spread of terrorism.
- Measures to prevent and combat terrorism.
- Measures to build states' capacity to prevent and combat terrorism and to strengthen the role of the United Nations system in that regard.
- Measures to ensure respect for human rights for all and the rule of law as the fundamental basis for the fight against terrorism.

In this, ongoing system artificial intelligence, machine learning and other software technology play a key role. Though fundamental remains the same- information. Correct information or accuracy and precision in the information is no doubt the game changer. As data science is now into everything for instance healthcare, finance, business and what now, its possible crucial contribution cannot be denied. This work contributes for the same to connect data mining and counters terrorism [6–8]. Data mining, text mining, sentiment analysis, machine learning systems, and predictive analytics are techniques which can be utilized to recognize and battle terrorism. Text mining helps us to decipher unknown information by extracting it automatically from different written sources [9–15].

2   Classification Techniques Used for Data Mining

Classification is a form of data analysis that extracts models describing important data classes. For example, we can build a classification model to categorize if the tumor is benign or malignant for patients going under treatment of possible breast cancer. The process is shown for the possible suspect data in a Figs. 1 and 2 in simplified way. Such analysis helps to provide us with a better understanding of the data at large scale. It is used for several research applications including fraud detection, terrorism prediction, target marketing, performance prediction, medical diagnosis, finance, weather prediction, business intelligence, homeland security. Several approaches are used for classification of datasets, as there are numerous techniques for classification and rule extraction. Classification algorithms can be categorized as probabilistic or non-probabilistic classifiers, Binary, and Multiclass classifier. Data classification is a two-step process namely learning step (where a classification model is constructed) and second step called classification step (where the model is deployed to predict the class labels of given data).

A sample, X is represented by an n-dimensional attribute vector, $X = (x_1, x_2, \ldots, x_n)$ depicting n measurements made on the sample from n attributes of database, $(A_1, A_2, \ldots, A_n)$. Each attribute represents a feature of X. The accuracy of a classifier for a given set of test samples is the percentage of test set samples that are correctly classified by the classifier. The associated class label of each test sample is compared with the learned classifier's class prediction for that sample [16–23]. A brief working insight of few prominent classifiers are presented below.

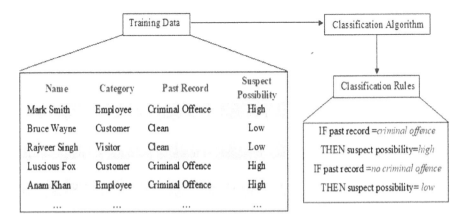

Fig. 1. Learning: training data are analyzed by a classification algorithm and here the class label attribute is suspect possibility, and the classifier is represented in the form of classification rules.

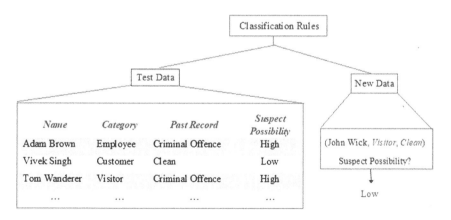

Fig. 2. Classification: test data are used to estimate the accuracy of the classification rules. If the accuracy is considered acceptable, the rules can be applied to the classification of new data samples.

2.1 Naïve Bayes

It is a classification technique based on Bayes' Theorem, in which it assumes that the presence of a particular feature in a class is unrelated to the presence of any other feature. This model is easy to build and particularly useful for very large data sets. Along with simplicity, Naive Bayes is known to outperform even highly sophisticated classification methods. It is easy to understand the algorithm in some steps. First, it converts the data set into a frequency table. After this create likelihood table by finding the probabilities. Lastly using the equation of Naïve Bayes, to calculate the posterior probability for each class. The outcome of prediction is the class with greatest posterior probability. In a probabilistic model, using Naïve Bayes theorem the conditional probability can be given by:

$$(C_K|x) = \frac{P(x|C_K)P(C_K)}{P(x)} \tag{1}$$

In other words, it can be written as Posterior = (prior × likelihood) | evidence. Under the conditional distribution over the class variable C, the equation can be given by:

$$P(C_K|x_1, \ldots, x_n) = \frac{1}{Z} (P(C_K) \prod_{i=1}^{n} P(|x_i|C_K) \tag{2}$$

where the evidence (Z) = P(x) is scaling factor, only dependent upon $x_1, \ldots, x_n$. Bayesian Classifier although have some complexities such as, it requires prior information of probabilities and in its absence, it is often predicted on the basis of background knowledge and earlier available data about original distributions.

2.2 Decision Tree

A decision tree is a flowchart-like tree structure, where each internal node (non-leaf node) denotes a test on an attribute, each branch represents an outcome of the test, and each leaf node (or terminal node) holds a class label. The topmost node in a tree is the root node. The construction of decision tree classifiers does not require any domain knowledge or parameter setting, and therefore is appropriate for exploratory knowledge discovery. Decision trees can handle multidimensional data. Their representation of acquired knowledge in tree form is intuitive and generally easy to assimilate by humans. The learning and classification steps of decision tree induction are simple and fast. In various computations, the characterization is executed recursively till every single leaf is immaculate, that is the data order which should be as flawless as would be prudent. The objective is a progressive observation of a decision tree until it gets adjusted of flexibility and precision. This method used the entropy that is the calculation of unstructured information. Here Entropy $\vec{X}$ is measured by:

$$\text{Entropy}\,(\vec{X}) = -\sum_{i=1}^{n} \frac{|Xi|}{|\vec{X}|} \log\left(\frac{|Xi|}{|\vec{X}|}\right) \tag{3}$$

$$\text{Entropy}\,(i|\vec{X}) = \frac{|X|}{|\vec{X}|} \log\left(\frac{|Xi|}{|\vec{X}|}\right) \tag{4}$$

Total gain can be represented by

$$\text{Total Gain} = \text{Entropy}\,(\vec{X}) - \text{Entropy}\,(i|\vec{X}) \tag{5}$$

2.3 Multilayer Perceptron

Multilayer Perceptron consists of a minimum of three layers of nodes and except the input nodes; each node is a neuron that uses a nonlinear activation function. It uses a supervised learning method known as back propagation for training. If a multilayer perceptron has a linear activation function in all neurons, then according to linear algebra any number of layers can be reduced to a two-layer input-output model. In perceptron learning happens by changing connection weights after each portion of data is processed, depending upon the amount of error in the output compared to the expected result. This is a supervised learning example, which is carried out by back propagation. The error in output node j in the nth data point is represented by

$$e_j(n) = d_j(n) - y_j(n) \tag{6}$$

where d is the target value and y is the value yielded by the perceptron. The node weights are managed on the basis of corrections that minimize the error in the entire output, given by

$$E(n) = \frac{1}{2} \sum_j e_j^2(n) \tag{7}$$

Using gradient descent, the change in each weight can be depicted as

$$\Delta \omega_{ji}(n) = -\eta \delta \frac{\delta E(n)}{\delta v_j(n)} y_i(n) \tag{8}$$

where y_i is the output of the previous neuron and η is the learning rate, which is selected to ensure that the weights swiftly converge to a response, without oscillations. The derivative depends on the induced local field v_j, which itself varies. For an output node the, derivative can be simplified to

$$-\frac{\delta E(n)}{\delta v_j(n)} = e_j(n) \emptyset'(v_j(n)) \tag{9}$$

where $\emptyset'$ is the derivative of the activation function previously mentioned, which does not vary by itself. The analysis is more tedious for the change in weights to a hidden node, although it can be shown that the required derivative is

$$-\frac{\delta E(n)}{\delta v_j(n)} = \emptyset'(v_j(n)) \sum_k \frac{\delta E(n)}{\delta v_k(n)} w_{kj}(n) \tag{10}$$

This depends on the change in weights of the nth nodes, which represent the output layer. In order to change the hidden layer weights, the output layer weights change according to the derivative of the activation function.

3 Methodology

The work itself starts from collection of data. All attempts were made to keep to data coherent and correct as much as possible. The second step is data preprocessing. It is a crucial undertaking and basic stride in text mining, and information retrieval. Data pre-processing is a data mining strategy that includes transforming raw data into a format suitable for experimental purposes. Data pre-processing stage is supposedly the most time-consuming of the whole knowledge discovery phase. The data is filtered after pre-processing by means of several filters such as linear and non-linear. All the attacks location with their corresponding latitude and longitude were embedded to the world map. The density plot snapshot is shown in Figs. 3 and 4 all attacks in the Word, Middle East and North Africa (40422 attacks) respectively. This research work focuses on classification based upon class – Attack Responsibility for the dataset comprised of reported terrorist attacks from 1970–2015. The classification algorithms applied for knowledge discovery for this data are Decision Tree random forest, Lazy classifier IBK linear NN, Lazy classifier IBK Filtered Neighbor Search, Lazy tree, IBK, Ball Tree, Lazy classifier K-star, Multilayer Perceptron, Multiclass Classifier and Naïve Bayes. So the class we have chosen for analysis for terrorist's attacks and causalities is based upon the fact if an organization has taken credit for it or not or whether it been an anonymous attack.

Fig. 3. Density distribution plot of all the terrorist attack (world) embedded with their latitude and longitude.

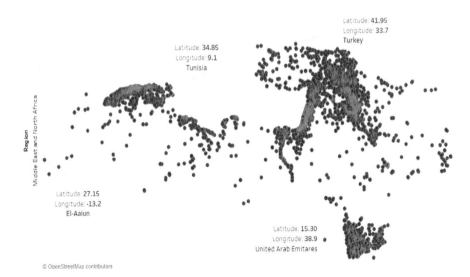

Fig. 4. Middle East & North Africa density distribution plot of 40422 terrorist attacks embedded with their latitude and longitude.

3.1   Description of Dataset

The terrorist attack data is obtained from the online data sources including sources released by the ministry of defense, government, private organizations and by collecting it individually. This data set is comprised of 156772 reported attacks all over the world, happened between the years 1970 to 2015. Analysis of the data after refining the

Table 1. Numerical distribution of experimental dataset.

Attack Type	AT-1	Armed Assault			
	AT-2	Assassination			
	AT-3	Bombing			
	AT-4	Facility/Infrastructure			
	AT-5	Hostage(Kidnapping)			
	AT-6	Hijacking			
	AT-7	Others			
Property Loss	S	Major			
	M	Moderate			
	L	Minor			
	U	Unknown			
Region	R1	Central America & Caribbean			
	R2	Central Asia			
	R3	East Asia			
	R4	Eastern Europe			
	R5	Middle East & North Africa			
	R6	North America			
	R7	Oceania			
	R8	South America			
	R9	Southeast Asia			
	R10	Sub-Saharan Africa			
	R11	South Asia			
	R12	Western Europe			
Weapon Type	WT-1	Explosives-Bombs			
	WT-2	Fake Weapons			
	WT-3	Firearms			
	WT-4	Incendiary			
	WT-5	Melee			
	WT-6	Miscellaneous			
	WT-7	Sabotage Equipment			
	WT-8	Unknown			
	WT-9	Vehicle			
Timeline	T-1	1970−75			
	T-2	1976−80			
	T-3	1981−85			
	T-4	1986−90			
	T-5	1991−95			
	T-6	1996−00			
	T-7	2001−05			
	T-8	2006−10			
	T-9	2011−15			

(continued)

Table 1. *(continued)*

			Claimed	No Claim	Anonymous
Target	TG-A	Airports & Aircraft			
	TG-B	Business			
	TG-C	Diplomats			
	TG-D	Educational Institutions			
	TG-E	Government Personal			
	TG-F	Journalists & Media			
	TG-G	Military			
	TG-H	Non-State Militia			
	TG-I	Others			
	TG-J	Police			
	TG-K	Private Citizens & Property			
	TG-L	Religious Figures /Institutions			
	TG-M	Telecommunication			
	TG-N	Tourists			
	TG-O	Transportation			
	TG-P	Utilities			

OK 20K 40K   2K 4K 6K 8K OK 20K 40K
Claimed   No Claim   Anonymous

raw data ends in nine attributes for categorizing data for this study. The attributes which were considered for this analysis were namely month of attack, year of the attack, Region, Weapon Type, Target, Attack Type, Data Source and Property Loss.

For this work "attack responsibility" was the class, spread over three categories which are Claimed, Not-Claimed and Anonymous. Further categorization of data is also done for ease of classification. Table 1 sums the entire data categorization. It depicts the numerical distribution of data in tabular form for a better understanding of class and attributes for the terrorist attack dataset. All the attributes their code, values, total and other factors are provided in the table. Division of total attacks has been done on the basis of the class attack responsibilities which are claimed, not-claimed and anonymous.

4   Results and Discussion

The classifiers used for this data are: Lazy classifier IBK linear NN, Lazy classifier IBK Filtered Neighbor Search, Lazy classifier IBK, Ball Tree, Lazy classifier K-star, Decision Tree Random Forest, Multilayer Perceptron, Multiclass Classifier and Naïve Bayes. We attained results with high-end accuracies ranging from 90–95%. Figure 5 shows the performance result of used classifiers. Tables 2, 3, 4, 5, 6 and 7 show the outputs achieved by Lazy Classifier IBK, Lazy Classifier K-star. Decision Tree Random Forest, Multilayer Perceptron, Multiclass Classifier and Naïve Bayes. This study endeavored to classify global terrorist attacks by utilizing different text mining classifiers such as Decision Tree random forest, Lazy classifier IBK linear NN, Lazy classifier IBK Filtered Neighbor Search and Lazy classifier K-star. We achieved fair accuracies in the context of classifiers performance. This analysis can be used to draw patterns for extracting information about the terror attacks. Some of the outcomes of the work are as follows:

- Out of total reported attacks from 1970 to 2015 which are 156772, 14664 are claimed, 75966 are Not-Claimed and 66142 are anonymous.
- The ratio of claiming the attacks by terrorist organizations varies from 10–15%. In this study, we have achieved a rate of 11%. (Although it is justified that more attacks happened from 2016 to present date and more responsibilities have been taken of those attacks so that is why the ratio is variable).
- The major stake is Asia in sustaining the attacks which are almost 45% of the total attacks.
- Government, Law Enforcement (Military and Police) bodies are the prime targets making up to 40% followed by Private Citizens and Property which is 22%.
- Losses occurred due to the terrorist attacks is highest for Moderate losses which are alone 64%.
- The types of attack such as bombing and armed assault are the most common which is 48.45% and 24.50% respectively.
- Explosives/Bombs and Firearms are the prominent weapon types deployed in these attacks with 50.47% and 33% respectively.

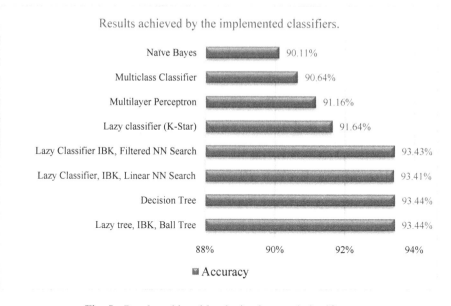

Fig. 5. Results achieved by the implemented classifiers.

Table 2. Result of Lazy Classifier IBK.

TP rate	FP rate	Precision	Recall	F-measure	MCC	ROC area	PRC area	Class
0.987	0.987	0.987	0.987	0.987	0.987	0.987	0.987	Claimed
0.114	0.114	0.114	0.114	0.114	0.114	0.114	0.114	No-claim
1.00	1.00	0.00	1.00	1.00	1.00	1.00	1.00	Anonymous

Table 3. Result of Lazy Classifier K-star.

TP rate	FP rate	Precision	Recall	F-measure	MCC	ROC area	PRC area	Class
0.999	0.161	0.853	0.999	0.921	0.845	0.972	0.962	Claimed
0.104	0.00	0.965	0.104	0.187	0.303	0.923	0.572	No-claim
1.00	1.00	0.00	1.00	1.00	1.00	1.00	1.00	Anonymous

Table 4. The result of Decision Tree.

TP rate	FP rate	Precision	Recall	F-measure	MCC	ROC area	PRC area	Class
0.987	0.114	0.891	0.987	0.936	0.875	0.983	0.979	Claimed
0.332	0.005	0.851	0.332	0.477	0.509	0.948	0.669	No-claim
1.00	1.00	0.00	1.00	1.00	1.00	1.00	1.00	Anonymous

Table 5. The result of Multilayer Perceptron.

TP rate	FP rate	Precision	Recall	F-measure	MCC	ROC area	PRC area	Class
0.984	0.156	0.856	0.984	0.916	0.834	0.954	0.929	Claimed
0.141	0.008	0.611	0.141	0.230	0.268	0.864	0.389	No-claim
1.00	1.00	0.00	1.00	1.00	1.00	1.00	1.00	Anonymous

Table 6. The result of Multiclass Classifier.

TP rate	FP rate	Precision	Recall	F-measure	MCC	ROC area	PRC area	Class
0.999	0.180	0.839	0.999	0.912	0.828	0.942	0.910	Claimed
0.009	0.001	0.511	0.009	0.018	0.059	0.826	0.269	No-claim
1.00	1.00	0.00	1.00	1.00	1.00	1.00	1.00	Anonymous

Table 7. The result of Naïve Bayes.

TP rate	FP rate	Precision	Recall	F-measure	MCC	ROC area	PRC area	Class
0.986	0.176	0.841	0.986	0.908	0.817	0.939	0.906	Claimed
0.010	0.002	0.358	0.010	0.019	0.048	0.818	0.246	No-claim
1.00	1.004	1.994	1.00	1.997	1.995	1.00	1.00	Anonymous

5 Conclusion and Future Work

As the analysis is user need based approach, several more patterns can be found. For future research work, there are several horizons to work upon with this analysis. First and foremost is achieving an accuracy of classifiers up to 99%. This can be done by genetic algorithms and deep neural networks along with the classifiers used in this work or a combination of the different classifier. In this study, we worked up to the tertiary classification of some of the attributes and on average up to secondary for each attribute. It has benefits and drawbacks also. Limiting the categorization of attributes reduces the computation complexity but for classes having a marginally low occurrence, it induces some biasing.

One of the objectives is to also increase the sub-classification layers and attributes both in order to find more useful trends. As with large data set expectation by which we mean Big Data: is to get rid of bias and variance in prediction. Secondly, using terrorist data it is possible to predict the organizations involved in the reported attacks. However, further study can be carried out to detect a terrorist group using historical data. In this lieu of work data mining, approaches can be deployed from finding the trends to counter terrorism by developing Terrorist Group Prediction Model. The parameters for the building could be web-based content, activities, social network data, phone calls, emails etc. by using the graph and pattern mining.

References

1. Tolan GM, Soliman OS (2015) An experimental study of classification algorithms for terrorism prediction. Int J Knowl Eng 1:107–112. https://doi.org/10.7763/ijke.2015.v1.18
2. Malathi A, Santhosh Baboo S (2011) Evolving data mining algorithms on the prevailing crime trend – an intelligent crime prediction model. Int J Sci Eng Res 2
3. Sachan A, Roy D (2012) TGPM: terrorist group prediction model of counter terrorism. Int J Comput Appl 44(10):49–52
4. Thuraisingham B (2003) Data mining, national security, privacy and civil liberties. SIGKDD Explor 4(2):1–5
5. Kumar V, Zinovyev R, Verma A, Tiwari P (2018) Performance evaluation of lazy and decision tree classifier: a data mining approach for global celebrity's death analysis. In: International conference on research in intelligent and computing in engineering (RICE). IEEE Xplore, pp 1–6. https://doi.org/10.1109/rice.2018.8509082
6. Bolz F et al (2001) The counterterrorism handbook: tactics, procedures, and techniques. CRC Press, Boca Raton
7. Ghosh A (1998) Ecommerce security: weak links and strong defenses. Wiley, New York
8. Thuraisingham B (2003) Web data mining technologies and their applications in business intelligence and counter-terrorism. CRC Press, Boca Raton
9. Hehenberger MR (2002) Text-based knowledge discovery: search and mining of life-science documents. Drug discovery today 7(11, Suppl):S89–S98
10. Navathe SB, Ramez E: Data warehousing and data mining. In: Fundamentals of database systems. Pearson Education Pvt Inc., Singapore, pp 841–872 (2000)
11. Michael WB (2004) Automatic discovery of similar words. In: Survey of text mining: clustering, classification and retrieval. Springer Verlag, New York, LLC, pp 24–43

12. Salvetti F, Lewis S, Reichenbach C (2004) Automatic opinion polarity classification of movie. Colo Res Linguist 17:2
13. Zhang D, Xu H, Su Z, Xu Y (2015) Chinese comments sentiment classification based on word2vec and SVM perf. Expert Syst Appl 42(4):1857–1863
14. Joachims T (2006) Training linear SVMs in linear time. In: Proceedings of the 12th ACM SIGKDD international conference on knowledge discovery and data mining. ACM, pp 217–226
15. Boger Z, Kuflik T, Shoval P, Shapira B (2001) Automatic keyword identification by artificial neural networks compared to manual identification by users of filtering systems. Inf Process Manag 37:187–198
16. Pierrea S, Kacanb C, Probstc W (2000) An agent-based approach for integrating user profile into a knowledge management process. Knowl-Based Syst 13:307–314
17. Provost F, Fawcett T (2001) Robust classification for imprecise environments. Mach Learn 42(3):203–231
18. Evfimievski A, Srikant R, Agrawal R, Gehrke J (2002) Privacy preserving mining of association rules. In: Proceedings of the eighth ACM SIGKDD international conference on knowledge discovery and data mining, Edmonton, Alberta, Canada
19. Kalpana R, Bansal KL (2014) A comparative study of data mining tools. Int J Adv Res Comput Sci Softw Eng 4:216–223
20. Kumar V, Tiwari P, Mishra BK, Kumar S (2017) Implementation of n-gram methodology for rotten tomatoes review dataset sentiment analysis. Int J Knowl Disc Bioinf (IJKDB) 7 (1):30–41. https://doi.org/10.4018/ijkdb.2017010103
21. Kumar V, Verma A, Mittal N, Gromov SV (2019) Anatomy of preprocessing of big data for monolingual corpora paraphrase extraction: source language sentence selection. In: Emerging technologies in data mining and information security. Advances in intelligent systems and computing, vol 814. Springer Nature, Singapore, pp 495–505. https://doi.org/10.1007/978-981-13-1501-5_43
22. Kumar, V, Kalitin D, Tiwari, P (2017) Unsupervised learning dimensionality reduction algorithm PCA for face recognition. In: International conference on computing, communication and automation (ICCCA). IEEE Xplore, pp 32–37. https://doi.org/10.1109/ccaa.2017.8229826
23. Han J, Kamber M, Pei J (2012) Data mining concepts and techniques, 3rd edn. Morgan Kaufmann, Burlington

Game Theoretic Framework to Mutation Testing

JooYoung Lee$^{(\boxtimes)}$ (iD)

Innopolis University, Innopolis, Russia
`j.lee@innopolis.ru`

Abstract. Mutation testing is an intuitive approach to test errors in software as well as to measure the quality of test suites. Due to its prohibitively high computation cost for generating all mutants, different approaches have been proposed to alleviate the cost. In this paper, first we introduce such efforts and then propose a new framework, *mutation game*. We formulate a game between Tester who wants to kill mutants and Demon who supports living mutants. We also propose strategies and an algorithm for the repeated game.

Keywords: Software testing · Mutation testing · Game theory

1 Introduction

Software testing is an essential process of software development. Typically, short term goals of software debugging includes bug discovery and bug prevention. While traditional debugging methods provide effective tools to remove bugs before publishing, evaluating the quality of software tests heavily relies on measuring the coverage of test cases. There are different types of coverage criteria including *node coverage, prime path coverage, all-uses coverage*, etc. It is shown that there is some statistically significant correlation between code coverage and bug kill effectiveness [4]. The coverage criteria, however, do not necessarily signify the quality of test suites. Also, various practical limitations exist in using code coverage for testing, such as finding an appropriate coverage tool, and the time overhead to check many coverage criteria.

Mutation testing generates *mutants*, mutated versions of the original program, and test cases are expected to 'kill' (detect and reject) the *mutants*. Therefore, from survived mutants we can locate weak points in test data and help testers to design new software tests. Also, *mutation score*, the number of killed mutants divided by the number of total mutants, can be a metric to evaluate the quality of existing software tests. Mutation testing is widely considered the strongest test criterion in terms of finding the most faults [8].

There are two fundamental hypotheses to support the theoretical validity of mutation testing [3]. The first is called *Competent Programmer Hypothesis (CPH)* which states that programmers are competent and therefore faults are

© Springer Nature Switzerland AG 2020
P. Ciancarini et al. (Eds.): SEDA 2018, AISC 925, pp. 159–164, 2020.
https://doi.org/10.1007/978-3-030-14687-0_14

merely a few simple faults which can be corrected by a few small syntactical changes [1,2]. The second is called *Coupling Effect* which states that "Test data that distinguishes all programs differing from a correct one by only simple errors is so sensitive that it also implicitly distinguishes more complex errors" [2].

2 Related Works

Mutation testing is considered one of the strongest criterion to find the most number of faults [8]. Due to its computational cost, however, it has little practical use. In this section, we introduce some of the efforts which strategically reduce the number of generated mutants.

Petrovic and Ivankovic proposed a probabilistic approach to limit the number of generated mutants to one per each line [8]. In this approach, mutants are only generated from interesting lines avoiding *arid* lines. A line or node from Abstract Syntax Tree (AST) is defined arid if it is associated with testing or it controls non-functional properties or it is axiomatic. Authors show that by restricting the generation of uninteresting mutants, the cost of computation is significantly lowered and at the same time, the quality of mutants improved. Reported through the developer tool, 75% of the findings were useful.

Mathur suggested to use less mutation operators by omitting two operators which produce the most number of mutants, namely SVR (scalar variable replacement) and ASR (array reference for scalar variable replacement), from 22 Mothra mutation operators [6]. As a result, *constrained mutation* achieves linear complexity in program size. In the following work [9], authors also showed that by applying *constrained mutation*, while reducing the size of the test suites, the fault detection effectiveness stayed almost the same compared to non-constrained mutation.

Offutt et al. [7] extended the idea of *constrained mutation* proposed by Mathur [6]. They called it *2-selective mutation* and extended the theory to *N-selective mutation* by omitting N most plentiful operators. In this work, authors presented results using *2-selective*, *4-selective*, and *6-selective* mutations. The obtained savings were 23.98%, 41.36%, and 60.56% respectively. Obviously, the more operators omitted, the more savings (less generated mutants) were obtained. In terms of mutation scores, each mutation set achieved 99.99%, 99.84%, and 99.71% respectively. The findings shed light on the coupling effect of the operators and the strength of each operator should be accounted a further elimination.

3 Mutation Game

In this section, we propose a *mutation game*. In the game, there are two players, namely Tester and Demon. Tester uses a set of test cases as its strategy, $S_T = \{t_1, t_2, \ldots, t_n\}$, and Demon has a set of mutants as its strategy, $S_D = \{m_1, m_2, \ldots m_m\}$. Note that a single test can kill multiple mutants and multiple tests can kill the same mutants. Mutants for which at least one test fails are killed. Consequently, mutants that are not detected by the test suite are called living mutants [8].

Each player takes a turn and plays an action and the game is repeated. Assume that after each round of the game, both players know which action was played by each other. Let us suggest a simple payoff matrix where payoffs for each player are either 1 or 0. If t_i can kill m_j, the payoffs for the players are 0 for Demon and 1 for Tester. We formally define the game as follows.

Definition 1 *(Mutation game). Let (S, f) be a game with 2 players, where S_i is the strategy set for player i, $S = S_T \times S_D$ is the set of strategy profiles and $f(x) = (f_T(x), f_D(x))$ is its payoff function evaluated at $x \in S$. Let x_i be a strategy profile of player i and x_{-i} be a strategy profile of all players except for player i. When each player $i \in \{T, D\}$ chooses strategy x_i resulting in strategy profile $x = (x_T, x_D)$ then player i obtains payoff $f_i(x)$.*

Tester

		t_1	t_2	...		t_n
	m_1	0, 1	0, 1			1, 0
	m_2	1, 0	1, 0			0, 1
Demon	...					
	m_m	1, 0	1, 0	0, 1	1, 0	1,0

Fig. 1. Example of a mutation game. In this game, a mutant m_1 can be killed by any of the tests t_1 and t_2. Also, m_2 can be killed by t_n.

Figure 1 shows payoffs for each player, Demon and Tester. Let us further assume that after each iteration, mutants are killed if mutants are detected by a test case but test cases are not removed.

3.1 Strategies

In a repeated game, players can adopt strategies. In this section, we propose several strategies and reason about the results of the game. We safely assume that there is no dominant strategy solution for this game since it is not realistic/useful if there exist a mutant or test case that is strictly dominated by another.

Strategy1: (Random). The first strategy is a random strategy where Demon and Tester randomly picks a mutant and a test case respectively. The game is repeatedly played until all the mutants are killed.

Strategy2: (Sequential). The second strategy is a sequential move. Both Demon and Tester picks the first possible action to start with, e.g., (m_1, t_1). If m_1 killed by t_1, Demon plays m_2 in the next round. Else, Demon plays m_1 again until it is killed. Similarly, if t_1 kills m_1, Tester plays t_1 again in the next round. When t_1 no longer kills a mutant, Tester moves on to the next test case, t_2.

Strategy3: (Probabilistic). The third strategy is based on probability. In this scenario, both players keep probabilities associated to the number of mutants and test cases. Probability of each mutant, p_{m_j} and probability of each test case, p_{t_k} is defined in Eqs. 1 and 2.

$$p_{m_j} = \frac{|m_j|}{\Sigma_{j \in M} |m_j|} \tag{1}$$

where $|m_j|$ is the number of test cases that can kill m_j.

$$p_{t_k} = \frac{|t_k|}{\Sigma_{k \in N} |t_k|} \tag{2}$$

where $|t_k|$ is the number of mutants that t_k can kill.

After each round of the game, the probabilities are updated accordingly. If a mutant is killed at any round, it is removed from the strategy immediately. On the other hand as long as a test case can potentially kill a mutant, it stays in the strategy.

> **Input:** $G = (S, f)$
> **Output:** $F = \Sigma f^i$
> i = 0;
> $j = \underset{j}{\operatorname{argmax}} \, p_{m_j}$;
> $k = \underset{k}{\operatorname{argmax}} \, p_{t_k}$;
> **while** $S_D \neq \emptyset$ **do**
> $\quad$ $x^i = (m_j, t_k)$;
> $\quad$ $F = F + f^i$;
> $\quad$ $i = i + 1$;
> $\quad$ **if** m_j is killed by t_k **then**
> $\quad\quad$ $S_D = S_D - \{m_j\}$;
> $\quad\quad$ Update p_{t_k} for all k;
> $\quad\quad$ $k = \underset{k}{\operatorname{argmax}} \, p_{t_k}$;
> $\quad\quad$ **if** t_k cannot kill more mutants **then**
> $\quad\quad\quad$ $S_T = S_T - \{t_k\}$;
> $\quad\quad\quad$ Update p_{m_j} for all j;
> $\quad\quad\quad$ $j = \underset{j}{\operatorname{argmax}} \, p_{m_j}$;
> $\quad\quad$ **end**
> $\quad$ **else**
> $\quad\quad$ $k = rand()\%N + 1$;
> $\quad$ **end**
> **end**

Algorithm 1. Repeated Mutation game using the probabilistic strategy. In theory, if any mutant survives, the game can be infinite. f^i is the payoff function at i^{th} iteration given strategies of both players. F is the accumulation of payoffs given from f at each iteration.

3.2 Nash Equilibrium

It is proved that every game with a finite number of players in which each player can choose from finitely many pure strategies has at least one mixed strategy Nash Equilibrium. We omit to present Nash Equilibrium of the proposed game since it is meaningless without a definite set of strategies. We plan to introduce a repeated Mutation game in which we will find Nash Equilibrium (generally different from that of a stage game) [5].

4 Discussion

The proposed game is simple but not realistic. We need to include the cost for each action, i.e., mutant and test case, to be able to find meaningful strategies and Nash Equilibrium. The game is also pointless if we had a test case which can kill all the mutant or a mutant which cannot be killed by any test case (since either of them will dominate all other strategies). Therefore, to include real scenarios to the game, we need to have definite action profiles for the game.

5 Conclusion

In this paper, we presented a framework for Mutation game and its strategy variants. To complete the game theoretic framework, we plan to provide the solution of the game, i.e., Nash Equilibrium, as well as the repeated version of the game and its convergence. Then, we can discuss its usefulness in eliminating unnecessary (soon to die) mutation variants as well as improving the quality of test cases by removing unused test cases.

Acknowledgments. The author would like to thank Dr. Bertrand Meyer for his insights, Dr. Jooyong Yi for his technical comments and Dr. Victor Rivera for his helpful advice. The author, however, bears full responsibility for the paper.

References

1. Acree AT, Budd TA, DeMillo RA, Lipton RJ, Sayward FG (1979) Mutation analysis. Technical report GIT-ICS-79/08, Georgia Institute of Technology, Atlanta, Georgia
2. DeMillo RA, Lipton RJ, Sayward FG (1978) Hints on test data selection: help for the practicing programmer. Computer 11(4):34–41
3. Johansson E (2016) Evaluating the effectiveness of test coverage criteria using mutation analysis: an evaluation of test coverage criteria in C
4. Kochhar PS, Thung F, Lo D (Mach 2015) Code coverage and test suite effectiveness: empirical study with real bugs in large systems. In: 2015 IEEE 22nd international conference on software analysis, evolution, and reengineering (SANER), pp 560–564. https://doi.org/10.1109/SANER.2015.7081877
5. Lee J, Oh JC (September 2014) Convergence of true cooperations in Bayesian reputation game. In: 2014 IEEE 13th international conference on trust, security and privacy in computing and communications, pp 487–494. https://doi.org/10.1109/TrustCom.2014.61

6. Mathur AP (September 1991) Performance, effectiveness, and reliability issues in software testing. In: [1991] Proceedings the fifteenth annual international computer software applications conference, pp 604–605. https://doi.org/10.1109/CMPSAC.1991.170248

7. Offutt AJ, Rothermel G, Zapf C (1993) An experimental evaluation of selective mutation. In: Proceedings of the 15th international conference on software engineering, Baltimore, Maryland, USA, 17–21 May 1993, pp 100–107. http://portal.acm.org/citation.cfm?id=257572.257597

8. Petrovic G, Ivankovic M (2018) State of mutation testing at Google. In: Proceedings of the 40th international conference on software engineering 2017 (SEIP)

9. Wong WE, Horgan JR, London S, Mathur AP (1995) Effect of test set minimization on fault detection effectiveness. In: Proceedings of the 17th international conference on software engineering. ICSE 1995. ACM, New York, pp 41–50. https://doi.org/10.1145/225014.225018

A Survey on Code Analysis Tools for Software Maintenance Prediction

Valentina Lenarduzzi[1], Alberto Sillitti[2(✉)], and Davide Taibi[1]

[1] Tampere University of Technology, Tampere, Finland
{valentina.lenarduzzi,davide.taibi}@tut.fi
[2] Innopolis University, Innopolis, Russian Federation
a.sillitti@innopolis.ru

Abstract. Software maintenance is a widely studied area of software engineering that it is particularly important in safety-critical and mission-critical applications where defects may have huge impact and code needs to be checked carefully through the analysis of data collected using a number of tools developed to investigate specific aspects. However, such tools are often not available to practitioners preventing them from applying the most recent and advanced approaches to industrial projects. This paper is an initial investigation about code analysis tools used to perform research studies on software maintenance prediction. We focus on the identification of tools that are available and can be used by practitioners to apply the same maintenance approaches described in published academic papers.

Keywords: Software maintenance · Tools · Software measurement · Software quality

1 Introduction

Software maintenance is a deeply studied area with hundreds of studies performed every year and published in top international conferences and journals. In many cases, these studies are empirical ones and based on data collected from a wide range of sources and then analyzed using a variety of mathematical techniques [5]. Moreover, most of the research activities and results presented in such studies are based on the usage of a wide range of software tools.

The code analysis tools used in these studies are very diverse and, in many cases, are ad-hoc developed prototypes used only to perform a single study [5]. In such cases, the developed tools are usually not maintained and are often not even released by the authors. This makes the replication of such studies very difficult and the adoption of the proposed approach nearly impossible for practitioners.

This is a clear limitation of the research activities in the software maintenance area that need to be more open to external validation of the performed researches to push ahead the state of the art and have a real impact on the practice.

As pointed out in a recent analysis of the evolution of the software maintenance research area [5], maintenance models have increased their complexity over the last 40 years using more sophisticated mathematical approaches for data analysis. However,

P. Ciancarini et al. (Eds.): SEDA 2018, AISC 925, pp. 165–175, 2020.
https://doi.org/10.1007/978-3-030-14687-0_15

this increase of complexity does not result in an improvement of performance of the developed models and the external validation of the approaches is almost inexistent, also due to the lack of availability of the tools used to extract and analyze the data. For such reasons, researchers are almost reinventing the wheel in each study and do not leverage on the work performed by others.

To help researchers and practitioners in the identification of available code analysis tools that have been used to publish scientific studies, we have classified them with regard to their support for software maintenance activities. We performed a review of the available tools, reporting both on commercial and open source products. Since the tools cannot be reliably retrieved from bibliographic sources, we adapted the systematic mapping approach defined by Petersen [1] by adding a step specifically intended for the identification of tools. We considered code analysis tools supporting maintenance activities as defined in the ISO/IEC 25010:2011 SQuaRE [4] (modularity, reusability, analyzability, modifiability, testability, reliability).

Only a few previous works have compared existing code analysis tools for specific software maintenance areas [6–8]. Unlike our work, [6] reports an overview on search-based optimization techniques, proposing four tools for software modularization support. Unfortunately, none of these tools are available anymore.

In [7], the authors compare three static code analysis tools (Fortify SCA, Splint, and Frama-C). However, the comparison focuses on the point of view of detecting security vulnerabilities.

In [8], the authors highlight the issues of detecting Java concurrency bugs using static code analysis tools (FindBugs, CheckThread, RacerX, and RELAY).

Moreover, to the best of our knowledge, no previous works exists that systematically identifies existing and available code analysis tools for software maintenance.

The remainder of this paper is structured as follows: in Sect. 2, we introduce the adopted methodology; in Sects. 3 and 4, we present and discuss the results; in Sect. 5, we point out the threats to the validity of this study; finally, in Sect. 6, we draw the conclusions and sketch possible future work.

2 Methodology

To perform the survey in a systematic way and achieve results that can be considered valid by the scientific community, we have analyzed in deep the most popular approaches for performing reviews in a coherent and replicable way. In particular, we have considered: Systematic Literature Review (SLR) [2], Multivocal Literature Review (MLR) [3], Systematic Mapping (SM) [1]. We have realized that none of them are suitable for our study for a number of reasons:

1. SLRs focus only on peer-reviewed publications. However, many tools are not described in usual academic papers that are subject to peer evaluations.
2. MLRs give the same level of importance to peer and non-peer reviewed sources. Even if they fit better our goals, they do not provide a specific procedure to integrate the results and weight the contributions.

3. SMs are very good for very initial studies and collect basic information in an unstructured way. However, they are too generic and do not provide a rigorous enough framework for the identification of the tools we are interested to include in this study.

Based on such considerations, we have realized that an extension of the SM approach was more suitable to address the problem we face in this investigation (Sect. 2.2).

2.1   Goal and Research Question

The goal of this study is to identify and classify code analysis tools for supporting software maintenance that have been adopted in industry and validated academically.
Therefore, we define our main Research Questions (RQs):

- RQ1: Which are the most commonly used code analysis tools to support software maintenance? (We aim at classifying the tools)
- RQ2: Have these tools been adopted in research? (We aim at identifying the most popular tools used to perform research activities).

2.2   Research Strategy

In our work, we follow a different procedure compared to the ones commonly adopted in systematic mappings, systematic literature reviews, and multivocal literature reviews since all of them have the limitations described at the beginning of Sect. 2.
Therefore, the process developed is based on the following steps:

Step 1: Tools Identification. We report the search strategy adopted for identifying the tools available on the web.

Step 1.1: Keywords definition. Based on our RQs, we define the search terms applied for the identification of the tools, as presented in Table 1. We adopted the PICO structure (Population, Intervention, Comparison, and Outcome), skipping the Outcome and Comparison terms, since the focus of our research is a general investigation, as suggested by Kitchenham and Charters [2]. To retrieve a reliable set of tools, we expanded the term "maintenance" with the terms included in the maintenance sub-characteristics defined in the standard ISO/IEC 25010:2011 SQuaRE [4].

Table 1. The search terms.

P: Software maintainability	**P1 terms:** "software", "maintainability", "maintenance", "modularity", "reusability", "analyzability", "modifiability", "testability", "reliability"
I: tool	**I1 terms:** "tool*", "static analysis", "dynamic analysis"

Step 1.2: Bibliographic Sources definition. Differently from the traditional approach for systematic review, instead of searching in bibliographic sources suggested by Kitchenham and Charters [2], we applied the search terms in google.com. This choice allowed us to broaden the search to non-scientific results to get the most comprehensive list of tools.

Step 1.3: Inclusion and exclusion criteria definition. We defined inclusion and exclusion criteria to identify the most relevant tools. We obtained the final criteria (Table 2) by means of refinements from an initial set of inclusion and exclusion criteria.

Table 2. The search terms.

Inclusion criteria	Exclusion criteria
Open source or commercial tool	Tool not available (e.g., not actively developed, never released, etc.)
	Tool does not directly support maintenance activities (e.g., IDEs)
	Tools that are not based on source code analysis (e.g., issue tracking)

Step 1.4: Search and Selection process. Two authors separately searched in the selected bibliographic source the tools according to the defined search keywords. They manually checked each retrieved tool by means of visiting the official web page and they applied the inclusion and exclusion criteria. In case of disagreement between the two authors, the third author was involved so as to apply the criteria. The search and selection process returned 25 tools.

From the retrieved 60 tools, we rejected 14 tools related to hardware maintenance, 18 research prototypes, including 14 not available anymore and 4 never mentioned in industrial case studies. Finally, we also rejected three IDEs tools.

Step 2: Popularity Assessment. The popularity of the tools identified in the previous step needs to be evaluated. This is an important step of the procedure, since it allows to consider differently tools that have been used or mentioned online very rarely. The results of the adoption are very important for practitioners: most of the companies cannot risk adopting a prototype or a tool with limited support and/or a narrow community.

For this purpose, we have considered the trend of searches in the last 5 years using Google Trends and the number of hits reported in google.com. Even if the numbers provided are not absolute, they help us in investigating the turnover.

These two values can be used as proxy for assessing the adoption of a tool by practitioners. Beside a number of Google results does not imply the same number of installations, we can assume that if tool "A" is mentioned 100K more than tool "B", the tool "A" is used and adopted much more than "B".

The trend reported in Google Trends, when paired with the number of results, can be a good sign of adoption trend of a tool.

The result of this step is the number of results reported in google.com and its 5-years trend reported in Google Trends (Decreasing, Increasing, Constant, No-Trend).

Step 3: Papers reporting experience using the selected tools. We identified papers reporting the usage of the identified tools to prove the adoption of the tools reported in the scientific literature. Moreover, we investigated existing case studies reporting the usage of the tools in industrial contexts.

To investigate the popularity of the tools in the research community, we adopted the traditional approach based on citations in scientific publications. We analyzed the tool adoption by identifying peer-reviewed publications reporting their usage. The number of reported citations of the tools can be considered as a proxy-measure for assessing the suitability of the tool to support repeatable research investigations that could be performed by advanced practitioners.

Step 3.1: Keywords Definition. In our case, the search terms needed to be defined by combining the search terms identified in Table 1 and the list of tools retrieved.

Step 3.2: Identification of bibliographic sources. A search process can be conducted automatically or manually across specific journals and conferences. To better address this step, we decided to combine both procedures.

Step 3.3: Inclusion and exclusion criteria definition. The results obtained from the search were then filtered by applying the inclusion and exclusion criteria to the title and the abstract. In this step, we excluded research prototype tools identified in the previous step that were never mentioned in the industrial case studies reported in the selected papers, even if they are available and even if they have existing citations in the scientific literature that does not include an industrial case study.

Step 3.4: Search and Selection process. Two authors separately searched in the selected bibliographic source the tools according to the defined search keywords. They manually evaluated each retrieved paper and they applied the inclusion and exclusion criteria. In case of disagreement between the two authors, the third author was involved so as to apply the criteria. The results of this step are reported in Table 3.

2.3   Data Extraction

In this step, we extract the relevant data from the tools that passed the inclusion and exclusion criteria.

3   Results

In the Step 1 of the Search Strategy approach (Tools identification), we retrieved 60 tools. Thanks to the inclusion and exclusion criteria we rejected 35 tools, resulting in 25 selected tools for this review (Table 3).

About the rejected tools, they are all research prototypes and in most of the cases executables or source code are no longer available. Unfortunately, some research prototypes that are also mentioned in industrial case studies are not available anymore (PROM [11, 13, 14], MacXim [17, 18] and RIGI [12]), while the four available tools are rarely mentioned in research works and never in industrial case studies.

Table 3. The tools classification.

Tool	Goal	Supported language	License	Cit.	Google results	Trend
CodeSonar	Code review	Java, C, C++	Comm.	347	56,500	=
Findbugs	Bug detection	Java	LGPL	346	656,000	↓
Coverity	Bug detection/testing	Java, C, C++	Comm.	315	235,000	=
PMD	Bug detection/testing	Java	LGPL	238	502,000	=
Polyspace	Run-time errors	C, C++	Comm.	212	110,000	=
Checkstyle	Coding standards	Java	LGPL	155	438,000	↓
Klocwork	Safety, reliability	Java, C, C++, C#	Comm.	132	79,400	↓
Parasoft Jtest	Testing	Java. C, C++, .NET	Comm.	114	548,000	=
Squale	Code review	Java, C, C++, .NET, PHP, Cobol	LGPL	110	291,000	↑
IBM AppScan	Testing, security flaws	Java	Comm.	86	364,000	=
JLint	Code review	Java	LGPL	86	141,000	=
SonarQube	Code review	All Comm. Lang.	LGPL + Comm.	74	417,000	↑
Lattix	Technical debt, modularity, reusability	Java, C, C++	Comm.	69	93,700	↓
Fortify Static Code Analyzer	Code analysis, vulnerability detection	All common lang.	Comm.	50	581,000	=
ConQAT	Code review	Java, C#, C++, ABAP, ADA	Comm.	37	25,800	=
Ndepend	Dependency analysis	.NET	Comm.	35	85,600	↓
CAST	Code review	All Comm. languages	Comm.	24	134,000	↑
Structure101	Architecture analysis	Java, .NET	Comm.	24	131,000	=
LDRA testbed	Dynamic code analysis	Java	Comm.	20	348,000	=
Axivion Bauhaus Suite	Reverse engineering and architecture recovery	Ada, C, C++, C# and Java	Comm.	3	1,710	=
Source meter	Code review	Java, C/C++, C#, Python and RPG	Freeware and Comm.	2	262,000	=
Jarchitect	Structural code analysis	Java	Comm.	2	65,700	=
Imagix	Structural code analysis	Java, C, C++	Comm.	1	399,000	=
Codacy	Code review	JavaScript, Scala, Java, PHP, Python, CoffeeScript, CSS, Ruby, Swift, C/C++	Comm.	1	83,100	↑
Parasoft dotTEST	Code review	C/C++, Java, .NET	Comm.	1	25,200	↓

3.1   RQ1. Which Are the Most Commonly Used Tools to Support Software Maintenance?

The most mentioned tools in google.com are Findbugs, Fortify, JTest, and PMD, all above 500K citations. However, analyzing the trend of the results in Google, no significant changes emerged or, at least, the tools with an increasing trend still have a very low number of results compared with the remaining tools. The only partial exception is SonarQube with over 400K citation and an increasing trend. This result also supports the importance of the tools reported.

To understand the purpose of this tools, we classified them based on the maintenance activities they can support.

Code Review is one of the most often considered activities by the selected tools, with 33% related tools, mainly for code understandability, support for code inspection, and error prediction. Reverse Engineering, Structural Code Analysis, and Testing are also popular activities, with 12.5% of the tools supporting each of them. Moreover, other tools focus on specialized activities such as Dynamic Code Analysis, Analysis of Technical Debt, Modularity and Reusability, and Security Flaws Analysis.

Considering the license of the selected tools, 83% of them are distributed with a commercial license, while the remaining ones are freely available with open source licenses (mainly LGPL).

Considering the supported programming languages, Java is the most frequently considered one, immediately followed by C and C++.

3.2   RQ2. Have These Tools Been Adopted in Research?

The selected tools have an average of 90 citations. However, the average is highly pushed up by the first three tools (CodeSonar, FindBugs, and Coverity) with more than 250 citations, which together account for nearly half of all citations. Fifteen tools have a number of citations below the average, with six tools having fewer than three citations.

Several academic works adopt tools with a very low number of results in google. com. As example CodeSonar and Polyspace are used by several research works but they are not well known or with a very low number of Google results. Moreover, these tools are mainly applied in non-industrial case studies.

This confirms the low level of adoption of several tools in academic works, including industrial case studies. In 60% of the cases, the interest in such tools (according to Google Trends) is almost constant, while in 16% of the cases is increasing and in 24% of the cases is decreasing. Therefore, we can conjecture that in most of the cases users could be satisfied with the tools they use or there are no better alternatives they can consider.

4 Discussion

Focusing on the two research questions, we have found out that the tools commonly used by practitioners and by researchers are very different, with only a single common tool among the most popular in the two cases: FindBugs.

We have also analyzed if there is a correlation between then citations in google.com and the ones in the scientific literature but the results were not significant. Even accepting an unusually low level of confidence, the correlation is very weak (0.35) showing that the two sets of citations are almost completely unrelated from each other.

These facts could be due to several reasons that need further investigation (e.g., difficulty of usage, different objectives, etc.). However, it is quite clear that the practitioners' and the researchers' communities value different tools to perform their work.

In any case, the most popular tools in research have a constant or negative trend for practitioners, the same happens for the most popular tools for practitioners. This is a clear evidence that neither set of tools are actually able to satisfy practitioners and there are no clear alternatives. The increasing trends are very limited and focused on tools with a low level of popularity, this could be related to the fact that they are being tested by practitioners, but such tools are not very satisfying and/or being able to be adopted by a large set of users.

We can also notice that almost all the tools identified are commercial and the availability of open source ones is very limited. In particular, the most used tools in research are available for free (as open source or with a free license for a subset of the features) while most of the tools used by practitioners are commercial. Since most researchers have limited budgets, it may happen that the features and the quality of the tools could be considered of less importance compared to the free availability. This may affect their usage in research projects and their citation in research papers. However, this is an aspect that needs further investigation. Moreover, tools are commonly used based on need and they are not used to continuously support the development process. In the selected literature, only two tools were integrated into the development pipeline (SonarQube and Cast). The integration into a DevOps pipeline, into a continuous monitor process, could provide higher benefit on their usage [15, 16].

Another aspect that needs to be considered is the lack of reliable maintenance models developed by researchers 5. Nearly all the models available are specifically developed to address a very limited set of projects (in many cases, such projects are proprietary and results cannot be replicated) and they lack of external validation, preventing practitioners from applying such models in their specific contexts.

Such lack of reliable results from the research produces also a lack of tools that practitioners can really use in their daily work. Looking at the excluded tools, we can notice that nearly all of them have a very low number of citations. This means that almost no research is based on them and they have been developed and used only for a limited number of projects (often just one) with the objective of writing a paper and not to create a tool able to support the adoption of a model by the community.

5 Threats to Validity

As suggested by Yin [9], we defined Internal validity, Construct validity and Reliability. Since we do not draw any conclusions about mapping studies or systematic literature reviews in general, external validity threats are not applicable.

Moreover, Petersen et al. [10] suggest assessing the quality of a study by means of profiling an objectively checklist. We reached an excellent score of 72%, higher than the average (33%–48%) of similar studies [10].

Internal Validity: Our study does not draw cause-effect relationships. Moreover, since our analysis only uses descriptive statistics and basic correlation, the threats are minimal. However, we understand that the identification of the citations based on scientific literature might only reflect a portion of the adopted tools, while the analysis of gray literature could have provided different results, somehow promoting more tools with a bigger community or with a longer history, without considering their quality.

Construct Validity: The measure of popularity could be of this threat. At the best of our knowledge there is no way to collect the exact number of users that use a specific tool, and therefore the measures we considered as proxy measures, could have affected the results. The terms adopted are well known and stable enough to be used as search strings.

Reliability: We defined search terms and applied procedures that can be replicated by others. However, we are aware that the same search string applied in google.com could return a slightly different set of results that can become relevant in a few years or even in a few months. It could be interesting to monitor the evolution over time.

6 Conclusions and Future Work

This paper provides an initial analysis and classification of the static and dynamic code analysis tools for supporting and predicting software maintenance that have been adopted in industry and validated academically.

The study focuses on tools that are not just research prototypes and that are available to support studies in academic and industrial settings. In particular, we have found out that most of the tools used in industrial case studies were no longer available at the time of writing, preventing other researchers or practitioners from easily adopting the same approaches and analyses by using already existing and tested tools.

Moreover, most of the widely-adopted tools are commercial ones and there are almost no open source communities that are able to build and support such kinds of tools. Research prototypes often have a large number of citations in scientific literature but are never adopted in industrial case studies. In most of the cases (60%), the interest in such tools is quite constant over the years providing a resistance in the dismissal of tools or in the adoption of new ones.

An interesting side result is that a lot of tool prototypes for reliability prediction (e.g. [19]) have been developed in the past by researchers, but their usage in industry is very limited. Moreover, no commercial tools for reliability prediction were identified from the selected toolset. The results of the classification of the selected tools reported in Table 3 can be used by researchers to select the most frequently used tools to obtain reliable results for their research, to conduct empirical studies, or to perform other work. Moreover, results can be beneficial for industry practitioners in to easily access to a classification of existing tools used to perform maintenance research studies.

Another aspect is that almost all the selected tools have been designed to support developers in performing some kind of activities (e.g., code reviews, testing, static analysis, etc.) but they do not include maintenance models to help users in making estimates/predictions. On the contrary, the rejected tools include several ones that implement such models. Moreover, the tools implementing a model are the least cited ones (almost no citations at all). This means that the research activities performed are not basing their work on existing results and tools, but new ones are developed. This is a sign of immaturity of the field and the need of further research in this area to create reusable models and tools that can be adopted by practitioners.

References

1. Petersen K, Feldt R, Mujtaba S, Mattsson M (2008) Systematic mapping studies in software engineering. In: International conference on evaluation and assessment in software engineering
2. Kitchenham B, Charters S (2007) Guidelines for performing systematic literature reviews in software engineering, version 2.3
3. Garousi V, Felderer M, Mäntylä MV (2016) The need for multivocal literature reviews in software engineering: complementing systematic literature reviews with grey literature. In: 20th international conference on evaluation and assessment in software engineering
4. ISO/IEC 25010:20111 SQuaRE. https://www.iso.org/standard/35733.html
5. Lenarduzzi V, Sillitti A, Taibi D (2017) Analyzing forty years of software maintenance models. In: 39th international conference on software engineering (ICSE 2017)
6. Bavota G, Di Penta M, Oliveto R (2013) Search based software maintenance: methods and tools. In: Evolving software systems
7. Mantere M, Uusitalo I, Röning J (2009) Comparison of static code analysis tools. In: 3rd international conference on emerging security information, systems and technologies
8. Manzoor N, Munir H, Moayyed M (2012) Comparison of static analysis tools for finding concurrency bugs. In: 23rd international symposium on software reliability engineering workshops
9. Yin RK (2009) Case study research: design and methods. SAGE Publications, Thousand Oaks
10. Petersen K, Vakkalanka S, Kuzniarz L (2015) Guidelines for conducting systematic mapping studies in software engineering: an update. Inf Softw Technol 64:1–18
11. Sillitti A, Janes A, Succi G, Vernazza T (2003) Collecting, integrating and analyzing software metrics and personal software process data. In: 29th Euromicro conference
12. Kleine HM, Muller HA (2010) Rigi - an environment for software reverse engineering, exploration, visualization, and redocumentation. Sci Comput Program 75(4):247–263
13. Coman I, Sillitti A (2007) An empirical exploratory study on inferring developers' activities from low-level data. In: International conference on software engineering and knowledge engineering (SEKE 2007), Boston, MA, USA, 9–11 July 2007

14. Coman I, Robillard PN, Sillitti A, Succi G (2014) Cooperation, collaboration and pair-programming: field studies on backup behavior. J Syst Softw 91(5):124–134
15. Janes A, Lenarduzzi V, Stan AC (2017) A continuous software quality monitoring approach for small and medium enterprises. In: 8th ACM/SPEC on international conference on performance engineering companion (ICPE 2017 Companion)
16. Lenarduzzi V, Stan AC, Taibi D, Tosi D, Venters G (2017) A dynamical quality model to continuously monitor software maintenance. In: 11th European conference on information systems management (ECISM 2017)
17. del Bianco V, Lavazza L, Morasca S, Taibi D, Tosi D (2010) The QualiSPo approach to OSS product quality evaluation. In: 3rd international workshop on emerging trends in free/libre/open source software research and development
18. del Bianco V, Lavazza L, Morasca S, Taibi D (2009) Quality of open source software: the QualiPSo trustworthiness model. In: OSS 2009. IFIP advances in information and communication technology, vol 299
19. Lavazza L, Morasca S, Taibi D, Tosi D (2012) An empirical investigation of perceived reliability of open source java programs. In: Proceedings of the ACM symposium on applied computing, pp 1109–1114. https://doi.org/10.1145/2245276.2231951

Stance Prediction for Russian: Data and Analysis

Nikita Lozhnikov[1(✉)], Leon Derczynski[2(✉)], and Manuel Mazzara[1(✉)]

[1] Innopolis University, Innopolis, Russian Federation
{n.lozhnikov,m.mazzara}@innopolis.ru
[2] ITU Copenhagen, Copenhagen, Denmark
leod@itu.dk

Abstract. Stance detection is a critical component of rumour and fake news identification. It involves the extraction of the stance a particular author takes related to a given claim, both expressed in text. This paper investigates stance classification for Russian. It introduces a new dataset, RuStance, of Russian tweets and news comments from multiple sources, covering multiple stories, as well as text classification approaches to stance detection as benchmarks over this data in this language. As well as presenting this openly-available dataset, the first of its kind for Russian, the paper presents a baseline for stance prediction in the language.

1 Introduction

The web is rife with half-truths, deception, and lies. The rapid spread of such information, facilitated and accelerated by social media can have immediate and serious effects. Indeed, such false information affects perception of events which can lead to behavioral manipulation [1]. The ability to identify this information is important, especially in the modern context of services and analyses that derive from claims on the web [2].

However, detecting these rumours is difficult for humans, let alone machines. Evaluating the veracity of a claim in for example social media conversations requires context – e.g. prior knowledge – and strong analytical skills [3]. One proxy is "stance". Stance is the kind of reaction that an author has to a claim. Measuring the stance of the crowd as they react to a claim on social media or other discussion fora acts as a reasonable proxy of claim veracity.

The problem of stance detection has only been addressed for a limited range of languages: English, Spanish and Catalan [4–6]. With these, there are several datasets and shared tasks. While adopting now more mature standards for describing the task and structuring data, RuStance enables stance prediction in a new context.

Debate about media control and veracity has a strong tradition in Russia; the populace can be vocal, for example overthrowing an unsatisfactory ruling empire in 1918. Veracity can also be questionable, for example in the case of radio

© Springer Nature Switzerland AG 2020
P. Ciancarini et al. (Eds.): SEDA 2018, AISC 925, pp. 176–186, 2020.
https://doi.org/10.1007/978-3-030-14687-0_16

transmitters in key areas being used to send targeted messages before the internet became the medium of choice [7]. Indeed, news on events and attitudes is Russia is often the focus of content in US and European media, with no transparent oversight or fact checking. The context is therefore one that may benefit greatly from, or least be highly engaged with, veracity and stance technology.

This paper relates the construction of a stance dataset for Russian, RuS-TANCE. The dataset is available openly for download, and accompanied by baselines for stance prediction on the data and analysis of the results.

2   Data and Resources

Before collecting data, we set the scope of the task and criteria for data likely to be "interesting". Collection centred around a hierarchical model of conversation, with stories at the top having source rumours/claims, which are referenced as "source tweets" in prior Twitter-centric work [8], that are the root of discussion threads, with responses potentially expressing a stance toward that source claim.

2.1   Requirements

The data was collected during manual observation in November 2017. Tweets that started a useful volume of conversational activity on a topic we had been observing were considered "interesting", and were added to the list of the potentially valuable sources.

For individual messages, we needed to determine the objective support towards a rumour, an entire statement, rather than individual target concepts. Moreover, we were to determine additional response types to the rumourous tweet that are relevant to the discourse, such as a request for more information (questioning) and making a comment (C), where the latter doesn't directly address support or denial towards the rumour, but provides an indication of the conversational context surrounding rumours. For example, certain patterns of comments and questions can be indicative of false rumours and others indicative of rumours that turn out to be true.

Following prior work [9,10], we define Stance Classification as predicting a class (label) given a text (features) and topic (the rumour). Classes are *Support, Deny, Query, Comment*, Examples are shown in Figs. 1 and 2.

- **Support:** the author of the response supports the veracity of the rumour, especially with facts, mentions, links, pictures, etc. For instance, "Yes, that's what BBC said."
- **Deny:** the author of the response denies the veracity of the rumour, the opposite case of the Support class. For instance, "Under the bed???? I've been there and there were not any monsters!"
- **Query:** the author of the response asks for additional evidence in relation to the veracity of the rumour. This one is usually is said in a questionable manner. For instance, "Could you provide any proof on that?"

– **Comment**: the author of the response makes their own comment without a clear contribution to assessing the veracity of the rumour. The most common class, but not the worst. The examples of the class usually contains a lot of emotions and personal opinions, for example, "Hate it. This is awesome!" - tells us nothing about the veracity.

Claim: The Ministry of Defense published irrefutable evidence of US help for ISIS.

– *Reply 1*: Come'on. This is a screenshot from ARMA. [**deny**].
– *Reply 2*: Good job! [**comment**]
– *Reply 3*: Is that for real? [**query**]
– *Reply 4*: That's it! RT also say so [**support**]

Fig. 1. A synthetic example of a claim and related reactions (English).

Claim: #СИРИЯ Минобороны России публикует неоспоримое подтверждение обеспечения Соединенными Штатами прикрытия боеспособны... https://t.co/auTz1EBYX0

– Reply 1: Это не фейк, просто перепутали фото. И, кстати, заметили и исправили. Это фото из других ново... https://t.co/YtBxWebenL [deny].
– Reply 2: Министерство фейков, пропаганды и отрицания. [comment]
– Reply 3: Что за картинки, на которых вообще не понятно - кто,где и когда? [query]
– Reply 4: Эту новость даже провластная Лента запостила, настолько она ржачная) [support]

Fig. 2. A real example of a claim and related reactions (Russian).

2.2 Sources

In order to create variety, the data is drawn from multiple sources. This increases the variety in the dataset, thus in turn aiding classifiers in generalizing well. For RuStance, the sources chosen were: Twitter, Meduza [11], Russia Today [RT] [12], and selected posts that had an active discussion.

– **Twitter:** is one of the most popular sources of claims and reactions. We paid attention to a well-known ambiguous and false claims, for example the one [13] mentioned – Russian Ministry Of Defense.

- **Meduza:** is an online and independent media located in Latvia and focused on Russian speaking audience. Meduza is known to be in opposition to Kremlin and Russian politicians. The media has ability for registered users to leave comments to a particular events. We collected comments on some popular political events that were discussed more than the others [14, 15].
- **Russia Today:** is an international media that targets world wide auditory. It is supposed that the editors policy of Russia Today is to support Russian government. Its main topics are politics and international relation which means there are always debates. We gathered some political and provocative publications with a lot of comments [16].

To capture claims and reactions in Twitter we used software developed as part of the PHEME project [17] which allows to download all of the threads and replies of a claim-tweet. For other sources we downloaded it by hand or copied.

The dataset sources are Twitter and Meduza with 700 and 200 entities respectively.

Firstly, Twitter is presented with over 700 interconnected replies, i.e. replies both to the claim and to other replies. The latter might be a cause of a large number of arguments and aggression (i.e. have high emotional tension) and as a result, the replies are poorly structured from the grammatical perspective, contain many non-vocabulary words in comparison with national or web corpora. Tweets that were labeled as "support" and "deny" tend to have links to related sources or mentions. Users of Twitter also use more multimedia, which brings in auditory content not included in this text corpus.

Secondly, Meduza comments are discovered to be more grammatically correct and less aggressive but still non-neutral and sarcastic. Meduza users are mostly deanonymized, but unfortunately, this is our empirical observation and not mentioned in the data. Comments on the articles vary in amounts of aggression, however still less aggressive than tweets. We hypothesize that this is caused by the fact that news articles provide more context and have teams of editors behind it. Users of Meduza tend to provide fewer links and other kinds of media, which may be due to the user interface of the site, or a factor of the different nature of social interactions on this platform.

Finally, Russia Today content is very noisy, and difficult to parse. This provided the smallest contribution to the dataset, and had the least structure and coherence in its commentary. Overall, the dataset contains both structured and grammatically correct comments and unstructured messy documents. This is indicative of a good sample; one hopes that a dataset for training tools that operate over social media will contain a lot of the noise characteristic of that text type, enabling models equipped to handle the noise in the wild.

Typical headlines for collection included (translated):

- The Ministry of Defense accused "a civil co-worker" in the publication of the screenshot from the game instead of the photo of the terrorists. And presented a new "indisputable evidence";
- The Ministry of Defense posted an "indisputable" evidence of US cooperation with ISIS: a screenshot from a mobile game;

- The Bell has published a wiretapping of Sechin and Ulyukaev phone calls;
- Navalny seized from "Life" 50 thousand rubles. He didn't receive this money as a "citizen journalist" for shooting himself;
- "If the commander in chief will call into the last fight, Uncle Vova, we are with you." Deputy of the Duma with the Cadets sang a song about Putin;
- "We are very proud of breaking this law." Representatives of VPN-services told "Medusa" why they are not going to cooperate with Russian authorities;
- Muslims of Russia suggested to teach "The basics of religious cultures" from 4 til 11 grade;
- "Auchan" (supermarket) will stop providing free plastic bags.

Table 1. Dataset class distribution

Support	Deny	Query	Comment
58 (6%)	46 (5%)	192 (20%)	662 (69%)

The most valuable classes – **Support** and **Deny** – are outnumbered by more general-purpose classes. This is similar to the class distribution that other stance classification datasets usually have [4–6]. Indeed, FNC-I, the Fake News Challenge dataset [18], also has a quite similar class distribution: 70% - comments, 20% - queries, 10% - support & deny (Table 1).

In the interests of describing the origins and potential biases in the dataset, a brief data statement [19] follows.

- **Curation rationale.** Text was drawn from sources likely to hold debate and discussion, incorporating many different viewpoints. The dataset should be useful to those building systems to be applied to commentary on Russian news.
- **Language variety.** The BCP-47 descriptors relevant are ru-Cyrl-RU and ru-Cyrl-LV.
- **Speaker demographic.** Most speakers are L1 with other information hidden. It is assumed that all have internet access and most are adept internet users.
- **Annotator demographic.** Annotators are males between 25–40 of European descent, with high degrees of education.
- **Speech situation.** Utterances are from November 2017, written, without editing, and spontaneous, in a public internet conversation context.

3   Implementation

As a baseline, and to provide a platform for analysis of the data, we built a pipeline of corpora preprocessing, feature extraction and classifier training.

3.1   Preprocessing

Dealing with natural languages is often a complicated task with many caveats; this is no better with social media. Phenomena prevalent in social media text include typos, acronyms, slang and another examples of non-dictionary and unexpected words. This can mean that finding representations for words can be very noisy, because the embedding models are usually trained using normal and grammatically correct corpora. In the case of RuStance, this is exacerbated by the paucity of large Russian datasets or embedding collections.

In order to be able to process in a form of vectors in a meaningful n-dimensional space, the texts of the dataset have to be converted from a human friendly representation to *Word Embeddings* [20]. First, a strict filtering step removes all of the social-media-specific entities like *-like* mentions, hashtags, URLs. We decided to proceed only with words and standard punctuation.

To convert a token to the expected format we had to pick a threshold that would cut out the outliers. Also, by having a fixed length input (array of tokens) it is possible to fit smaller texts into the input by substituting the missing tokens with zero-tokens.

[24] provide a pretrained model of word embeddings for Russian called RusVectores. Since our dataset covers mostly web and social conversations, we used a model that was trained on a web corpora and Wikipedia. This hopefully increases the probability that words from our dataset will be in the vocabulary.

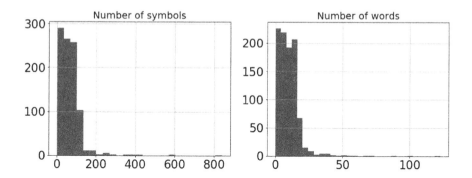

Fig. 3. Dataset symbols/words length distribution.

To maintain a uniform feature representation, we set the input length to 25 words. We set the length of the input to be 25 words because over 95% of the records would pass the threshold (Fig. 3) and the outliers will be abandoned. For the records that have less than 25 words we used pad sequence from Keras [22] which appends empty word embeddings in order to equalize lengths. Unknown tokens (the tokens that are not in the vocabulary) are substituted with zero-vectors.

So far an entry of the dataset for the classifier has 25 features = word embeddings and a class label. Then features can be used as an input for both Bayesian

Table 2. Evaluation metrics after cross-validation, k = 5

	Bagging	AdaBoost	Boosting	SGD	Logistic regression
F1	**0.832**	0.530	**0.865**	0.266	0.259
Accuracy	**0.925**	0.766	**0.925**	0.582	0.678

and Deep Learning models. At this step we have prepared dataset that later will be split for delayed cross-validation, model training and evaluation.

After the tokenization Keras Tokenizer is next to be trained on the texts to substitute words with Term Frequency - Inverse Document Frequency [TFIDF] indices. Using that indices and Gensim package [27] we create Embedding layer of the Deep Learning model [24].

It is known from the documentation that the models are initialized with the most common defaults, thus we expect our models to perform at a non-zero level.

4   Evaluation and Discussion

The dataset is split into train and test partitions, and then perform cross-validation (see Table 2).

We evaluate model performance using *accuracy* and *f1-measure*. The dataset consists of short messages, so accuracy will be more representative in the context of overall analysis, whereas f1-measure will partially compensate the imbalance of classes. In this case, accuracy tends to be more effective, since we want to exclude as many false positives as possible in order to not to call arbitrary media to be fake.

With trained split stratified validation and precision as a metric we built *Confusion Matrices* (Fig. 4) for top-scoring classifiers. The fit-predict part was performed with *GridSearch* cross-validation with K equals 5 and train/test split coefficient equals 0.1.

4.1   Analysis

To analyse misclassification errors, we generated confusion matrices. These enable visual comparison of accurate and error predictions for each class. The ideal outcome of such a visualization would be a unit matrix since the main diagonal would contain ones which means that every class was predicted 100% correct. In practice, the stronger (close to 1) the main diagonal – the better the classifier performs.

According to the confusion matrices one can infer that models tend to overfitting by predicting *Comment*-class all of the time, which would result in around 70% of true positives. However, the models that use Bagging and Boosting are more reliable and predict all of the classes more or less equally.

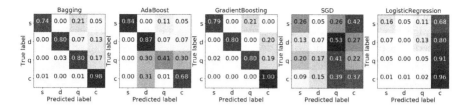

Fig. 4. Confusion matrices for SQDC classification of stance on RuStance.

5 Related Work

Evaluation of the reliability of the information is a difficult and time consuming task, even for trained professionals [25]. Fortunately, the process can be divided into steps or stages, some of which can later be automated.

The first step towards classifying claims as either fake or trustworthy is to find out how others react to said claims. This process of Stance Detection plays significant role in fact-checking pipelines [25,26,29].

Since initial work on determining veracity of social media [9], powerful systems and annotation schemes have been developed to support the analysis of rumours and misinformation in text. In this paper, the authors introduced methods to enhance the performance of classifiers trained on relatively long-term rumours in tweets. The idea was later extended by [23] who, in addition, tracked the presence of positive or negative markers in a tweet. Later [4] provides a novel hand crafted dataset of rumoured claims. Our work is to provide a similar dataset for Russian tweets.

[10] exploited the temporal sequence of tweets, although the conversational structure was ignored and each tweet was treated as a separate unit. In other domains where debates or conversations are involved, the sequence of responses has been exploited to make the most of the evolving discourse and perform an improved classification of each individual post after learning the structure and dynamics of the conversation as a whole.

Classification of stance towards a claim on Twitter has been mentioned in SemEval-2016 task 6 [29]. Subtask B tested stance detection towards an unlabelled target, which required a weakly supervised or unsupervised approach. The dataset of this competition was not related to rumours or breaking news, it only considered a 3-way classification and did not provide any relations between tweets, which were treated as individual instances.

Recent state-of-the-art work on stance classification includes the top-scoring system in the RumourEval exercise [28], which decomposed conversational branches into lines for use as context input to an LSTM [33], and another approach based on proximity to English words representative of certain stance classes [30]. Stance detection has also been used to develop datasets of diverse stance, enabling construction of balanced summaries and examination of argumentation and counter-argumentation [31]. Finally, while RumourEval, the Fake News Challenge, FEVER [32] and others have provided datasets, and some have developed creative uses of this stance data, all of these resources are in English; RuStance is the first such dataset for Russian, a language hosting an active, nuanced and passionate political debate.

6   Conclusion

RuStance [21] is the first dataset for stance classification in Russian. We think it is a highly area to study, collecting data for future researchers with tools and data, and opening up the arena of fake news in Russia to global researchers. It comprises a dataset from multiple sources, including many conversation threads, and a mixture of social interaction.

A baseline is included. As we assumed, a thousand examples is definitely not enough to fit the LSTM layer. Nevertheless, we achieved accuracy over 90% using classifiers without any tuning whatsoever. Confusion matrices suggest that even if a single precision rate is high the class imbalance is still a huge issue and the bottleneck that stops us from fitting a really accurate classifiers.

New metrics are also to be considered and developed. Russian tweets seem to be as reliable as English in terms of consistency and class distribution. This dataset and baseline provide first steps into analyzing fake news spread and generation among Russian speakers, and we hope, with further work, multilingually.

References

1. Rapoza K (2017) These two Russian 'fake news' outfits get billions of hits on Facebook. https://www.forbes.com/sites/kenrapoza/2017/09/22/these-two-russian-fake-news-outfits-get-billions-of-hits-on-facebook
2. Zubiaga A, Aker A, Bontcheva K, Liakata M, Procter R (April 2017) Detection and resolution of rumours in social media: a survey. ArXiv e-prints
3. Mrowca D, Wang E, Kosson A (2017) Stance detection for fake news identification
4. Ferreira W, Vlachos A (2016) Emergent: a novel data-set for stance classification. In: Proceedings of the 2016 conference of the North American chapter of the Association for Computational Linguistics: human language technologies. ACL
5. Anta AF, Chiroque LN, Morere P, Santos A (2013) Sentiment analysis and topic detection of spanish tweets: a comparative study of of NLP techniques. Procesamiento del lenguaje natural 50:45–52
6. Taulé M, Martí MA, Rangel FM, Rosso P, Bosco C, Patti V et al (2017) Overview of the task on stance and gender detection in tweets on Catalan independence at IberEval 2017. In: 2nd workshop on evaluation of human language technologies for Iberian languages, IberEval 2017, vol 1881, CEUR-WS, pp 157–177
7. Enikolopov R, Petrova M, Zhuravskaya E (2011) Media and political persuasion: evidence from Russia. Am Econ Rev 101(7):3253–85
8. Zubiaga A, Kochkina E, Liakata M, Procter R, Lukasik M (2016) Stance classification in rumours as a sequential task exploiting the tree structure of social media conversations. arXiv preprint arXiv:1609.09028
9. Qazvinian V, Rosengren E, Radev DR, Mei Q (2011) Rumor has it: identifying misinformation in microblogs. In: Proceedings of the conference on empirical methods in natural language processing. Association for Computational Linguistics, pp 1589–1599

10. Lukasik M, Srijith P, Vu D, Bontcheva K, Zubiaga A, Cohn T (2016) Hawkes processes for continuous time sequence classification: an application to rumour stance classification in Twitter. In: Proceedings of 54th annual meeting of the Association for Computational Linguistics. Association for Computational Linguistics, pp 393–398

11. Meduza (2018) http://meduza.io

12. Channel RT TV (2018) Russia Today. https://rt.com

13. The Guardian (2017) Russia's 'irrefutable evidence' of us help for ISIS appears to be video game still. https://www.theguardian.com/world/2017/nov/14/russia-us-isis-syria-video-game-still

14. Meduza (2017) Meduza.io: on fake evidence. https://meduza.io/shapito/2017/11/14/minoborony-vylozhilo-neosporimoe-dokazatelstvo-sotrudnichestva-ssha-i-ig-skrinshot-iz-mobilnoy-igry

15. Meduza (2017) Meduza.io: on Ministry of Defense. https://meduza.io/shapito/2017/11/14/minoborony-vylozhilo-neosporimoe-dokazatelstvo-sotrudnichestva-ssha-i-ig-skrinshot-iz-mobilnoy-igry

16. Channel RT TV (2017) Russia Today: on Russian president candidates 2018. https://russian.rt.com/inotv/2017-11-14/Rukovoditel-internet-kampanii-Sobchak-eyo-uchastie

17. Derczynski L, Bontcheva K (2014) PHEME: veracity in digital social networks. In: UMAP workshops

18. Pomerleau D, Rao D (2017) Fake news challenge. http://www.fakenewschallenge.org

19. Anonymous (2018) Data statements for NLP: toward mitigating system bias and enabling better science. OpenReview.net

20. Mikolov T, Sutskever I, Chen K, Corrado GS, Dean J (2013) Distributed representations of words and phrases and their compositionality. In: Advances in neural information processing systems, pp 3111–3119

21. Rustance. https://figshare.com/articles/dataset_csv/7151906/2

22. Chollet F et al (2015) Keras. https://github.com/keras-team/keras

23. Liu X, Nourbakhsh A, Li Q, Fang R, Shah S (2015) Real-time rumor debunking on Twitter. In: Proceedings of the 24th ACM international on conference on information and knowledge management. ACM, pp 1867–1870

24. Kutuzov A, Kuzmenko E (2017) WebVectors: a toolkit for building web interfaces for vector semantic models. Springer, Cham, pp 155–161

25. Ghulati D (2016) Introducing factmata—artificial intelligence for political fact-checking. https://medium.com/factmata/introducing-factmata-artificial-intelligence-for-political-fact-checking-db8acdbf4cf1

26. Baird YPS, Sibley D (2017) Talos targets disinformation with fake news challenge victory. http://blog.talosintelligence.com/2017/06/talos-fake-news-challenge.html

27. Řehůřek R, Sojka P (May 2010) Software framework for topic modelling with large corpora. In: Proceedings of the LREC 2010 workshop on new challenges for NLP frameworks. ELRA, Valletta, pp 45–50. http://is.muni.cz/publication/884893/en

28. Derczynski L, Bontcheva K, Liakata M, Procter R, Hoi GWS, Zubiaga A (2017) SemEval-2017 task 8: RumourEval: determining rumour veracity and support for rumours. arXiv preprint arXiv:1704.05972

29. Mohammad SM, Sobhani P, Kiritchenko S (2017) Stance and sentiment in tweets. ACM Trans. Internet Technol. (TOIT) 17(3):26

30. Aker A, Derczynski L, Bontcheva K (2017) Simple open stance classification for rumour analysis. In: Proceedings of RANLP

31. Ruder S, Glover J, Mehrabani A, Ghaffari P (2018) 360° stance detection. In: Proceedings of the 2018 conference of the North American chapter of the Association for Computational Linguistics: demonstrations. Association for Computational Linguistics, pp 31–35
32. Thorne J, Vlachos A, Christodoulopoulos C, Mittal A (2018) Fever: a large-scale dataset for fact extraction and verification. In: Proceedings of the 2018 conference of the North American chapter of the Association for Computational Linguistics: human language technologies (long papers), vol 1. Association for Computational Linguistics, pp 809–819
33. Kochkina E, Liakata M, Augenstein I (2017) Turing at SemEval-2017 task 8: sequential approach to rumour stance classification with branch-LSTM. In: Proceedings of SemEval (2017)

Software Requirements Complexity Analysis to Support the "ADVISORY NETWORK IN TO THE NATION FORCES BUILD-UP"

Luca Galantini[1]([⊠]), Angelo Messina[2], and Mario Ruggiero[3]

[1] Università Cattolica del Sacro Cuore, Milan, Italy
luca.galantini@unicatt.it
[2] University of Innopolis, Innopolis, Russian Federation
segreteria@dssea.eu
[3] Defense & Security Software Engineers Association, Rome, Italy
marius.rogerius@gmail.com

Abstract. In the modern strategic scenarios, military operations are more and more complex, in particular when compliance with the so called "soft power" (DIME – Diplomatic, Information, Military and Economic) is required. Because of this, military doctrine and, in particular, the support at the Command and Control are obliged to a continuous evolution, trying to match the military requirements with the operation's reality.

Inside the complexity of the modern scenarios, the operations to support a failed/failing states must be quoted, where the military commanders are in charge to support and sustain, frequently, the full range of a national domains mostly "civilian". The most exhaustive example of that is the Operations Resolute Support (RS) in Afghanistan, where the RS Commander is responsible to support not only the Defense and Interior Affairs Ministry and Minister, but also the Finance, Public Health and others Departments, together with the highest state's institutions represented by the National Security Authority and the Republic's Presidency.

In this context, current Military Command and Control Systems are unable to correlate and analyze the daily activity performed by thousand of Alliance's Advisors. NATO, with the support of the US Defense Digital Service developed an Agile Project, delivering a first response tool in 14 weeks. This was possible also because of the former experience developed in the LC2Evo Italian Army C4I demonstrator.

Taking into account the relevance of security assistance in the modern scenarios, it's important to implement and further develop the above mentioned experience at national level, in order to be prepared to support the national contingents in the future operations.

Keywords: Military operations · Software requirement analysis · Software complexity · Agile

© Springer Nature Switzerland AG 2020
P. Ciancarini et al. (Eds.): SEDA 2018, AISC 925, pp. 187–197, 2020.
https://doi.org/10.1007/978-3-030-14687-0_17

1 Evolution of the Military Operations

1.1 Trends [1]

After the Second World War, the international geostrategic framework was characterized by the prevalent application of indirect strategy techniques, in order to avoid direct confrontation between superpowers. With the changes that took place following the breakup of the Soviet Union, there was a reawakening of planned interventions and conducted in the field of direct strategy techniques, at least up to the present day. The call to the direct strategy has involved not only the deployment in numerous Operational Theaters of US and NATO military units, but also the need to recall a unitary use, in time and space, of the so-called national powers, namely the Economic, Diplomatic and Military under the institutional direction of Political power. This, obviously not only at the national level, but also and above all at the international level.

The most evident result, even in countries with a very structured foreign and defense policy, such as the USA, was to see a stronger interaction and synergy between the "foreign" and the "defense" world. At a multinational level, such as the European Union and NATO, it was necessary to develop and implement a comprehensive approach to international crises, condensed in the directives on the "Comprehensive Approach" that these organizations have developed taking into account the different institutional approach.

1.2 History and Legacy

The European Union has been shaped throughout time, particularly since the disintegration of USSR and Yugoslavia. By expanding its membership and developing its own means and tools of managing and preventing conflicts, the EU aims at becoming a global actor with civilian and military capacities. In this way it shapes its neighbourhood as its sphere of influence by displaying its new means and tools, namely the CSDP operations. The European Security and defence legal environment has constantly changed over the past years. Hence it's important to deeply understand its origins with the European Defence Community, and further developments as the Common Security and Defence Policy, and the intergovernmental pillar of the European Security and Defence Policy. The origins of the European Defence Community (to be followed by EDC) can be found just after the first attempts to the creation of establishment of the first supranational project the European Coal and Steel Community, on the 27 May 1952 [2], with Germany, France, Belgium, Netherlands, Italy and Luxembourg signing it. Regrettably the EDC failed to succeed and come to end on the 1954 [3], rejected by the French Government. This Treaty constituted the first common initiative on ensuring the maintenance of peace, with a clear defence purpose of the Western Europe "against any aggression by participating in the western defence within the framework of the North Atlantic Treaty and by accomplishing the integration of the defence forces of the Member States and the rational and economic utilization of their resources" [4]. This principle made a clear reference to the principle of mutual defence, present at the 1949' Washington Treaty. Thus, this treaty had a clear impetus of taking a step forward the formation of an integrated and united Europe, with the creation of common institutions, common armed forces and a common budget for the community defence [5]. Indeed, it's important to emphasize that the treaty was born with a clearly supranational character, in opposition to all the future defence related treaties, heavily anchored in intergovernmental rules. Although NATO's essential role, as set in the preamble of the 1949's Washington Treaty, is to "safeguard the freedom, common heritage and civilisation of their peoples, founded on the principles of democracy, individual liberty and the rule of law". They [Parties to this Treaty] seek to promote stability and well-being in the North Atlantic area. They [Parties to this Treaty] are resolved to unite their efforts for collective defence and for the preservation of peace and security. They therefore agree to this North Atlantic Treaty" [6], the post-cold war international security scenario has shifted this view [7]. In accordance with the article 5 of its foundation treaty, NATO's mandate is to provide collective self-defence to its members, as following:

"The Parties agree that an armed attack against one or more of them in Europe or North America shall be considered an attack against them all and consequently they agree that, if such an armed attack occurs, each of them, in exercise of the right of individual or collective self-defence recognised by Article 51 of the Charter of the United Nations, will assist the Party or Parties so attacked by taking forthwith, individually and in concert with the other Parties, such action as it deems necessary, including the use of armed force, to restore and maintain the security of the North Atlantic area. Any such armed attack and all measures taken as a result thereof shall immediately be reported to the Security Council. Such measures shall be terminated when the Security Council has taken the measures necessary to restore and maintain international peace and security". Furthermore, as per article 2 of the North Atlantic Treaty, NATO has to "contribute toward the further development of peaceful and friendly international relations by strengthening their free institutions, by bringing about a better understanding of the principles upon which these institutions are founded, and by promoting conditions of stability and well-being".

In this way, the Alliance's strategic purposes shall be constantly adapted to the natural development of its security challenges[1]. The current strategic concept[2], published at the Lisbon Summit in November 2010, highlighted the collective defence (accordingly with article 5 of Washington Treaty)[3], crisis management and cooperative

---

[1] Terrorism, WMD, cyber security insecurity, socio-economic problems, transnational organized crime, ethnic tensions, fragile and failed states, environmental destruction, climate change, competition over resources, etc.

[2] The 2010 Strategic Concept 'Active Engagement, Modern Defence' is an official document that outlines NATO's enduring purpose and nature and its fundamental security task, describing NATO's values and strategic objectives for the next decade.

[3] Art. 5 of the Washington Treaty states "*The Parties agree that an armed attack against one or more of them in Europe or North America shall be considered an attack against them all and consequently they agree that, if such an armed attack occurs, each of them, in exercise of the right of individual or collective self-defence recognised by Article 51 of the Charter of the United Nations, will assist the Party or Parties so attacked by taking forthwith, individually and in concert with the other Parties, such action as it deems necessary, including the use of armed force, to restore and maintain the security of the North Atlantic area. Any such armed attack and all measures taken as a result thereof shall immediately be reported to the Security Council. Such measures shall be terminated when the Security Council has taken the measures necessary to restore and maintain international peace and security*".

security as its main core tasks in this new international security arena. Noteworthy is the acknowledgement and reconnaissance of the Alliance's strategic partners, such as relevant countries and other international organisations, through establishment of strategic partnerships in supporting tasks needed for enhancing the security and stability of the Euro-Atlantic area. Noteworthy is that that EU and NATO have now 22 Member States in common.

This particular security landscape reinforces mutual interests to be carried on, in accordance with the principles shared by both organisations[4], and allows for more flexibility while conducting crisis management operations and most importantly, empowering the Union to have greater capacity to respond to international crises, without undermining the essential role of the NATO, remaining the hinge of the collective defence of its members as well as its territorial integrity[5].

1.3   Current Operation[6]

The recent military operations conducted by the NATO countries, while preserving some phases in which classical military activities played a major role, have been characterized by being focused, in global terms, on the reconstruction of so-called "failed states" or to prevent their collapse (failing states).

The NATO countries Armed Forces had to carry out large-scale activities relegated for many decades, within the Armed Forces themselves, to a minor role or within the strict competence of Special Forces, such as training and monitoring in the field of Security and Defense Forces. To this has also been added a completely new role in the modern military panorama of almost all NATO countries, that is, the function of "adviser" at the strategic level of top management and politicians of the organization of a state. This particular function was carried out not only in favor of the Departments of Defense and the Interior of the countries concerned, but often also in support of the highest levels of the state, as well as departments far from the defense and security sector, such as Health, Finance, Justice.

Generally speaking, the EU-NATO permanent arrangements on crisis management, the so called "Berlin Plus" arrangements, historically built upon pre-existing arrangements between NATO and WEU, is a gathering of several documents of different nature, concluded in December 2002 and finalized in March 2003[7]. This range included the NATO-EU Agreement on Security of Information, assured EU access to NATO's planning capabilities for actual use in the military planning of EU-led crisis management operations, presumed availability of NATO capabilities and common

---

[4] Partnership, Mutual consultation and transparency, equality and decision-making autonomy and respect for the interest of Members of EU and NATO are among the shared values and principles described in the 2002's EU-NATO Declaration.

[5] Despite the creation of mutual assistance clause (art. 42 (7) of TEU), NATO remains responsible for collective defence of its members. In no case these provisions will undermine NATO defensive role in the international security sphere (art. 42 (2) of TEU).

[6] Op. Cit. [1].

[7] Only the NATO-EU Agreement on Security of Information and the EU-NATO Declaration of 2002 are available to the public.

assets, such as communication units and headquarters for EU-led crisis management operations, procedures for release, monitoring, return and recall of NATO assets and capabilities, terms of reference for NATO's Deputy SACEUR (who in principle will be the operation commander of an EU- led operation under the "Berlin Plus" arrangements and who is always a European, and European Command Options for NATO) and finally the NATO-EU Declaration on ESDP of the 16 December 2002 on the EU access to NATO's Planning capabilities in the context of an EU-led crisis management operation [8, 9].

1.4   Nato-EU Evolution

The NATO-EU Declaration on ESDP of the 16 December 2002 is at the heart of the Framework agreement of Berlin Plus, being concluded by High Representative for the CFSP at the time Javier Solana, also former Secretary General of NATO, and the NATO Secretary General, George Robertson, the two heads of organisations. At this point, is worth to recall these arrangements only became effective in case the EU wants to rely upon NATO capabilities. Besides the shared principles of cooperation already emphasized, the EU-NATO declaration of 2002 consists in crisis consultations arrangements following a non-committal approach, addressing in which ways NATO could further contribute to the EU-led operations: (1) NATO assured the EU access to NATO planning; (2) The availability of NATO capabilities and common assets to the EU (listed by NATO as possibly available to the EU, after NATO's formal approval) and (3) NATO can render available its European Command elements options upon EU's request. In the declaration there is a clear formulation of a set of obligations: (a) The EU is "ensuring the fullest possible involvement of non-EU European Members of NATO within ESDP, implementing the relevant Representative on 13 December 2002" and (b) NATO "is supporting ESDP in accordance with the relevant Washington Summit decisions, and is giving the European Union, inter alia and in particular, assured access to NATO's planning capabilities, as set out in the NAC decisions on 13 December 2002". In light of these provisions, contrary to the wording used in the Security of Information Agreement, it can be understood from the weak wording as well as its vagueness of its used terms such as the wording "is supporting", "is ensuring" or "is giving", the Parties to this text appear to don't have the intent to bind effect[8]. Additionally, that fact that the Berlin Plus Agreement was concluded by the High Representative for CFSP and not by the Council, authorized by the Presidency, pinpoints the non binding character of the framework[9]. Thus in this way and with a positive inference, both organisations have regulated a particular subject, crisis management, skirting issues of political domain. The EU has accepted to involve NATO to the fullest extend, while NATO has agreed to support the EU with its own capabilities

---

[8] Is also important to stress that calling an agreement " a declaration" often indicates the parties intent to don't be vinculated to that text.

[9] Following the strict sense of the article 24 of TEU *"The Council, acting unanimously, may authorize the Presidency, assisted by the Commission as appropriate, to open negotiations to that effect. Such agreements shall be concluded by the Council acting unanimously on a recommendation from the Presidency"*.

and assets, however this last formulation is imprecise in juridical terms and do not impose any obligations to NATO of providing planning capabilities as well as other military assets. Those assets are most likely to be given to the EU[10], however they are not hundred percent ensured since the agreement clearly stipulates that in case of unexpected conditions NATO might want to recur to them in case where an attack against an NATO member can happen.

EU-NATO Civilian Missions comprehends the whole spectrum of instruments that help to restore the civil government. In a nutshell, civilian missions are all these missions absent at its core of military objectives and instruments [10]. These types of missions are at the bulk of the efforts in crisis management, being a fundamental part of the EU-NATO priorities on the CFSP missions. The Civilian Headline goal of 2010 was built upon an extensive experience in civilian crisis management of the EU and therefore should be considered a living process, open to adjustment on lessons learnt from operations in different contexts.

The 2010 Headline goal, like the 2008 Headline goal, aimed at achieving the EU's civilian capacity to respond effectively to crisis management operations by 2010, by deploying civilian crisis management of high quality, with the support functions and the equipment required in a short time-span and in sufficient quantity in line with the ESS. The new Headline goal set a clear illustrative framework of planning and development of the civilian operations, taking into account the previous experiences and assessments and setting its targets accordingly[11]. This Headline goal had a great focus on Civil-Military cooperation along with the continuous focus on improving its readiness and deployability.

1.5   Scenarios Evolution

The creation of future scenarios (see footnote 6), to be used in the process of building future capabilities, always creates strong frictions in the force development and force planning bodies. Despite this, however, it is necessary to have the strength to pursue this path which is the only one that allows us to create balanced capabilities with a vision that takes into account past experiences but allows us to correctly interpret future trends.

---

[10] NATO has included the reference to the procedures for release, monitoring, return and recall of NATO assets and capabilities, as previously eluded.

[11] More specifically: *"Sufficient numbers of well-qualified personnel are available across the civilian ESDP priority areas and for mission support, to enable the EU to establish a coherent civilian presence on the ground where crisis situations require it to do so; ESDP capabilities such as planning and conduct capabilities, equipment, procedures, training and concepts are developed and strengthened according to need. One of the results will be that missions have adequate equipment and logistics and other enabling capabilities, including for effective procurement procedures; The EU is able to use all its available means, including civilian and military ESDP, European Community instruments and synergies with the third pillar, to respond coherently to the whole spectrum of crisis management tasks; The development of civilian capabilities is given increased political visibility at EU as well as at Member States' level; The EU strengthens its co-ordination and co-operation with external actors as appropriate".*

As a result, we can say that in the near future, the ability to operate in environments where the use of direct strategy will prevail, with the consequent need to operate in synergy with the Diplomatic and Economic power remains fundamental for the fulfilment of the mission following the "Comprehensive Approach" methodology.

As a result of what has just been said, the operating environment will almost certainly be characterized by an extensive use of asymmetric techniques and tactics, even with the presence of time windows or areas in which symmetrical or dissymmetric activities will be developed.

Finally, the presence of hybrid entities, with presence at a political and economic level, will contribute to increase the complexity of the operational environment. This will require, perhaps even more than today, the ability to the Armed Forces to be able to develop activities of advising and mentoring (sometimes) at the highest levels of the countries concerned, increasing the capacities currently expressed, for instance, in Afghanistan.

2 The Scenario Complexity

2.1 Evident Complexity

Evident complexity
Starting from the experience gained by the NATO countries in the intervention in Afghanistan which is the most complete example of support for the Nation Building, we can highlight the following prominent elements:

- the military level of Theater is the Operative one, with reference to the definition of the levels of responsibility in the conduct of military operations. It is therefore intended, in doctrinal terms, to integrate and develop parallel and sequential campaigns and multiple operations in order to achieve the strategic objectives received. This level, structured to generate influence on the adverse forces, is in the support of the Nation Building to simultaneously support the strategic, military strategic and operational levels of the Failed/Failing State, issuing directives and simultaneously coordinating the tactical level. In this way, a case of asymmetry of the levels of responsibility is obtained due to the fact that a Commander of the Operative level acts as advisor of a strategic manager;
- the canonical intervention sectors, with the use of thousands of advisors and tens of thousands of contractors, include not only the recruitment, training, operations and logistics of the security and defense forces, but they must be pushed into the micro-management, such as supervision and sometimes management of the entire cycle of human resources, including the definition and management of the staff payments cycle, not losing sight of the entire procurement cycle such as the definition of present and future capabilities of the Defense and Security Forces;
- the advising activity is also required at the level of the definition of the national strategy, the definition of the descending strategies of the sector with particular impact on that of security and defense such as energy, as well as support for the construction and development of the system health, tax and judiciary;

- data collected by advisors, normally on unclassified networks, operating in Failed/Failing State infrastructures, must be able to be quickly inserted in the classified network of the mission, in order to allow the definition of the interrelations and actions to be developed.

This articulation of the advising function involves the drafting of thousands of daily and/or periodic reports in order to define the progress of the various lines of action defined for the achievement of the capabilities objectives defined in the relevant sectors. The reports, focused on the Advisor - Advisee report, also highlight the problems encountered and the suggestions proposed. Even if the synchronization of these activities is also supported by specific meetings, it is clear that the day-to-day staffing activity is based on the analysis of the different reports, intercepting the elements of interest of each individual branch. In this way, however, the coordination, coordination and updating of each individual action line and activity becomes extremely difficult to perform and the possibility of loss of precious data/input becomes high.

- the frequent rotation of Advisors (shifts of six months or one year) leads to the loss of important data if not properly cataloged and made available for subsequent rotations.

2.2   Embedded Complexity

Embedded complexity is stemming mainly from the following:
requirements concerning the activity which the application software is required to support are very often described in large and articulated formal documents called Concept of Operations (CONOPS). These documents are the result of a compromise between pure operational needs and very strict legal constraints mandated by the specific Operation rules. Moreover this documents often call for "ad hoc" ground rules to be defined during the operation. ....

The language used to draft the CONOPS and the related documents are a mixture of formal bureaucratic expression and domain based terminology which makes it very difficult to extract the underlying functional requirement.

The large organizations such as NATO, UN, EU have their own requirement change management procedures which they expect to follow which are normally far away from "agile" principles.

3   Emerging Requirements and Their Volatility

3.1   Traditional vs "Agile" Requirements Elicitation

In spring 2016, inside the NATO - Resolute Support HQ (RS-HQ) emerged the need to develop a software to support the activity of advisors that was more adherent to specific needs than the use of common Office tools, even if customized, until then used. For this reason, a Crisis Response Urgent Requirement was defined and approved at local level, aimed at filling this capacity gap.

In order to be able to adhere to the Theater requirements in a short time, the USA sent a Defense Digital Service Team to the Operative Theater in order to develop the software tool in close synergy with the staff involved. It is worthwhile to remember that the problem of adapting traditional procurement procedure to Agile is still unsolved and widely debated [11].

It was a 14-week Agile development that led to the definition of a valid product to address the primary needs of the advisors community and the support needs of the staff and the RS team.

The product has been installed in the RS classified network and NATO, through NCIA, has been entrusted with the maintenance of the product as well as the further development and refinement of the software to the evolving needs.

In short, this is a further demonstration of how with the classic software development process we can not meet rapidly evolving needs, not only for indecision or imprecision of the military component in the definition of the requirements linked to future needs, but for the deviations imposed by the adverse party that often has reaction times of the order of days or weeks. As demonstrated by the US Defense Digital Service, the problems related to the use of "agile" software on protected networks must be dealt with from a systemic perspective, taking into account the consequences and losses deriving from not making the right software available at the right time compared to an absolutistic vision of security.

3.2 iAgile for Mission Critical Application: Italian Army LC2Evo Approach [12]

The acronym stands for Land Command & Control Evolution and this is a successful effort the Italian Army General Staff made to device a features & technology demonstrator that could help identifying a way ahead for the future of the Command & Control support software.

The main scope, related to the software engineering paradigm change in the effort, was to demonstrate that a credible, innovative and effective software development methodology could be applied to complex user domains even in the case of rapidly changing user requirements. The software project was embedded in a more ambitious and global effort in the frame of the Italian Defence procurement innovations process aimed at implementing the Concept Development & Experimentation (NATO CD&E) which was initially started by the Centro Innovazione Difesa (CID).

As clearly demonstrated at Ch.1 and 2, the Military operations are characterized by huge complexity in the support software requirements and at the time of LC2Evo start, the experience in Iraq and Afghanistan had clearly demonstrated the operating scenario was changing an a few months cycle and the most required characteristic for a C2 system by the user was flexibility. The possibility of adapting the software functions to an asymmetric dynamically changing environment seemed to be largely incompatible with the linear development lifecycle normally used for mission critical software in the Defence & Security area. As a matter of fact the USA DoD issued a recommendation (D.o.D. 5000.2) at the end of 2013 inviting contractors to switch to a production cycle based on short deliveries of working software increments. Even if the word "agile" was not mentioned in the quoted document, it is easy to recognize in it the whole pattern of the agile software development lifecycle.

In the framework of a paramount coordination effort led by The Italian Army COMFORDOT (three star level Command in charge, among other things of the Army Operational Doctrine) the Army General Staff Logistic Department got full delegation to lead, with the help of Finmeccanica (after changed into Leonardo), a software development project using agile methodology (initially Scrum, then ITA Army Agile and finally DSSEA ® iAgile) aimed at the production of a technology demonstrator capable of implementing some of the Functional Area Services of a typical C2 Software.

Strictly speaking of software engineering, one of the key issues was providing the users with a common graphical interface on any available device in garrison (static office operation) in national operations (i.e. Strade sicure) or international operations. The device type could vary from desk top computers to mobile phones [13].

The development was supposed to last for 6 to 8 months at the Army premises to facilitate the build-up of a user community network and to maximize the availability of user domain experts, both key features of the new agile approach. In the second phase the initial team was supposed to move to the contractor premises and serve as an incubator to generate more teams to work in parallel.

The first team outcome was so surprisingly good and the contractor software analysts and engineers developed such an excellent mix with the army ones that, both parts agreed to continue phase two (multiple teams) still at the Army premises.

The effort reached the peak activity after 18 months from start when 5 teams were active at the same time operating in parallel (The first synchronized "scrum of scrum like" reported in the mission critical software area).

More that 30 Basic production cycles (Sprints of 4–5 weeks) were performed, all of them delivered a working software increment valuable for the user. The delivered FAS Software tested in real exercises and some components deployed in operations. One of the initial tests was performed during a NATO CWIX exercises and concerned cyber security. The product, still in a very initial status, was able to resist more than 48 h to the penetration attempt by a very good team of Hackers.

More than a million equivalent line of software were developed at a unit cost of less than 10 Euros, with an overall cost reduction of 90% reaching a customer satisfaction exceeding 90%. One of developed FAS is still deployed in Afghanistan at the multi-national Command.

The details of this project methodology and technical outcomes have been released during the conferences of the SEDA series and published in many technical articles (cfr my linkedin list.).

4 Conclusion and Future Work

Both the quoted experiences of developing mission critical software applications in the operation theatre and homeland have clearly demonstrated that a very relevant part of the design uncertainty resides in the complexity and the volatility of the user requirement. The structure of the current military operation impose an "embedded complexity" which is due to the number and variety of agents involved in operation. In recent operation the number of agents included in the Command & Control loop has grown to include civilian organization and NGOs and the "rules" to be supported now are not "purely" military doctrine but include complex sets of international rules imposed by treaties and agreements. The usual complexity imposed by the traditional Command & Control functions are nowadays even more complex and volatile. In this paper a touch of the real complexity of the support software application user requirement was given based on the real experience of the authors in operation. This is a point of view which is seldom considered and, according to the authors, is heavily connected the most important element of the software lifecycle: the requirement analysis. Using a layer of expert translators (as in the software linear development) tends to freeze the requirement losing the capability to evolve and adapt the software development as the operation progresses.

Agile methods can be a valuable option to tackle the above described complexity but the implementation of the agile development frame in the described scenario implies multiple coordinated teams and the inclusion of many domain experts in the very core of the teams (as in the reported examples).

It is necessary to develop a new Software Engineering Paradigm to cope with the complexity imposed by the described scenarios [14].

References

1. Italian Defence General Staff (2008) JIC 007 Asimmetria e Dissimmetria dei conflitti
2. Treaty of European Defence Community (1952)
3. Wikipedia. https://en.wikipedia.org/wiki/Treaty_establishing_the_European_Defence_Community. Accessed 24 Oct 2018
4. European Defence Community, Treaty of EDC, Art. 2
5. European Defence Community, Treaty of EDC, Art. 1
6. NATO, The North Atlantic Treaty (1949)
7. Yost D (2010) NATO's evolving purposes and the next strategic concept. Int Aff 86(2):489–522
8. Reichard M (2006) The EU-NATO relationship: a legal and political perspective. Ashgate, pp 273–276
9. Reichard M (2008) The European union and crisis management - policy and legal aspects. Blockmans, pp 233–253
10. Koukratos P (2013) The international responsibility of the European union. Hart Publishing
11. Messina A, Modigliani P, Chang S How agile development can transform defense IT acquisition. In: Proceedings of 4th international conference in software engineering for defence applications, 978-3-319-27894-0, 339732_1_EN
12. Messina A, Cotugno F (2014) Adapting SCRUM to the Italian army: methods and (open) tools. The 10th international conference on open source systems San Jose, Costa Rica, 6–9 May 2014
13. Ventrelli C, Trenta D, Dettori D, Sanzari V, Salomoni S ITA army agile software implementation of the LC2EVO army infrastructure strategic management tool. In: Proceedings of 4th international conference in software engineering for defence applications, 978-3-319-27894-0, 339732_1_EN
14. Messina A, Ciancarini P, Ruggiero M, Russo D (2016) A new agile paradigm for mission-critical software development. CrossTalk 29(6):25–30
15. di Bella E, Fronza I, Phaphoom N, Sillitti A, Succi G, Vlasenko J (2013) Pair programming and software defects–a large, industrial case study. IEEE Trans Softw Eng 39(7):930–953

Hybrid Agile Software Development for Smart Farming Application

Angelo Messina[1] and Ilya Voloshanovskiy[2(✉)]

[1] University of Innopolis, Innopolis, Russian Federation
segreteria@dssea.eu
[2] MTS Company, Innopolis, Russian Federation
i.voloshanovskiy@innopolis.ru

Abstract. Agile methodology for software development has nowadays become the most used ones by Industry and start-ups. Notwithstanding the still persistent general perception of "agile" being free and unregulated, most of the agile methods are very demanding on rituals and on building an effective and working relationship with their Customers/Stakeholders. Some projects, as the one presented in this paper, are characterized by a huge area of uncertainty where neither the customers nor the developers have a consistent knowledge of the domain. In such cases the agile rituals cannot be initiated properly until a knowledge acquisition phase is performed.

In this paper the peculiar "research project" approach used to solve the domain knowledge gap is presented and the subsequent "hybrid agile" method is discussed. A peculiar mix of domain specific research, technology survey and rethinking the agile scrum frame has made it possible to deliver a successful application working in the Smart Farming domain.

Keywords: Hybrid agile · Improved agile · Research and study

1 Introducing the Innopolis University MTS Master Project

1.1 General Introduction of the MSIT Project

The MSIT Project is the most important component of the Master of Science in Information Technology – Software Engineering (MSIT-SE) degree which is designed for junior professionals with one to two years of work experience in software development and who want to boost their career and become technical leaders, software architects or project managers. The program was developed on the basis of the Carnegie Mellon University (CMU) program in Software Engineering and is delivered face-to-face locally at Innopolis by instructors with a long record of industry experience and selected external experts in specific areas that complement the MSIT-SE core courses. The program runs for 12 months and three semesters starting from the middle of August.

The MSIT project can be considered as a Magisterial Theses where candidates work in teams to solve a real industrial problem developing a working and documented software application. Proposal coming from Industry (some are large entities) go

P. Ciancarini et al. (Eds.): SEDA 2018, AISC 925, pp. 198–205, 2020.
https://doi.org/10.1007/978-3-030-14687-0_18

through a selection process performed by the IU Master Program Faculty members and the shortlisted ones are assigned to the master students for full development. The results are estimated and graded accordingly to level of achievements of the most important parameters: Customer Satisfaction, quality of the delivered software, clarity of the software engineering process used for the development. A Mentor from the IU Faculty Staff is assigned to every team to supervise the effort.

1.2 The MTS Project

The goal of the Smart Farming application is to provide to farmers the ability to effectively manage their farm, concentrating on breeding cows. Cow is an expensive resource for farmers in the contemporary world and the loss of this resource due to a disease or other reasons can be financially damaging. Smart Farming application is supposed to help farmers to know about non-standard situations promptly, to track their cows' health, and improve farm operation by analyzing cows' metrics. The main goal of the industrial project is to develop an MVP of the described application to be able to demonstrate the ideas behind it to potential investors.

1.3 "Agile" at IU

The University of Innopolis offers many courses specifically addressing the area of the application of the agile methodologies in the software development process. There are specific elective courses addressing the evolution of the agile methodology and the application to specific mission critical domains requiring the implementation of high reliability and security attributes. As in the DSSEA ® iAgile methodology [1] the accent is on the practical application and verification of the process implementation through class-time exercises which are very close to "real" software development cycles.

2 The MTS Project

2.1 Initial Definition and Design Choices

Initially it was decided to implement the most interesting functions considering that there is only one farm is served by our web-service so that to avoid spending time on the things, which we considered "routine" and not really important for the core part of the prototype, such as complicated access control and multiple farms management, this strategy has been proven effective in the implementation of Scrum even in particularly structured domains [2]. However, when we started to implement the functionality for other roles, we've seen that a certain amount of properly designed access control was necessary to properly model the work of other roles with the system. Moreover, we realized that the case where - one farmer can have several farms was a mandatory feature of our model. This episode clearly shows how knowledge acquisition has been an integral part of the design process as well as the architectural design and the coding. The specificity of the domain could be a powerful drive to modify the requirement acquisition strategy even for larger scale programs [3]. The above mentioned repositioning of the model led to the redesign of the system's architecture and to a certain amount of refactoring on the later stages.

2.2 The Requirement Analysis "Crises"

- When it became clear that we were facing a very particular situation when both the customers and the team lack the core part of domain knowledge and had only vague understanding of what should be done to bring the value to the potential end-users, it was decided to introduce an "inception phase", which was supposed to be conducted at Sprint 0. At inception phase, the team was tasked to investigate the domain and come up with the ideas of basic functionality needed for the MVP of the product [4]. This phase was not structured in a particular way but relied on individual initiative.
- In the end of the inception phase a user stories workshop has been conducted to integrate the individually acquired knowledge into the User Stories and generate a new base set of US to be inserted in the product backlog.
- We tried to follow the devised requirements first couple of sprints, thinking that we had caught at least the essence of the users' needs. However, the further we were proceeding into the sprints, the more we realized that the acquired knowledge through the inception phase was not enough. At this phase of the development, it was also decided to engage a real farmer into the process, who could help us to properly address the risks of misunderstanding of the users' needs.

2.3 Solving the Knowledge Gap

It was then decided to add structure to the knowledge acquisition phase using a Research and Study approach [5]. The challenge was to include the R&S activities into the agile process without compromise for the "true" software engineering activities.

The role of the Scrum PO was enlarged to address the following "non-software development" activities to fill the knowledge gap:

- scientific articles selection and analysis;
- assessment of farmers specialized resources in the Internet;
- communication and interaction with a real farmer;
- comparative investigation concerning the competitors and their consumers;
- business case investigation and outline of a potential application buyer - agro holding.

3 Re-shaping Scrum for the MTS Pilot Project

3.1 Integration of Non-design Activity and Customer Involvement

After several initial sprints in the try to adapt to the specifics of the project with its lack of understanding, we come up with the changes in the initial process, to which we decided to stick for the future.

- PO is an active part of the team.
- During each sprint the PO leads and organizes the process of the domain specific knowledge acquisition.
- Based on the knowledge acquired, PO generates with the Team the ideas for the changes or new functionality, which can be implemented in the future sprints (*there is a similarity here with process implemented in the ITA Army agile methodology during the production of the LC2Evo Command and Control Software* [6]
- Each sprint the PO meets personally (usually in the middle of a sprint) the customers and explains his ideas, the customers give their feedback, and they, together with PO, decide what to include in the Product Backlog. [7] *PO Board in DSSEA iAgile*)
- PO is responsible for the preliminary acceptance of the tasks in Product Backlog, which was reflected in the Definition of Done.
- PO is responsible to explain to the team how the functionality will be used by the users and to write scenarios of usage, which then will be used as acceptance criteria.
- Customers give their feedback at each meeting with PO and at each Sprint Review.

3.2   Hybrid Sprints

Differently from the traditional agile development in this project execution the number of "knowledge acquisition" stories were very relevant and sometimes larger than the usual User Stories generated by the user requirements analysis or by non-functional requirements.

The process of defining the application domain was achieved step-by-step by analyzing and discussing the results of the scientific works considered. As said before, the inclusion of this knowledge (ideas on how to proceed) was in integral part of the Sprints even if the time spent cannot be accounted for as a traditional Software Engineering activity such as coding or testing.

If we evaluate the effectiveness of the Sprints only considering the code or the function points produced, we observe an abnormal distribution of resources (hours) between "management" and "development". On the other end the peculiarity of this project was the particular way the requirements were generated by the insertion of a research study at the very core of the software development process without changing the essence of the scrum-like framework adopted.

For the above reasons and for the successful outcome, the applied methodology can be considered as a hybridization of the general agile paradigm where a considerable part of the requirement analysis has been made through teamwork elaboration of scientific research papers addressing the particular domain of "Farming".

This implementation has a strange logic parallelism with the V-Model for software development where in the effort of overcoming the unsuccessful features on the "Waterfall" methodology more than 50% of the requirement analysis activity was devoted to prepare testing specification far before any line of code was written. In the case we discuss, placing so much effort in the generation of the requirements has generated an increment of the process effectiveness which is testified but the low level of rework on the User Stories (Table 1).

Table 1. Time spent on the rework (refactoring).

Sprint	Time spent	Comment
0	0:00:00	Inception phase
1	0:00:00	
2	0:00:00	
3	0:00:00	
4	12:05:00	New team member joined in the previous sprint took a sprint to refactor the backend and use all the abilities of Spring framework, which will allow save time in the future on the backend development
5	0:00:00	
6	0:00:00	
7	2:39:10	Due to the implementation of new features
8	9:42:34	Due to the new requirement (multi farming) the significant changes in the architecture were needed to be done
9	13:31:01	Due to the new requirement (multi farming) the significant changes in the architecture were needed to be done
10	02:40:00	Due to the implementation of new features

Diagram of hybrid sprints

3.3 Metrics

Besides the usual metrics used in the frame of agile such as velocity, delivered LOCs or FPs, Rework %, etc. it was necessary to use a different one to measure the quality of the process integrating new knowledge into the Sprints. For this purpose, the number of user stories originated by the specific knowledge acquisition story was used (Tables 2 and 3).

Table 2. Number of user stories originated by knowledge acquisition story.

Sprint	Number of US	Comment
0	17	A sprint dedicated primarily to knowledge acquisition
1	0	Started doing things
2	0	Continue going further
3	0	Concentrated on the shaping the process, defining quality, and configuration management; were redoing front-end because the customers did not like it at all - together with them it was decided to move to Material design
4	5	Experiments with the parallel knowledge acquisition process; new team member joined in the previous sprint took a sprint to refactor the backend and use all the abilities of Spring framework, which will allow save time in the future on the backend development

(continued)

Table 2. (*continued*)

Sprint	Number of US	Comment
5	1	Increasing the knowledge base of Aggregate platform; the customers are not happy with the quality of the product - writing tests to increase the coverage, applying the developed quality plan
6	3	Concentrating on the preparation for end-of-semester presentation; continuing extending the tests base
7	4	PO started to do knowledge acquisition activities and meet at sprint basis with the customers (as it described above) - the idea to conduct such meetings belongs to the customers and helped to improve the quality of knowledge acquisition activities
8	4	Working with the established knowledge acquisition process
9	5	Working with the established knowledge acquisition process
10	0	Wrapping things up, preparing to the last presentation; no knowledge acquisition activities are conducted

4 Phases of the Project and Artefacts

Table 3. Time distribution between different activities.

Sprint	Total time	Management	Quality	Development	LOC added	LOC deleted
1	99:11:48	68:47:24	0:00:00	30:24:24	958	28
2	78:46:07	43:32:13	0:00:00	35:13:54	734	52
3	126:59:18	40:23:58	35:34:59	51:00:21	2631	257
4	109:18:51	25:00:00	3:33:54	80:44:57	3368	2476
5	107:12:54	9:19:02	41:48:16	56:05:36	972	151
6	121:51:09	39:07:39	18:43:09	64:00:21	3403	406
7	136:40:11	12:36:04	27:33:22	96:30:45	12595	1754
8	185:21:15	34:04:06	52:43:50	98:33:19	5219	2314
9	201:36:40	12:25:00	89:12:57	99:58:43	8966	5064
10	171:05:18	80:56:29	30:32:00	59:36:49	13206	349

4.1 Initial

This phase is characterized by low LOCs production and very moderate refactoring. This was mainly due to the domain knowledge gap.

4.2 Middle

Phase characterized by evident increase in the production as the knowledge gap is filled while the refactoring keeps at the same level of the initial phase.

4.3   Final

In this last phase the LOCs production increases regularly while the refactoring keeps initially almost at the average value while the acquired domain knowledge gives confidence in the quality of the application to deliver in the final version (Fig. 1).

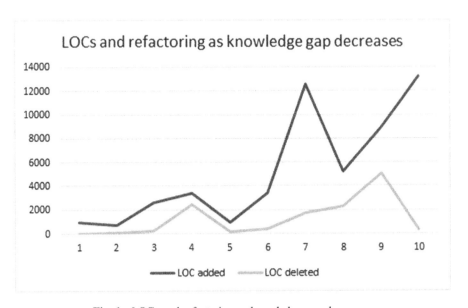

Fig. 1. LOCs and refactoring as knowledge gap decreases.

5   Conclusion and Future Work

It appears that the continuous search for the correct implementation of the agile methodology in the production of software application cannot avoid the inclusion of hybrid solution in the new paradigm. With "hybrid" we mean a software development cycle which is including at the very core of the production activity something radically different from the usual software engineering activities. Even if "hybrid" in conjunction with agile is a relatively new concept, the need of including ad-hoc strategies in the implementation of agile are present in literature [6]. In the presented case, the usual requirement elicitation activity was almost completely substituted by a Research & Study one. In a traditional approach the software lifecycle would have been interrupted to take care of the relevant knowledge gap in a preliminary phase possibly run by specific experts. This eventuality, disconnecting for a period of many weeks the stakeholders and the development team could have led to shifting or even to failure of the project.

The Team was able to structure the R&S phase according to the agile methodology and "knowledge gap" was treated as "functional gap" which required proper skill acquisition and specific knowledge acquisition. Treating this knowledge gap issues as "User Stories" the team was able to continue the agile lifecycle without interruption.

The poor production of working code in the beginning is widely compensated by the high quality of the software delivered at the end and the decrease of refactoring which indicates the level of confidence of the developers in the Product Backlog.

More investigation has to be conducted as far the initial phase of a generic hybrid agile software development cycle might be structured upfront using this R&S-SDP integrated architectural pattern leveraging from the traditional R&S activities. The whole area of the CASE tools to be used in support of this new agile methodologies remains open and widely untouched [7].

References

1. Messina A (2014) Adopting Agile methodology in mission critical software production. In: Consultation on Cloud Computing and Software closing workshop. Bruxelles
2. Messina A Cotugno F (2014) Adapting SCRUM to the Italian army: methods and (open) tools. In: The 10th international conference on open source systems, San Jose, Costa Rica
3. Cotugno F, Messina A (2014) Implementing SCRUM in the army general staff environment. In: The 3rd international conference in software engineering for defense applications – SEDA, Roma, Italy, 22–23 September 2014
4. Mauro V, Messina A (2016) AMINSEP – agile methodology implementation for a new software engineering paradigm definition – a research project proposal. In: Proceedings of 4th international conference in software engineering for defense applications. ISBN: 978-3-319-27894-0 (Print) 978-3-319-27896-4
5. Ciancarini P, Messina A, Poggi F, Russo D (2018) Agile knowledge engineering for mission critical software requirements. In: Synergies between knowledge engineering and software engineering projects. Advances in intelligent systems and computing. https://doi.org/10.1007/978-3-319-64161-4_8
6. Messina A, Modigliani P, Chang S (2015) How agile development can transform defense IT acquisition. In: Proceedings of 4th international conference in software engineering for defense applications. ISBN: 978-3-319-27894-0 (Print) 978-3-319-27896-4
7. Ciancarini P, Messina A, Silliti A, Succi G (2015) A-CASE "Agile" computer aided software engineering environment. In: International CAE conference. Innopolis University, 19–20 October 2015. http://proceedings2015.caeconference.com/speakers/messina.html

Detection of Inconsistent Contracts Through Modular Verification

Alexandr Naumchev[1,2(✉)]

[1] Innopolis University, Innopolis 420500, Russian Federation
a.naumchev@innopolis.ru
[2] Toulouse University, Toulouse, France

Abstract. Existing techniques of Design by Contract do not prevent developers from specifying inconsistent contracts. Any attempt to write a program to meet an inconsistent contract will fail, leading to wasted resources. The present article describes a technique for catching inconsistent contracts in the development time. Applying the technique may save projects' resources and lower the likelihood of failure.

Keywords: Inconsistent contract · Eiffel · Design by Contract · AutoProof · Specification drivers

1 Introduction

In the world of program correctness, it takes two to tango: a specification and implementation. A program is correct if the implementation satisfies the specification. If they do not match, the program is incorrect. In general, work on program verification takes the specification for granted and blames any fault on the implementation. But it is possible to write a specification that no implementation can satisfy. Given the routine contract

```
require
  a > 0
ensure
  b > old a
  b < 0
```

One cannot implement the routine, since it would have to yield a value of b that is both negative and greater than the positive initial value of a. Little work has addressed the issue of "wrong" specifications, perhaps because the general notion of"wrong" is difficult to define and assess: wrong with respect to what? Most likely to another, higher-level specification, but this is just an escalation of the problem. As the example suggests, however, a specific case of "wrong" does not raise this problem: a specification can be inconsistent, hence impossible to implement. Then we want to know right away; and even if we have written an implementation and the verification process – inevitably – cannot prove it correct, it should direct us to looking for the bug where it lies: in the

P. Ciancarini et al. (Eds.): SEDA 2018, AISC 925, pp. 206–220, 2020.
https://doi.org/10.1007/978-3-030-14687-0_19

specification. This article presents a technique to find out automatically that a specification is inconsistent.

Empirical studies of contracted programs reveal that the problem is not limited to artificial examples such as this one, but in fact arises widely in practice. Ciupa et al. [CPO+11] (also [CMOP08], [MCL+]), in their studies of bugs in contracted programs, found that an astounding 62.42% of contract violations during random testing of their program sample resulted from incorrect specifications (rather than incorrect implementations), although they do not state which ones are inconsistencies.

The technique presented here, enjoying automatic tool support thanks to the AutoProof verification environment [TFNP15], is powerful enough to catch the following inconsistencies in classes with a contract:

- Inconsistency of the invariant, which results in impossibility to have instances of the class (Sect. 3).
- Inconsistency of a routine's postcondition, which invalidates the client's state after calling a routine (Sect. 4).
- Inconsistency of a routine's precondition, which makes calling the routine of the class impossible (Sect. 5).

The approach also handles some nuances related to non-exported routines, which may be called only in the non-qualified way (Sect. 6).

2   Why Detect?

If a contract is inconsistent, this will eventually become apparent in any case. For example, if a class has an inconsistent invariant, it will not be possible to develop a correct creation procedure: all creation procedures must establish the class invariant on their completion [Mey92]. If a routine's precondition is inconsistent, no client of the class will be able to call the routine. If the precondition is satisfiable, but the postcondition is not, the outcome will be like the one for an inconsistent invariant: no one will be able to implement the routine correctly.

The biggest problem with this trial and error approach is the waste of resources: it may take multiple man-hours before the developer understands that the specification is not implementable at all. The present article describes an alternative approach, capable of catching inconsistencies before they turn into problems.

2.1   Example

To illustrate our approach, we use a class that describes an ordered triple of integers (Fig. 1), in which the order is represented by the class invariant. From the description of the approach it will be visible that it scales to classes of unlimited complexity, which is why it does not seem bad to pick an artificial and simplified example. There are no inconsistencies in the INTEGER_TRIPLE class' contract, which consists only of the invariant yet. Throughout the article we

```
class INTEGER_TRIPLE
feature
  a, b, c: INTEGER
invariant
  a > b
  b > c
end
```

Fig. 1. Example: an ordered triple of integers.

extend the example, intentionally introduce various inconsistencies to it and show how to detect them.

All experiments from the article are reproducible in the Eiffel verification environment [eve].

2.2 The Basic Idea

This approach relies on the presence of a static program verifier capable of checking routines of the following form:

$$Routine(ARGS) \; \{Pre(ARGS)\} \; Execution(ARGS) \; \{Post(ARGS)\} \qquad (1)$$

where $ARGS$ is a list of formal arguments, Pre and $Post$ are Boolean expressions, and $Execution$ is a sequence of command calls. All these components depend on, and only on, the list $ARGS$. Later, the article uses the term "specification drivers" for referring to routines of this form [NM16].

If there is an inherent inconsistency in either the signature, the precondition, or the execution part, it should be possible to prove the following specification driver:

$$Routine(ARGS) \; \{Pre(ARGS)\} \; Execution(ARGS) \; \{False\} \qquad (2)$$

This equation basically encodes a proof by contradiction of a potential inconsistency: prove false, assuming the possibility to use $Routine$, Pre and $Execution$ together. If the assumption is a logical contradiction, the prover will accept the proof.

This article successively examines how to apply this general form to express and prove inconsistency of invariants (Sect. 3), postconditions (Sect. 4) and preconditions (Sect. 5). The examples are written in Eiffel and checked with Auto-Proof.

3   Class Invariants

A class invariant is a property that applies to all instances of the class, transcending its routines [Mey92]. From this definition, an immediate conclusion follows: if the class invariant is inconsistent, then no objects can have it as their property. This conclusion leads us to the following definition.

Definition 1. *Class* TARGET_CLASS *has an inconsistent invariant, if, and only if, the following specification driver is provable:*

$$class\_invariant(n : TARGET\_CLASS)\ \{True\}\ \{False\} \tag{3}$$

In Eiffel `class_invariant` takes the following form:

```
class_invariant (n: TARGET_CLASS)
  require
    True
  do
  ensure
    False
  end
```

The `class_invariant` routine represents a proof by contradiction, in which the proof assumption is that there can be an object of `TARGET_CLASS`. If its invariant is inconsistent, then existence of such an object is not possible in principle, and assuming the opposite should lead to a contradiction. If AutoProof accepts the `class_invariant` specification driver, then `TARGET_CLASS` has an inconsistent invariant.

```
class INTEGER_TRIPLE
feature
  a, b, c: INTEGER
invariant
  a > b
  b > c
  c > a -- This is not possible
end
```

Fig. 2. Example of an inconsistent invariant

Assume an artificial inconsistency in the invariant of the `INTEGER_TRIPLE` class (Fig. 2). Following from the transitivity of the $>$ relation on integers, the last assertion is inconsistent with the first two. According to the Definition 1, it is necessary to encode the corresponding specification driver and submit it to AutoProof; but a specification driver must exist in some class, which is a minimal compilable program construct in Eiffel. Assume there is such a class, `INTEGER_TRIPLE_CONTRADICTIONS` (Fig. 3). If the class compiles, the next step is to submit the proof to AutoProof.[1] AutoProof accepts the proof (Fig. 4), from

---

[1] The **note explicit: wrapping** expression in the first line of the class is a verification annotation for AutoProof [apm]; its meaning is not related to the ideas under the discussion.

```
note explicit: wrapping
deferred class INTEGER_TRIPLE_CONTRADICTIONS
feature
  class_invariant (n: INTEGER_TRIPLE)
    require
      True
    do
    ensure
      False
    end
end
```

Fig. 3. Proof of the `INTEGER_TRIPLE` invariant inconsistency.

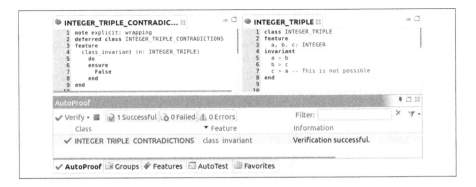

Fig. 4. Proving an inconsistency of the invariant.

which we conclude the existence of an inconsistency in the invariant of the `INTEGER_TRIPLE` class. Removal of the problematic assertion from the invariant makes AutoProof reject the `class_invariant` proof (Fig. 5). The rest of the article does not include any screenshots of Eve because one can easily download it and check the examples locally.

4   Postconditions

According to the principles of Design by Contract [Mey92], a routine will never complete its execution, if it fails to assert its postcondition; consequently, to express the contradiction, the corresponding specification driver needs to assume the termination and assert `False` in its postcondition. Two definitions follow for commands (Sect. 4.1) and functions (Sect. 4.2); the definitions differ according to the ways in which clients use commands and functions.

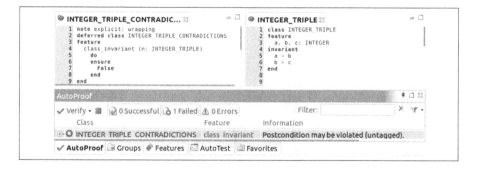

Fig. 5. Failure to find an inconsistency in the invariant.

4.1 Commands

Commands are state-changing routines, which is why clients can use command calls only in routines' bodies, not in contracts. To prove the inconsistency of a command's postcondition, it is necessary to assume that it is possible to call the command and continue execution of the program.

Definition 2. *An exported command* c *with a precondition* pre *and a list of formal arguments* ARGS *from class* TARGET_CLASS *has a contradictory postcondition, if, and only if, the following specification driver is provable:*

$$c\_post(t : TARGET\_CLASS; ARGS)\ \{t.pre(ARGS)\}\ t.c(ARGS)\ \{False\}$$
$$(4)$$

In Eiffel c_post takes the following form:

```
c_post (t: TARGET_CLASS; ARGS)
  require
    t.pre (ARGS)
  do
    t.c (ARGS)
  ensure
    False
  end
```

This is a proof by contradiction in which the assumption is the possibility to call the c command so that the execution reaches checking the postcondition of c_post. If the postcondition of c is inconsistent alone or is not consistent with the invariant of TARGET_CLASS, the execution will stop right after the call, and the outer postcondition will never be checked (Fig. 6).

Assume the task is to implement command move_c that should somehow change the value of the c attribute in the INTEGER_TRIPLE class:

The last line in the postcondition of the move_c command makes the value of c bigger than that of a, which is not consistent with the invariant.

```
class INTEGER_TRIPLE
feature
  a, b, c: INTEGER
  move_c
    do
    ensure
      a = old a
      b = old b
      c = 2 * a - old c -- Inconsistent with the invariant
    end
invariant
  a > b
  b > c
end
```

Fig. 6. A command with an inconsistent postcondition.

```
note explicit: wrapping
deferred class INTEGER_TRIPLE_CONTRADICTIONS
feature
  move_c_post (n: INTEGER_TRIPLE)
    require
      modify (n)
    do
      n.move_c
    ensure
      False
    end
end
```

Fig. 7. Specification driver for detection of the move_c command's inconsistent post-condition.

The move_c_post specification driver (Fig. 7) reflects a proof by contradiction of the inconsistency.[2]

AutoProof accepts the move_c_post specification driver, from which one can see the presence of an inconsistency in the postcondition of move_c; removal of its last line will make AutoProof rejecting the proof.

4.2 Functions

Functions are state-preserving value-returning routines, which may be used in other routines' pre- and postconditions. To prove by contradiction inconsistency of a function's postcondition, it is necessary to assume that the function can produce some value.

---

[2] The modify(n) expression inside the **require** block is a frame specification for Auto-Proof [PTFM14].

```
class INTEGER_TRIPLE
feature
  a, b, c: INTEGER
  diff_ab: INTEGER
    do
    ensure
      Result = b − a
      Result > 0 -- Inconsistent with the invariant
    end
invariant
  a > b
  b > c
end
```

Fig. 8. A function with an inconsistent postcondition.

```
note explicit: wrapping
deferred class INTEGER_TRIPLE_CONTRADICTIONS
feature
  diff_ab_post (n: INTEGER_TRIPLE; diff: INTEGER)
    require
      n.diff_ab = diff
    do
    ensure
      False
    end
end
```

Fig. 9. Specification driver for detection of a function with an inconsistent postcondition.

Definition 3. *An exported function f with a return type T, precondition* `pre`*, and a list of formal arguments* `ARGS` *from class* `TARGET_CLASS` *has a contradictory postcondition, if, and only if, the following specification driver is provable:*

$$f\_post(t : TARGET\_CLASS; ARGS; res : T) \{t.f(ARGS) = res\} \{False\} \tag{5}$$

In Eiffel `f_post` takes the following form:

```
f_post (t: TARGET_CLASS; ARGS; res: T)
  require
    t.f (ARGS) = res
  do
  ensure
    False
  end
```

If the postcondition of `f` is inconsistent alone, or is not consistent with the class invariant, it will never return any result. The `require` block in the Definition 3 states the opposite: there is some value `res` of type `T`, such that it equals the value of the function; this statement is the assumption of the proof by contradiction.

Assume the task is to implement a function `diff_ab` that returns the difference `b - a` between `a` and `b`. From the invariant of `INTEGER_TRIPLE`, one can see that this difference should always be negative, but the developer may confuse operators `>` and `<`, in which case the postcondition of `diff_ab` becomes inconsistent (Fig. 8).

Specification driver `diff_ab_post` (Fig. 9) reflects the proof by contradiction corresponding to the given example. AutoProof accepts `diff_ab_post`, thus disclosing the presence of an inconsistency.

5   Preconditions

Precondition of a routine constitutes requirements that every client has to meet to call the routine. If a precondition is inconsistent, no client will be able to meet it.

Definition 4. *An exported routine* `callable` *with precondition* `pre` *and list of formal arguments* `ARGS` *from class* `TARGET_CLASS` *has an inconsistent precondition, if, and only if, the following specification driver is provable:*

$$callable\_pre(t : TARGET\_CLASS; ARGS) \ \{t.pre(ARGS)\} \ \{False\} \qquad (6)$$

In Eiffel `callable_pre` takes the following form:

```
callable_pre (t: TARGET_CLASS; ARGS)
  require
    t.pre (ARGS)
  do
  ensure
    False
  end
```

Assume the `move_c` command requires the result of the `diff_ab` function to be greater than 0, which is not consistent with the class invariant, according to the postcondition of `diff_ab` (Fig. 10).

The `move_c_pre` specification driver reflects the Definition 4 as applied to the precondition of the `move_c` command. It has the same precondition as does the `move_c` command, where every non-qualified call is replaced with its qualified counterpart; the target for the call comes from the `move_c_pre`'s list of formal arguments.

Note that the `move_c_post` (Fig. 7) specification driver needs to be updated: the `move_c` command now has a precondition that has to be guaranteed by all its callers.

AutoProof discloses the presence of a contradiction by accepting the `move_c_pre` specification driver (Fig. 11).

```
class INTEGER_TRIPLE
feature                                                 diff_ab: INTEGER
  a, b, c: INTEGER                                        do
  move_c                                                  ensure
    require                                                 Result = b − a
      diff_ab > 0 -- Inconsistent with the invariant       end
    do                                                   invariant
    ensure                                                 a > b
      a = old a                                            b > c
      b = old b                                          end
    end
```

Fig. 10. The move_c command with an inconsistent precondition.

6 Non-exported Routines

A non-exported routine is a routine that cannot be invoked using a qualified call [Mey09]. Consequently, the definitions, which were presented so far, are not applicable to non-exported routines: those definitions rely on the ability to do qualified calls. The present section gives definitions applicable to non-exported routines.

Definition 5. *The non-exported command* c *with precondition* pre *and list of formal arguments* ARGS *has an inconsistent postcondition, if, and only if, the following specification driver is provable:*

```
note explicit: wrapping
deferred class INTEGER_TRIPLE_CONTRADICTIONS
feature
  move_c_pre (n: INTEGER_TRIPLE)
    require
      n.diff_ab > 0
    do
    ensure
      False
    end
end
```

Fig. 11. Specification driver for catching the inconsistent precondition.

$$c\_post(ARGS) \ \{pre(ARGS)\} \ c(ARGS) \ \{False\} \qquad (7)$$

In Eiffel `c_post` takes the following form:

```
c_post (ARGS)
  require
    pre (ARGS)
  do
    c (ARGS)
  ensure
    False
  end
```

Definition 6. *The non-exported function f with return type T, precondition* `pre`, *and list of formal arguments* `ARGS`, *has an inconsistent postcondition, if, and only if, the following specification driver is correct:*

$$f\_post(ARGS; res : T) \ \{f(ARGS) = res\} \ \{False\} \qquad (8)$$

In Eiffel `f_post` takes the following form:

```
f_post (ARGS; res: T)
  require
    f (ARGS) = res
  do
  ensure
    False
  end
```

Definition 7. *The non-exported routine r with precondition* `pre` *and list of formal arguments* `ARGS` *has an inconsistent precondition, if, and only if, the following specification driver is correct:*

$$r\_pre(ARGS) \ \{pre(ARGS)\} \ \{False\} \qquad (9)$$

In Eiffel `r_pre` takes the following form:

```
r_pre (ARGS)
  require
    pre (ARGS)
  do
  ensure
    False
  end
```

In Definitions 5, 6, and 7 the routine calls do not have targets, which means the calls can occur only in the class where the routines are defined or in one of its descendants [Mey09].

```
class INTEGER_TRIPLE
feature {NONE}
  a, b, c: INTEGER
  move_c
    require
      diff_ab > 0
    do
    end
  diff_ab: INTEGER
    do
    end
end
```

Fig. 12. The INTEGER_TRIPLE class with all the features non-exported.

Assume the INTEGER_TRIPLE class with all its routines non-exported (Fig. 12), which is denoted by the {NONE} specifier next to the feature keyword.

For such an example, the specification drivers class may be a descendant of the INTEGER_TRIPLE class so that it will be able to call its routines in the unqualified way (Fig. 13).

```
note explicit: wrapping
deferred class INTEGER_TRIPLE_CONTRADICTIONS
inherit INTEGER_TRIPLE
feature {NONE}
  move_c_post
    require
      diff_ab > 0
    do
      move_c
    ensure
      False
    end

  diff_ab_post (res: INTEGER)
    require
      diff_ab = res
    do
    ensure
      False
    end
  move_c_pre
    require
      diff_ab > 0
    do
    ensure
      False
    end
end
```

Fig. 13. Specification drivers for detection of contradictions in the non-exported routines.

7   Related Work

The problem of inconsistent specifications receives noteworthy attention in Z [WD96]. Without an explicit syntactical separation of Z assertions into pre- and postconditions and in the absence of an imperative layer, it is not clear how to apply the techniques from the present article. Detection of inconsistencies in Z may occasionally lead to development of complicated theories and tools [Mia02]. We are not aware of any work specifically targeting detection of inconsistencies in Design by Contract.

The problem of inconsistent contracts may also be viewed through the prism of liveness properties in concurrency [MK99]:

- An inconsistent class invariant makes the class "non-alive": it is not even possible to instantiate an object from the class.
- An inconsistent routine precondition makes the routine never callable.
- An inconsistent routine postcondition leads to its clients' always crashing after calling the routine.

8   Summary

A strength of the approach is the possibility to employ it for real-time detection of inconsistencies. Once generated, the specification driver for the invariant (Sect. 3) never changes; consequently, it is enough to recheck it whenever the invariant changes and display a warning in the event of successful checking. The same applies to detection of inconsistent pre-/postconditions, with the only difference that it will be necessary to update the preconditions of the corresponding specification drivers in the event of modifying the routine's precondition. In any case, such an update amounts to copying the precondition with possibly adding targets in front of the class' queries (Sect. 5).

Another strength of the approach is its applicability. Eve is not the only environment in which it is possible to write and statically check contracts: there is a similar environment for .net developers [Mic15], in which the techniques presented here are directly applicable. There are several programming languages that natively support contracts, for which the present approach is applicable conceptually, but still needs development of a verifier capable of checking specification drivers.

Other applications of specification drivers include seamless requirements specification [NM17] and its narrower application in the context of embedded systems [NMM+17]. These approaches contribute to seamless development [WN95], [Mey13], [NMR16] – a research program pursuing unification of requirements and code.

8.1   Limitations of the Approach

Results Interpretation. In the present approach, a positive response from the prover means something bad, which is detection of an inconsistency. This may be misleading: the developer may think, instead, that everything is correct. This requires fixing, possibly by development of a separate working mode in AutoProof.

Precision. The approach shows the presence of a contradiction but does not show its location. This is not a problem when developing from scratch: background verification may catch the contradiction as soon as it is introduced. However, if the task is to check an existing codebase, the only way to locate origins of contradictions seem to be in commenting/uncommenting specific lines of the contracts.

Frozen Classes. The approach for non-exported routines relies on the ability to inherit from the supplier class. It is not possible to inherit from a class, if it is declared with the `frozen` specifier [Mey09]. Nevertheless, it is always possible to apply the technique to exported routines of the supplier class.

8.2   Future Work

The present article describes the approach conceptually, yet no tools exist that could generate and check the necessary proofs automatically. Two main possibilities exist in this area:

- Build a contradiction detection functionality directly into AutoProof, without letting developers see the proofs.
- Develop a preprocessing engine on the level of Eiffel code that would generate classes with proofs for checking them with AutoProof in its current state.

Apart from automating the approach, it seems reasonable to investigate see whether the proof by contradiction technique may be of any help with other problems of program verification.

Acknowledgement. The authors are grateful to the administration of Innopolis University for the funding that made this work possible.

References

[apm] AutoProof manual: annotations. http://se.inf.ethz.ch/research/autoproof/manual/#annotations

[CMOP08] Ciupa I, Meyer B, Oriol M, Pretschner A (2008) Finding faults: manual testing vs. random+ testing vs. user reports. In: 2008 19th International Symposium on Software Reliability Engineering (ISSRE). IEEE, pp 157–166

[CPO+11] Ciupa I, Pretschner A, Oriol M, Leitner A, Meyer B (2011) On the number and nature of faults found by random testing. Softw Test Verif Reliab 21(1):3–28

[eve] Eve: Eiffel verification environment. http://se.inf.ethz.ch/research/eve/

[MCL+] Meyer B, Ciupa I, Liu LL, Oriol M, Leitner A, Borca-Muresan R (2007) Systematic evaluation of test failure results

[Mey92] Meyer B (1992) Applying 'design by contract'. Computer 25(10):40–51

[Mey09] Meyar B (2009) Touch of class: learning to program well with objects and contracts. Springer, Heidelberg

[Mey13] Meyer B (2013) Multirequirements. In: Seyff N, Koziolek A (eds) Modelling and quality in requirements engineering (Martin Glinz Festscrhift). MV Wissenschaft

[Mia02] Miarka R (2002) Inconsistency and underdefinedness in Z specifications. Ph.D. thesis, University of Kent

[Mic15] Microsoft (2015) Code contracts for .net. https://visualstudiogallery. msdn.microsoft.com/1ec7db13-3363-46c9-851f-1ce455f66970

[MK99] Magee J, Kramer J (1999) State models and Java programs. Wiley, Hoboken

[NM16] Naumchev A, Meyer B (2016) Complete contracts through specification drivers. In: 2016 10th international symposium on theoretical aspects of software engineering (TASE) (2016). IEEE Computer Society Press, pp 160–167

[NM17] Naumchev A, Meyer B (2017) Seamless requirements. Comput Lang Syst Struct 49:119–132

[NMM+17] Naumchev A, Meyer B, Mazzara M, Galinier F, Bruel J-M, Ebersold S (2017) Expressing and verifying embedded software requirements. arXiv preprint arXiv:1710.02801

[NMR16] Naumchev A, Meyer B, Rivera V (2016) Unifying requirements and code: an example. Lecture notes in computer science (including subseries lecture notes in artificial intelligence and lecture notes in bioinformatics), vol 9609, pp 233–244

[PTFM14] Polikarpova N, Tschannen J, Furia CA, Meyer B (2014) Flexible invariants through semantic collaboration. In: FM 2014: formal methods. Springer, Heidelberg, pp 514–530

[TFNP15] Tschannen J, Furia CA, Nordio M, Polikarpova N (2015) AutoProof: autoactive functional verification of object-oriented programs. arXiv preprint arXiv:1501.03063

[WD96] Woodcock J, Davies J (1996) Using Z: specification, refinement, and proof, vol 39. Prentice Hall, Englewood Cliffs

[WN95] Waldén K, Nerson JM (1995) Seamless object-oriented software architecture. Prentice-Hall, Englewood Cliffs

Balanced Map Generation Using Genetic Algorithms in the Siphon Board-Game

Jonas Juhl Nielsen and Marco Scirea[✉]

Maersk Mc-Kinney Moller Institute, University of Southern Denmark,
Odense, Denmark
msc@mmmi.sdu.dk

Abstract. This paper describes an evolutionary system for the generation of balance maps for board games. The system is designed to work with the original game Siphon, but works as a proof of concept for the usage of such systems to create maps for other board games as well. Four heuristics and a constraint, developed in collaboration with the game designer, are used to evaluate the generated boards, by analyzing properties such as: symmetry, distribution of resources, and points of interest. We show how the system is able to create diverse maps that are able to display balanced qualities.

1 Introduction

Board games are increasing in popularity, and successfully publishing a board game is becoming increasingly challenging [1]. Hence it is important to have a solid game design, so that this specific game will stand out of the other thousands that are developed. An important game design element in board games is balance. If a game is completely unbalanced, it gives an almost guaranteed victory for the same player every time, which makes it unpleasant to play. In a balanced game the initial player situation (e.g. the starting country in Risk) should not ensure victory or defeat. As an example of an advantage a player might have, consider having the first move, a balanced game should allow the player skills (and sometimes luck) to determine the progression of the game regardless of such advantages. This problem relates to all board games and is consequently an area that needs focus. Board games are becoming more and more advanced, which makes it complex to ensure balance. The incomplete information, randomness, large search space and branching factor makes it almost certain that the designer will miss substantial balance elements. When introducing procedural elements to the game – as the multiple scenarios for victory and dynamic map of Betrayal at House on the Hill – it becomes even harder to control the scope of all possible actions that the players can take. Artificial intelligence (AI) techniques can be employed to explore the game space and provide evidence that a certain rule in the game must be removed or revised. To acquire the balanced experience, play testing is essential. After the AI has generated various setups for the game, they can be tested by human players to evaluate if it still feels fun to play [2].

© Springer Nature Switzerland AG 2020
P. Ciancarini et al. (Eds.): SEDA 2018, AISC 925, pp. 221–231, 2020.
https://doi.org/10.1007/978-3-030-14687-0_20

This paper highlights the exploration of how to make a board game balanced using artificial intelligence techniques, specifically with procedural content generation (PCG) applied to board generation using a genetic algorithm (GA). This will be applied on a self-developed board game called Siphon. This study is based on a limited version of Siphon and concern itself with balancing game maps. In the future the project will consider other game play elements that also affect game balance. The simplified version of the game will be described in Sect. 3. The genetic algorithm uses several heuristics in the fitness functions that describe different desirable features to create interesting and balanced maps.

2   Background

Video Games drags our attention and makes us able to sit in front of the screen for hours and hours, just to progress a virtual character in games like World of Warcraft. Some board games and card games can inspire our strategic minds, to find optimized ways to beat friends and family in our most favorite games. Sometimes we are killing the greater evil together to survive in the darkness. What is it with all these games that drags our attention? A part of the answer can be simplified to "A great game design" [3].

This project introduces the game design element "balance", with the focus at board games. To accomplish balance in board games, it is important first to locate the areas, which can be unbalanced and then figure out a plan to make it balanced and keep the fun in playing the game. By introducing AI in game design, it provides the opportunity to explore the game space much quicker than a human being would ever be able to do. They can be used to discover unbalanced strategies or even rules that are not covered in the rule-book.

2.1   Designing Balance

Game Design is a huge area of study, which is about figuring the tools of play, the rules, the story plot and line, the possible strategies, etc. As this project focuses at the balance aspect of game design, we will look at internal, external and positional balance, where some good balanced games will be used as examples, to understand what a good design is [4].

One of the primary concerns of a game designer is to generate a game with long replay ability which is caused by e.g. varied experiences and removing unfairness. Internal balance has the focus of eliminating false decisions and regulating dominant strategies. Eliminating false decisions means that every action that has been chosen can lead to a high scoring weight when combined with other actions, which means that they are all valid for the chosen strategy. This means that no choices are considered as a false decision. If a game implements false decision making, the experienced players will have a huge advantage because of his/her knowledge of the game and not the ability of outsmarting the other. The board game Stratego uses internal balance, which introduces the false decision making. It uses two frameworks that are made to equalize the output of different game choices. Specifically, Stratego uses the intransitive relationship framework which means that there is always a counter (as also in Rock-Paper-Scissors) Stratego uses ranks to determine who beats who. The higher ranks beat the lower ranks with one exception: The highest rank is beaten by the lowest rank.

Another framework is the transitive relationship, which uses a cost-benefit curve to compare objects. If the benefit is higher than the cost, the object will be categorized as overpowered and if it is lower as under-powered. An example of a game that uses this, is Magic the Gathering. It is a card game, where each card has a cost to use and has an effect when it is used. Ticket to Ride is also an example of a game that is very popular and uses the transitive framework. Players are rewarded for selecting and completing tickets. The harder the ticket is, the more rewarding it is to complete.

Another kind of balancing is the external balance. This is the primary focus of this project, since it encompasses the designers' choices of how the game is built (e.g. being the starting player or the positions of resources gives a player huge advantage). Symmetry in games are often appealing, since humans feel comfortable by seeking patterns. It is one of the easiest ways of making balance, since it gives the same position and options for each player. The reason not to make everything symmetric is that it can appear less interesting and subtle imbalance can easily be overseen. An example of a mirror symmetric game is Chess. The Fritz database, which contains more than eight million chess games, shows that the average score is 0.55 points, where 1 is the white winning, 0 is loosing and 0.5 is a draw. This shows that the starting player, using the white pieces, has an advantage in chess. Asymmetrical game play often provides more interesting scenarios, but it is harder to make balanced. Also, it might be harder for new players to discover a good action in a given situation, which will widen the skill level between newcomers and experienced players. The art is finding a combination between the symmetry and asymmetry to have an interesting but clear design for all kind of players and still have a balanced game.

2.2   Using AI for Balance

A previous study [5] uses AI-based play testing on the board game Ticket to Ride, to detect loop holes and unbalanced strategies. The AI's plays the game using different evolutionary algorithms and evaluating each play through. By playing the game thousands of times with different maps, some strategies will be evaluated high. They might as well discover failure cases where the agents found game states that are not covered in the rules. These are the areas in the game design that needs focus. A high evaluation might be close to an optimal strategy, which might be overpowered. Evolutionary algorithm is the super-set of the genetic algorithm, which is used for Siphon (see Sect. 4).

The popular board game Settlers of Catan has multiple options of setting up the game board when playing. One is randomization which can provide very unbalanced boards for some players. Another way is following the setup that is displayed in the rule-book [6]. A third way is using "BetterSettlers".[1]

Civilization is a famous video game, which uses random map generation every time you play a new custom map. A study [7] describes how Civilization uses a real world map to discover types of environment by color, to gain the types of resources. After this the information from the map, (latitude and longitude) are used to generate the maps. At last genetic algorithms calculates the initial position, where the fitness function takes fairness and exploration into account.

Another study [8] uses a GA for generating new board games, by using simple and existing ones: Checkers, Tic-Tac-Toe and Reversi. Finding the game elements that they all have in common, creates the search-space for the algorithm and makes it able to generate 144 new games. Adding some nontraditional game elements, increases the search-space to 5616 games. After generating the games, they are played hundreds of times with a simulator. The games are then evaluated with the fitness function based on diversity and balance. By doing this, new and balanced board games are created, where the rules are defined by the AI. This is an example of how powerful GAs can be.

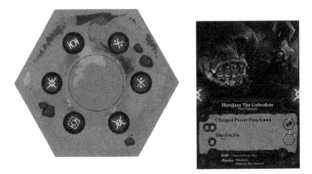

Fig. 1. Example of tiles and heroes cards in Siphon. Each tile can have up to 6 runes which can be used by heroes standing on them to activate abilities.

3   Game Design of Siphon

In this report, a simplified version of a self-developed board game called Siphon are used to demonstrate how GAs can be used to balance a board game.

Siphon is a competitive two-player game. Each player has 5 heroes, which they have picked in turns from a hero pool. The heroes are summoned in an arena which is a board consisting of 37 hexagon tiles, placed as a rectangle. A tile has up to four different rune types and can have up to six of these in any

---

[1] BetterSettlers (www.bettersettlers.com), uses an algorithm that provides the fairest distribution of starting setup.

combination. The runes count as resources. The system creates balanced boards by deciding where to place these runes. See Fig. 1.

A hero has various abilities and has two split runes on each side of the card. These are the same types that can appear on the tiles. To cast abilities, it is a requirement that the runes appearing on the abilities are met. See Fig. 1. This means that the split runes displayed on the hero cards, must combine a full rune on the board tiles, by rotating the hero card. See the rotating hero card on Fig. 2. Since the heroes can both move around on the tiles and rotate, and each hero are different, the branching factor is huge since there is many different actions available. It would be very difficult for a designer to find a board that is balanced just by trial-and-error, and the odds of finding a balanced board are minimal. As such the game present a good framework for the usage of AI techniques for balancing.

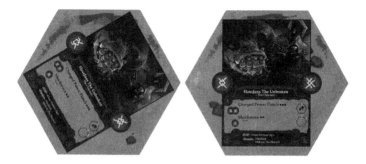

Fig. 2. The hero cards can rotate over a tile, and need to be align the correct runes to activate abilities.

4   Methods

The algorithm that we employed to balance Siphon is a genetic algorithm (GA), which is a subset of evolutionary algorithms [9]. GAs are inspired by evolution theory, which is the process of natural selection and survival of the fittest. These algorithms are used to solve problems where "brute force" algorithms would take too long. GAs generate solutions to optimize and search for problems that uses bio-inspired operators. These operators are *selection*, *mutation* and *crossover*.

Before going in depth with the before-mentioned operators, Darwinian evolution theory will be explained [10]. He uses three elements that describes the natural selection that occurs in our nature.

I. Heredity - The children of an evolution need to receive the properties of their parents. If a creature reproduces, it needs to pass down its traits to the next generation.

II. Variation - The traits that are present in the population need to vary, so that all creatures in it reproduce a variety of traits. If this was not the case, all children will possess the same genome as their ancestors.

III. Selection - The fittest creatures of a population need to be able to pass down their genes, so that the stronger survives. This is the evolution and is commonly referred as "survival of the fittest". A fit creature does not necessarily mean "physically fit" but is based on its likelihood of being good at a specific task. The selection operator uses this principle.

In GAs the creatures of a population are often referred to as "individuals", where the traits of the creature are the "genes". In practice, the individuals are the solutions we want to assess and the genes are the characteristics that define a solution. From the Darwinian principle of variation populations are usually initialized randomly. After creation, each individual of the population is evaluated by using an evaluation (fitness) function. The fitness values are then used by the *selection operator*. The operator is used to determine which individuals are going to be chosen to become the parents for the next generation. In this implementation all individuals of a population have been evaluated the most fit individuals are placed in a mating pool. For the reproduction step, a simple *one-point crossover operator* is used. This crossover operator consists of: (i) choosing two random individuals from the mating pool (parents), (ii) picking a random index to split the genomes, (iii) creating two new individuals (offspring) by combining the first part of the genome of one parent with the second part of the other (and vice-versa). After the crossover step, it is usually necessary to use a *mutation operator*. This operator is used to introduce random changes in the genes of the offspring, that would otherwise have only a combination of the genes of its parents. The mutation operator can be seen as a random local search mechanism, as it should never result in very big variations in the genome. As GAs contain many stochastic elements to the optimization process, there is a chance for evolution to get stuck in a local optimum, mutation used to limit this shortcoming of the algorithm family. An example of when this problem might appear occurs when the starting population does not have the genes that are required to get to the optimal fitness score. With mutation, it is possible to change a gene that is not a part of the ancestors and access the genes that allow for the possibility of acquiring the optimal fitness score.

4.1   Fitness Function

When creating a population, all individuals have to be evaluated in order to determine the "fitness" of the solutions they represent. Fitness functions are usually very domain-dependent and are in fact the driving force of the evolutionary process. z These heuristics have been defined in collaboration with the designers of the game (one of which is an author of this paper). One constraint and four heuristics have been defined. The constraint has to be fulfilled before an individual is evaluated on the other heuristics. In practice, that means that all generated boards **have to** satisfy the constraint, but might satisfy in different amounts (possibly not optimally) the other heuristics. The fitness scoring are as follows:

$$f(sol) \quad = \quad \begin{cases} Constr(sol) & \text{if } Constr(sol) < 1 \\ w_1 Even(sol) + w_2 Symm(sol) \\ + w_3 Rune(sol) + w_4 Mono(sol) & \text{if } Constr(sol) = 1 \end{cases}$$

The possible fitness values are between [0,2]. While the constraint has not been met this value is restricted between [0,1], effectively forcing evolution to first create individuals that satisfy constraints. Each of the other four heuristics is assigned a weight. A heuristic with a higher weight will be prioritized over the others. In Siphon, an individual is a complete board setup and the genes are the tiles.

The constraint heuristic (*Constr*) is based on the sums for each type of rune that is allocated on the hero cards. The percentage of these sums are then calculated from the total amount of runes on the hero cards. An individual (a board setup) is then evaluated with a higher score the closer it gets to these percentages by placing the runes on the tiles.

The even-distribution heuristic (*Even*) is evaluating how evenly distributed runes are all over the board (e.g. a board with x tiles and x runes of a type has 1 of these runes on each tile).

The symmetry heuristic (*Symm*) is based on symmetry, as that is often a quality that is desirable for balance. The designer felt it was too static to aim for complete mirror symmetry (which is also a trivial problem that does not require evolution to solve), so it was decided that a potentially imperfect diagonal symmetry was more desirable. With the weight parameter, it is possible to adjust the amount of symmetry, since the GA usually does not get a perfect score in this metric.

The rune-count heuristic (*Rune*) is evaluating that the number of runes on the tiles are below four, with the exception of the center tiles, which have six. This heuristic is used to create dominant tiles, which generates high risk - high reward areas on the board.

The mono-rune heuristic (*Mono*) is aiming toward having tiles with the same rune type. This heuristic's purpose is to create interesting territories (biomes) which might high rewards, so that the players have an incentive to want to capture it, which generates more action in the game.

Considering all the different heuristics that are part of the fitness function, individuals might never get to a perfect score, since satisfying a condition might break another. Nonetheless, the GA is able to create boards that have a close to perfect (if not optimal) score. These individuals are still valid candidates for a balanced board, and can be play tested by human players and compared to evaluate, which one that should be the chosen one. This might require many candidates, but the GA are able to decrease the number of candidates significantly. Figure 3 shows a board generated by the GA, with a fitness score of 1.92. This can be compared with other generated boards and get play tested by human players.

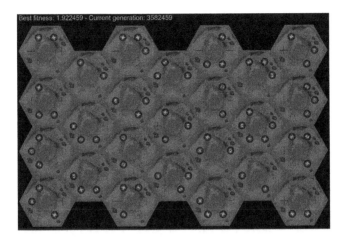

Fig. 3. An example of a generated map with fitness quite close to the optimal value (1.92).

5 Analysis

There are many parameters of the GA that can be adjusted. These include: the population size, the mutation rate, heuristic weights, and heuristic-specific parameters. By finding the right adjustment, it might be possible to get closer to optimal solutions, but finding these can be difficult and require a large amount of testing. In this section, the results from testing the GA used in Siphon are displayed and described.

5.1 Effect of Population Size

Figure 4 display the best fitness and the average fitness in a population over generations for two data sets. The results will be used to analyze what to optimize and to see after how many generations the GA stops finding new solutions. The only parameter that will be changed is the population size, which are displayed on the graphs. The static parameters are set to: Mutation Rate: 0.2, Heuristic Weights (w_1, w_2, w_3, w_4): 0.25, 0.25, 0.25, 0.25.

The best fitness curve are the same for both data sets and reach a plateau around 7–8000 generations. The two average fitness from the data sets are instead quite different. The data set with the population parameter set to 400, rarely satisfies the constraints and therefore has an average fitness lower than 1 (the score indicating the constraint satisfaction). The other data-set rises after there are enough individuals that satisfy the constraint in the mating pool. That is why there is a delay of the average fitness.

Figure 5 shows the amount of individuals in a population which satisfy the constraint over generations. Three data sets are used with same static parameters previously defined. The population parameter are set to 100, 200 and 400 for

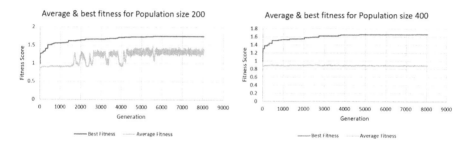

Fig. 4. Best and average fitness in the setup with 200 and 400 individuals in the population

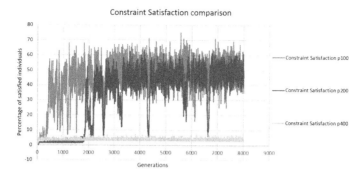

Fig. 5. Amount of individuals that satisfy the constraint as evolution progresses.

the data sets. It appears that for this problem a relatively low population size (around 150) makes it easier for the GA to satisfy the constraint (see Table 1). When the mating pool starts having individuals which satisfy the constraint, it will be move likely that following generations will as well.

Table 1. Population size effect

Population size	Fitness score
100	1.72
150	1.755
200	1.75
400	1.655
800	1.58

5.2 Individual Heuristic Effect

A population size of 150 will be a static parameter for the next data set. As previously tested, the fitness score rarely gets above 1, 8. This could be caused by the heuristics working against each other, or an optimal solution might not exist. Table 2 compares the maximum fitness scores found by running evolution when removing each heuristic one at a time. The ones that are still active are assigned identical weights.

Table 2. Heuristic effect

Parameters	Values
Population size	150
Mutation	0.2
Even-distribution	20
Symmetry	10
Rune-count	40
Mono-rune	30

By removing single heuristics one by one and evaluating the scores we wanted to investigate which heuristic are most difficult to satisfy. As the table shows, the symmetry heuristic are the most difficult, since the fitness function evaluates the score highest when it is removed. With these results the designer can consider lowering the weight of the more difficult heuristics or readjust the heuristic algorithm.

6   Conclusion

The purpose of this project was to explore the balancing of board games using a genetic algorithm. A simplified version of the self-developed game Siphon was used as a case-study. It presents a map that has a big impact on how the game plays out and consequently that is where the algorithm has been applied. A genetic algorithm uses a fitness function to evaluate how "good" a board is. The performance is dependent on the heuristics chosen, and as such these were developed together with the designer of Siphon. The heuristics were defined by analyzing and trying different strategies that seems to make an interesting game.

To provide a proper evaluation of the system, we plan to conduct an experiment where players will be asked to play generated boards, evaluate them, and compare them with human-designed ones. One of the advantages of using GAs for these generative tasks is that, given the stochastic elements of the technique, it allows for the generation of diverse maps, which still fulfill (to some degree) the requirements set by the heuristics. The heuristics do not necessarily need to be perfectly satisfied in order to make a balanced and interesting map. Another possible improvement to the system would be to replace the heuristic with a simulation: to evaluate if a map is balanced we can simulate a number of games and analyze the results (e.g. how many times did player one win? How long did the games last in average?). A possible candidate for the implementing the player controller is Monte Carlo Tree Search, which is domain independent (does not need heuristics). While this paper focuses on creating balanced maps, it would be interesting in future work to reverse this concept to create very unbalanced (unfair) maps. This could be used to create extra challenge for the players, or as a game mechanic if you assume the players would be able to act on the map layout. Since the algorithm is entirely based on the heuristics, they can be modified to any needs. For example, as wars are rarely balanced, a use case could be having the enemy part of the map as static, so that the system could create optimal positioning of resources for the player/user. That said, we believe that this system would be of more use in creating training simulations than optimizing positions for real-world usage.

To conclude, this paper presents an exploration of the usage of GAs for the generation of balanced board-game maps. The results, while preliminary, seem to show potential for creating diverse and interesting maps.

References

1. Adkins S (2017) The 2017–2022 global game-based learning market. In: Serious Play Conference
2. Yannakakis GN, Togelius J (2017) Artificial intelligence and games. Springer
3. Fullerton T (2008) Game design workshop: a playcentric approach to creating innovative games. CRC Press, Boca Raton
4. Harkey A (2014) Balance
5. de Mesentier Silva F, Lee S, Togelius J, Nealen A (2017) Ai-based playtesting of contemporary board games. In: Proceedings of the 12th international conference on the foundations of digital games. ACM, p 13
6. Teuber K (1995) The settlers of catan - game rules and almanac
7. Barros GA, Togelius J (2015) Balanced civilization map generation based on open data. In: 2015 IEEE congress on evolutionary computation (CEC). IEEE, pp 1482–1489
8. Hom V, Marks J (2007) Automatic design of balanced board games. In: Proceedings of the AAAI conference on artificial intelligence and interactive digital entertainment (AIIDE), pp 25–30
9. Mitchell M (1998) An introduction to genetic algorithms. MIT Press, Cambridge
10. Shiffman D (2012) The nature of code. Free Software Foundation. http://natureofcode.com/book/chapter-9-the-evolution-of-code/. Accessed 12 Apr 2018

Distributed Composition of Highly-Collaborative Services and Sensors in Tactical Domains

Alexander Perucci[✉], Marco Autili, Massimo Tivoli, Alessandro Aloisio, and Paola Inverardi

Department of Information Engineering, Computer Science and Mathematics, University of L'Aquila, L'Aquila, Italy
{alexander.perucci,marco.autili,massimo.tivoli, alessandro.aloisio,paola.inverardi}@univaq.it

Abstract. Software systems are often built by composing services distributed over the network. Choreographies are a form of decentralized composition that models the external interaction of the services by specifying peer-to-peer message exchanges from a global perspective. When third-party services are involved, usually black-box services to be reused, actually realizing choreographies calls for exogenous coordination of their interaction. Nowadays, very few approaches address the problem of actually realizing choreographies in an automatic way. These approaches are rather static and are poorly suited to the need of tactical domains, which are highly-dynamic networking environments that bring together services and sensors over military radio networks. In this paper, we describe a method to employ service choreographies in tactical environments, and apply it to a case study in the military domain.

1 Introduction

In the last decade, the Service Oriented Architecture (SOA) paradigm has been highly developed and applied in very different Information Technologies (IT) areas. SOA promotes the construction of systems by composing software services deployed on different devices, connected to the network.

Today's service composition mechanisms are based mostly on service orchestration, a centralized approach to the composition of multiple services into a larger application. Orchestration works well in rather static environments with predefined services and minimal environment changes. These assumptions are inadequate in tactical domains, which are characterized by highly-dynamic networking environments that brings together services and sensors over low-availability military radio networks with low-performing execution nodes. In contrast, service choreography is a decentralized composition approach that was recognized in the BPMN2 (Business Process Model and Notation Version 2.0[1] – the de facto standard for specifying choreographies), which introduced dedicated choreography-modeling constructs through *Choreography Diagrams*. These diagrams model

---

[1] www.omg.org/spec/BPMN/2.0.

© Springer Nature Switzerland AG 2020
P. Ciancarini et al. (Eds.): SEDA 2018, AISC 925, pp. 232–244, 2020.
https://doi.org/10.1007/978-3-030-14687-0_21

peer-to-peer communication by defining a multiparty protocol that, when put in place by the cooperating parties, allows reaching the overall choreography objectives in a fully distributed way. In this sense, service choreographies differ significantly from service orchestrations, in which only one entity is in charge of centrally determining how to reach the overall objective through centralized coordination of all the involved services.

Recent studies [1,12,17,19–21] have challenged the problem to apply SOA in a tactical domain, providing soldiers with services and sensors in order to better face common difficulties of a mission. Two main aspects are concerned: (i) disadvantaged networks, and (ii) the distributed nature of the tactical field. For example, networks available for land-based military operations suffer from problems like lack of connectivity guarantee, changing network topology that implies no guaranteed service delivery, and radio silence that implies no guaranteed data delivery. A fully distributed service composition approach appears then fitting the tactical environment. Hence, choreographies are a good candidate in that distributed deployment and peer-to-peer communication are appropriate for low-availability networks with low-performing execution nodes.

In this paper, we describe a method to employ service choreographies in tactical environments. Specifically, we report on the extension to the Tactical Service Infrastructure (TSI) that we proposed in the TACTICS project[2] with the approach to the automatic synthesis of choreographies that we proposed in the CHOReVOLUTION project.[3] A case study[4] in the military domain is used to describes our method at work.

The paper is structured as follows. Section 2 sets the context and Sect. 3 introduces a case study in the military domain. Section 4 describes our method at work on the case study. Section 5 discusses related work, and conclusions and future work are given in Sect. 6.

2   Setting the Context

This section sets the context of our work by introducing the projects TACTICS (Sect. 2.1) and CHOReVOLUTION (Sect. 2.2). Then, Sect. 2.3 describes the extension to the TACTICS TSI node that we propose in order to achieve tactical choreographies.

2.1   TACTICS

The TACTICS project studied a way to apply the SOA paradigm over tactical networks, in order to allow information exchange and integration with Command and Control (C2) systems [3,11], and Command, Control, Communication, Computers and Intelligence (C4I) systems [10] for land based military operations.

---

[2] https://www.eda.europa.eu/info-hub/press-centre/latest-news/2017/06/20/tactics-project-completed.
[3] www.chorevolution.eu.
[4] https://github.com/sesygroup/tactical-choreographies.

It addressed this problem through the definition and the experimental demonstration of the TSI enabling tactical radio networks (without any modifications of the radio part of those networks) to participate in SOA infrastructures and provide, as well as consume, services to and from the strategic domain independently of the user's location. The TSI provides efficient information transport to and from the tactical domain, applies appropriate security mechanisms, and develops robust disruption- as well as delay-tolerant schemes. This includes the identification of essential services for providing both a basic (core) service infrastructure and enabling useful operational (functional) services at the application level. As just mentioned, an interesting goal of the project was to provide an experimental setup (including a complete network setup using real radio communication, and running exemplary applications and operational services) to demonstrate the feasibility of the concepts in a real-life military scenario.

2.2 CHOReVOLUTION

The CHOReVOLUTION project developed the technologies required to implement dynamic choreographies via the dynamic and distributed coordination of (possibly, existing third-party) services. The overall goal was to implement a complete choreography engineering process covering all the development activities, from specification to code synthesis, to automatic deployment and enactment on the Cloud. The outcome was the CHOReVOLUTION Platform[5] together with an integrated development environment realized as a customization of the Eclipse platform called CHOReVOLUTION Studio.[6]

2.3 Tactical Choreographies

As already anticipated, the CHOReVOLUTION results were leveraged to extend the TACTICS TSI node. The TSI node is the core of the TSI. It is essentially a middleware that supports the consumption of remotely provided functional services, interoperating with existing Information Systems and Bearer Systems (i.e., radios). The main goal of the TSI node is to allow connection among services over disadvantaged tactical networks. A service can send a message to another service being agnostic of the problems (i.e., constraints) typical for this kind of networks.

The main functionalities of the TSI node are depicted in Fig. 1. The eight functionalities on the left-hand side belong to three middleware layers: *Service Mediator*, *Message Handler*, and *Packed Handler*. The seven functionalities on the right-hand side are cross-layer, so they can be used in every part of the TSI node. Briefly, the *Service Mediator* manages the user's session, adapts messages to the network bandwidth, and deals with the awareness of the context. The

---

[5] www.chorevolution.eu/bin/view/Documentation/.

[6] www.chorevolution.eu/bin/view/Documentation/Studio.

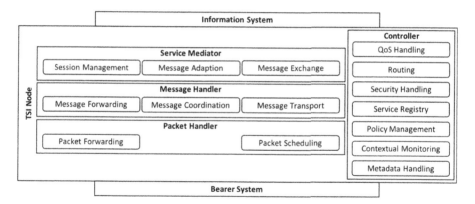

Fig. 1. TSI node

Message Handler stores and/or forwards a message to the next node of a route. Messages are stored in order to cope with network faults. If a message does not reach the destination, the *Message Handler* can resend it. Finally, it deletes a stored message once it receives a receipt confirmation. The *Packet Handler* is mainly used to forward IP packets between different radio networks. Providing detailed description of all the TSI node layers is outside of the scope of this paper. Interested readers can refer to [1,12] for details. The focus of this paper is on the *Message Handler* layer that was extended by adding the new functionality *Message Coordinator* in order to support choreographies. This new functionality is offered by a set of software entities called *Coordination Delegates* (CDs). These entities permit to handle the coordination of messages among the services running on TSI nodes in a way that the desired choreography is fulfilled. When interposed among the services according to a predefined architectural style (a sample instance of which is shown in Fig. 2), CDs proxify the participant services to coordinate their interaction, when needed.

As detailed in Sect. 4, CDs are automatically synthesized, out of a BPMN2 choreography specification, by exploiting the CHOReVOLUTION methodology. CDs guarantee the collaboration specified by the choreography specification through distributed protocol coordination [7]. CDs perform pure coordination of the services' interaction (i.e., *standard communication* in the figure) in a way that the resulting collaboration realizes the specified choreography. To this purpose, the coordination logic is extracted from the BPMN2 choreography diagram and is distributed among a set of *Coordination Models* (CMs) that codify coordination information. Then, at run time, the CDs manage their CMs and exchange this coordination information (i.e., *additional communication*) to prevent possible *undesired interactions*, i.e., those interactions that prevent the choreography realizability [4–7]. The coordination logic embedded in CDs is obtained by a distributed coordination algorithm implemented in Java; each CD runs its own instance of the algorithm. Once deployed on TSI nodes, CDs support the correct execution of the choreography by realizing the required distributed coordination

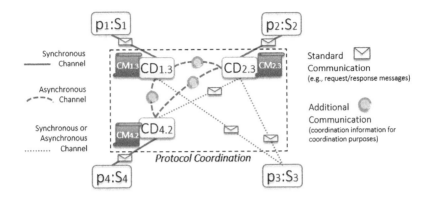

Fig. 2. Architectural style (a sample instance of)

logic among the participant services. Formal foundations of the whole approach and details of the distributed coordination algorithm can be found in [6,7].

3   Case Study

This section introduces a case study in the military domain inspired by the scenario in [2,18]. It concerns an instance of an Emergency Deployment System (EDS) [22], which supports the distributed management and deployment of personnel in cases of natural disasters, search-and-rescue efforts, and military crises. The system gathers information from the tactical environment and gets knowledge of the current status, e.g., the locations and status of the personnel, vehicles, and obstacles.

Considering the scenario represented in Fig. 3, our instance of EDS involves three main subsystems, i.e., Headquarter (HQ), Team Leader (TL), and Responder (RE):

- **HQ** is in charge of both planning and executing operations, in constant communication with TLs in its area and with other HQs in different areas of the same conflict. By gathering information from the field, HQ is fully aware of the current field status and displays on a map the locations and the status of TLs, troops, vehicles and obstacles of interest (both sensed from the environment or pointed out by personnel on the field). Enemy troops and vehicles locations are also displayed as communicated by REs to TLs. That is, HQ is able to send orders to and receive reports from the TLs and to analyze a log of the battlefield status.
- **TLs** are responsible for a smaller part of the field in which they lead the deployment of the REs. They gather information from the REs so to maintain and analyze the current status of the field they are responsible for, and forward it to the HQ in command and to the other TLs in the same area.
- **REs** receive direct orders from TLs. Their main function is to sense the environment and report the status of the location they are deployed in.

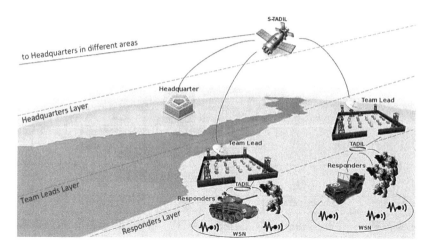

Fig. 3. EDS scenario

Conventional military operations are usually composed of three main activities:

- Reconnaissance – to sense the environment and report to the HQ;
- Command and Control – to plan and organize a strategy by delivering orders to assigned forces;
- Effect – to execute orders in the field and deliver reports to the HD.

Figure 4 shows the portion of the overall system architecture that concerns the services involved in a reconnaissance activity, according to the tasks flow specified by the choreography in Fig. 5.

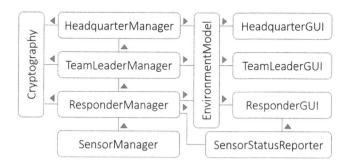

Fig. 4. System architecture related to the choreography in Fig. 5

In BPMN2, a choreography *task* is an atomic activity that represents an interaction by means of one or two (request and optionally response) message exchanges between two participants. Graphically, BPMN2 choreography diagrams use rounded-corner boxes to denote choreography tasks. Each of them

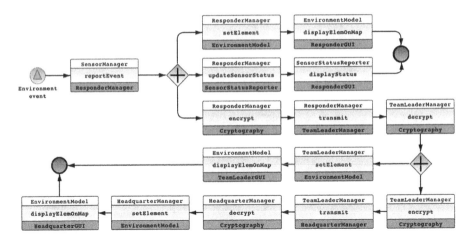

Fig. 5. BPMN2 choreography specification

is labeled with the roles of the two participants involved in the task (e.g., *SensorManager* and *ResponderManager*), and the name of the service operation (e.g., *reportEvent*) performed by the initiating participant and provided by the other one. A role contained in the white box denotes the initiating participant (e.g., *SensorManager* of the first task on the left). In particular, the BPMN2 standard specification employs the theoretical concept of a token that, traversing the sequence flows and passing through the elements in a process, aids to define its behavior. The start event generates the token that must eventually be consumed at an end event. For the sake of presentation, in Fig. 5, input/output messages are not shown although they are internally handled by the synthesis processor according to their XML schema type definition. The tactical choreography is triggered by events associated to variation of elements in the environment. Upon receiving an event, the `SensorManager` reports it to the `ResponderManager` by invoking the *reportEvent* operation. The `ResponderManager` reacts by creating three parallel flows in order to:

1. display on the `ResponderGUI` dedicated graphical markers to locate on a map the elements detected in the environment. For this purpose, the `ResponderManager` calls the *setElement* operation provided by the `EnvironmentModel` service, which in turn calls the *dispalyElemOnMap* operation provided by `ResponderGUI` to show the updated map on the RE's device.
2. display the current status of the sensors on the `ResponderGUI`. In order to accomplish this task, the `ResponderManager` calls the *updateSensorStatus* operation provided by the `SensorStatusReporter` service, which in turn calls *displayStatus* to update the sensors status on the RE's device.
3. transmit the detected elements to its TL – see the *transmit* operation. Before transmitting the message containing the information of the detected elements, the `ResponderManager` encrypts the message by using the `Cryptography`

service. Upon receiving the message, the TL decrypts it, and creates two parallel flows to:

3.1 update the map on the TL device: after the detected elements have been set through the *setElement* operation, the `EnvironmentModel` calls the *dispalyElemOnMap* provided by the `TeamLeaderGUI` service;

3.2 transmit the encrypted element updates to its HQ: upon receiving the updates, in order to display them on the `HeadquarterGUI`, the `HeadquarterManager` follows the same steps as the ones performed by the TLS and REs.

4   Method at Work

The CHOReVOLUTION methodology uses a synthesis processor that takes as input the BPMN2 choreography diagram in Fig. 5, and derives the intermediate automata-based model in Fig. 6, called Choreography-explicit Flow Model (CeFM). This model makes explicit coordination-related information that in BPMN2 is implicit, hence representing an intermediate model suitable for the automated synthesis of the CDs and their CMs. For instance, the state S14 models the fork state in Fig. 5 creating the two parallel flows described by the items 3.1 and 3.2 in Sect. 3.

As already done for the choreography diagram, we do not show the input/output messages of the operations, and make use of abbreviations for role names by showing only capital letters.

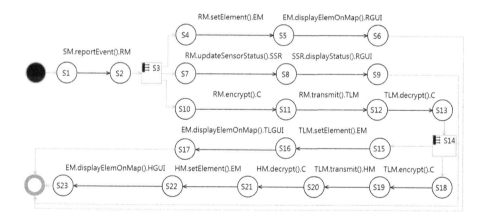

Fig. 6. CeFM derived from the choreography in Fig. 5

Table 1 shows the CMs synthesized from the CeFM of Fig. 6. We refer to [7] for a formal definition of the CeFM construction process and CMs synthesis. For the purposes of our case study, it is sufficient to say that coordination information are codified into the CMs as a set of tuples. For each tuple $\langle s, t, s', CD_{s'}, \rho, Notify_s, Wait_{siblings(s)} \rangle$:

Table 1. CMs tuples derived from the CeFM in Fig. 6

$CM_{EM,HGUI}$	$CM_{EM,RGUI}$
$\langle S22, displayElemOnMap(), S23, \{\}, true, \{\}, \{\}\rangle$	$\langle S5, displayElemOnMap(), S6, \{\}, true, \{\}, \{\}\rangle$
$\langle S23, \varepsilon, FINAL, \{\}, true, \{\}, \{\}\rangle$	$\langle S6, \varepsilon, FINAL, \{\}, true, \{\}, \{\}\rangle$
$CM_{EM,TLGUI}$	$CM_{SSR,RGUI}$
$\langle S16, displayElemOnMap(), S17, \{\}, true, \{\}, \{\}\rangle$	$\langle S8, displayStatus(), S9, \{\}, true, \{\}, \{\}\rangle$
$\langle S17, \varepsilon, FINAL, \{\}, true, \{\}, \{\}\rangle$	$\langle S9, \varepsilon, FINAL, \{\}, true, \{\}, \{\}\rangle$
$CM_{HM,C}$	$CM_{HM,EM}$
$\langle S20, decrypt(), S21, \{(HM, EM)\}, true, \{\}, \{\}\rangle$	$\langle S21, setElement(), S22, \{(EM, HGUI)\}, true, \{\}, \{\}\rangle$
$CM_{RM,C}$	$CM_{RM,EM}$
$\langle S10, encrypt(), S11, \{(RM, TLM)\}, true, \{\}, \{\}\rangle$	$\langle S4, setElement(), S5, \{(EM, RGUI)\}, true, \{\}, \{\}\rangle$
$CM_{RM,SSR}$	$CM_{RM,TLM}$
$\langle S7, updateSensorStatus(), S8, \{(SSR, RGUI)\}, true, \{\}, \{\}\rangle$	$\langle S11, transmit(), S12, \{(TLM, C)\}, true, \{\}, \{\}\rangle$
$CM_{SM,RM}$	$CM_{TLM,C}$
$\langle S1, reportEvent(), S2, \{\}, true, \{\}, \{\}\rangle$	$\langle S12, decrypt(), S13, \{\}, true, \{\}, \{\}\rangle$
$\langle S2, \varepsilon, S3, \{\}, true, \{\}, \{\}\rangle$	$\langle S13, \varepsilon, S14, \{\}, true, \{\}, \{\}\rangle$
$\langle S3, \varepsilon, S10, \{(RM, C)\}, true, \{\}, \{\}\rangle$	$\langle S14, \varepsilon, S15, \{(TLM, EM)\}, true, \{\}, \{\}\rangle$
$\langle S3, \varepsilon, S4, \{(RM, EM)\}, true, \{\}, \{\}\rangle$	$\langle S14, \varepsilon, S18, \{(TLM, C)\}, true, \{\}, \{\}\rangle$
$\langle S3, \varepsilon, S7, \{(RM, SSR)\}, true, \{\}, \{\}\rangle$	$\langle S18, encrypt(), S19, \{(TLM, HM)\}, true, \{\}, \{\}\rangle$
$CM_{TLM,EM}$	$CM_{TLM,HM}$
$\langle S15, setElement(), S16, \{(EM, TLGUI)\}, true, \{\}, \{\}\rangle$	$\langle S19, transmit(), S20, \{(HM, C)\}, true, \{\}, \{\}\rangle$

- s denotes the CeFM source state from which the related CD can either perform the operation t or take a move without performing any operation (i.e., the CD can step over unlabelled arrows). In both cases, s' denotes the reached target state;
- $CD_{s'}$ contains the set of (identifiers of) those CDs whose supervised services became active in s', i.e., the ones that will be allowed to require/provide some operation from s'. This information is used by the "currently active" CDs to inform the set of "to be activated" CDs (in the target state) about the changing global state;
- ρ is a boolean condition whose validity has to be checked to select the correct tuple, and hence the correct flow(s) in the CeFM; if no condition is specified, ρ is always set to *true*.
- $Notify_s$ contains the predecessor of a join state that a CD, when reaching it, must notify to the other CDs in the parallel flow(s) of the same originating fork. Complementary, $Wait_{siblings(s)}$ contains the predecessors of join states that must be waited for.

For instance, the first tuple in $CM_{EM,HGUI}$ specifies that $CD_{EM,HGUI}$ can perform the operation *displayElemOnMap* from the source state $S22$ to the target state $S23$; whereas, the second tuple specifies that $CD_{EM,HGUI}$ can step over $S23$ and reach the final state. The first tuple in $CM_{HM,C}$ specifies that $CD_{HM,C}$ can perform the operation *decrypt* from the source state $S20$ to the target state $S21$, and afterwards activates $CD_{HM,EM}$ on the new global state $S21$. At this point, as specified by the corresponding tuple, $CD_{HM,EM}$ can perform the operation *setElement* from $S21$.

The CDs are deployed on the TSI nodes, and at runtime they use the synthesized CMs to control the service interactions enforcing the realizability of the specified choreography.

5 Related Work

The work described in this paper is related to approaches and tools for automated choreography realization.

In [14], the authors propose an approach to enforce synchronizability and realizability of a choreography. The method implementing the approach is able to automatically generate monitors, which act as local controllers interacting with their peers and the rest of the system in order to make the peers respect the choreography specification. Our notion of CD is "similar" to the notion of monitor used in [14], since CDs are able to interact with the choreography participants in order to fulfill the prescribed global collaboration. However, the two synthesis methods are different. In [14], the monitors are generated through an iterative process, automatically refining their behavior.

In [13], the authors address the realizability problem based on a priori verification techniques, using refinement and proof-based formal methods. They consider asynchronous systems where peers communicate via possibly unbounded FIFO buffers. The obtained asynchronous system is correct by construction, i.e., it realizes the choreography specification. With respect to our method and other methods discussed in this section, this method is more scalable in terms of number of involved peers and exchanged messages. However, our approach focuses on realizing a choreography specification by reusing third-party peers (possibly black-box), rather than generating the correct peers from scratch. This is why we cannot avoid to deal with exogenous coordination by means of additional software entities such as the CDs.

The approach in [16] checks the conformance between the choreography specification and the composition of participant implementations. This approach permits to characterize relevant properties to check a certain degree of realizability. The described framework can model and analyze compositions in which the interactions can be asynchronous and the messages can be stored in unbounded queues and reordered if needed. Following this line of research, the authors provided a hierarchy of realizability notions that forms the basis for a more flexible analysis regarding classic realizability checks [15]. The approach statically checks realizability but does not automatically enforce it at run time as we do in our approach.

In [8], the authors identify a class of systems where choreography conformance can be efficiently checked even in the presence of asynchronous communication. This is done by checking choreography synchronizability. Differently from us, the approach in [8] does not aim at synthesizing the coordination logic, which is needed whenever the collaboration among the participants leads to global interactions that violate the choreography specification.

The ASTRO toolset supports automated composition of Web services and the monitoring of their execution [23]. ASTRO deals with centralized orchestration-based processes rather than fully decentralized choreography-based ones.

In [9], the authors present a unified programming framework for developing choreographies that are correct by construction in the sense that, e.g., they ensure deadlock freedom and communication safety. Developers can design both protocols and implementation from a global perspective and, then, correct endpoint implementations are automatically generated. Similarly to us, in [9] the authors consider the notion of multiparty choreography and defines choreography projections. Differently from us, the approach in [9] is not reuse-oriented in that the goal is the generation of the service endpoints' code, rather than the generation of the code for coordinating existing endpoints.

6   Conclusions and Future Work

In this paper, we exploited the results of both the TACTICS project and the CHOReVOLUTION project in order to employ choreographies in a tactical domain. Tactical environments are characterized by low-availability networks with low-performing execution nodes. Preliminary results show that choreographies are a good candidate in that distributed deployment and peer-to-peer communication specifically suite these environments. As future work, we plan to extend the approach to deal with choreography evolution, so to also support choreographies that can vary depending on context changes, intermitting services availability, and user preferences.

Acknowledgment. This research work has been supported by the EU's H2020 Programme, GA No. 644178 (project CHOReVOLUTION - Automated Synthesis of Dynamic and Secured Choreographies for the Future Internet), by the Ministry of Economy and Finance, Cipe resolution No. 135/2012 (project INCIPICT - INnovating CIty Planning through Information and Communication Technologies), and by the EDefence Agency, contract No. B 0980 IAP4 GP (project TACTICS - TACTICal Service Oriented Architecture).

References

1. Aloisio A, Autili M, D'Angelo A, Viidanoja A, Leguay J, Ginzler T, Lampe T, Spagnolo L, Wolthusen SD, Flizikowski A, Sliwa J (2015) TACTICS: tactical service oriented architecture. In: 3rd international conference in software engineering for defence applications (SEDA), pp 1–9
2. Andersson J, de Lemos R, Malek S, Weyns D (2009) Modeling dimensions of self-adaptive software systems. In: Software engineering for self-adaptive systems, volume 5525 of LNCS. Springer, Heidelberg, pp 27–47
3. Australian Defence Force Warfare Centre (2009) ADDP 00.1 Command and Control - Department of Defence, CANBERRA ACT 2600

4. Autili M, Di Ruscio D, Di Salle A, Inverardi P, Tivoli M (2013) A model-based synthesis process for choreography realizability enforcement. In: Fundamental approaches to software engineering, volume 7793 of LNCS. Springer, Heidelberg, pp 37–52

5. Autili M, Inverardi P, Tivoli M (2015) Automated synthesis of service choreographies. IEEE Softw 32(1):50–57

6. Autili M, Inverardi P, Tivoli M (2018) Choreography realizability enforcement through the automatic synthesis of distributed coordination delegates. Sci Comput Program 160:3–29

7. Autili M, Tivoli M (2014) Distributed enforcement of service choreographies. In: Proceedings of the 13th international workshop on foundations of coordination languages and self-adaptive systems (FOCLASA), pp 18–35

8. Basu S, Bultan T (2011) Choreography conformance via synchronizability. In: Proceedings of the 20th international conference on World Wide Web (WWW), pp 795–804

9. Carbone M, Montesi F (2013) Deadlock-freedom-by-design: multiparty asynchronous global programming. In: Proceedings of the 40th annual ACM SIGPLAN-sigact symposium on principles of programming languages, pp 263–274

10. National Research Council (1999) Realizing the Potential of C4I: Fundamental Challenges. The National Academies Press

11. Department of Defense (2010) Department of Defense Dictionary of Military and Associated Terms - Joint Publication 1-02 - as amended through 15 December 2014

12. Diefenbach A, Ginzler T, McLaughlin S, Sliwa J, Lampe TA, Prasse C (2016) TACTICS TSI architecture: a European reference architecture for tactical SOA. In: International conference on military communications and information systems (ICMCIS), pp 1–8

13. Farah Z, Ait-Ameur Y, Ouederni M, Tari K (2017) A correct-by-construction model for asynchronously communicating systems. Int J Softw Tools Technol Transf 19(4):465–485

14. Güdemann M, Salaün G, Ouederni M (2012) Counterexample guided synthesis of monitors for realizability enforcement. In: Automated technology for verification and analysis, volume 7561 of LNCS. Springer, Heidelberg, pp 238–253

15. Kazhamiakin R, Pistore M (2006) Analysis of realizability conditions for web service choreographies. In: Formal techniques for networked and distributed systems (FORTE). Springer, Heidelberg, pp 61–76

16. Kazhamiakin R, Pistore M (2006) Choreography conformance analysis: asynchronous communications and information alignment. In: WebServices and formal methods, volume 4184 of LNCS. Springer, Heidelberg, pp 227–241

17. Lopes RRF, Wolthusen SD (2015) Distributed security policies for service-oriented architectures over tactical networks. In: IEEE military communications conference (MILCOM), pp 1548–1553

18. Malek S, Beckman N, Mikic-Rakic M, Medvidovic N (2005) A framework for ensuring and improving dependability in highly distributed systems. In: Architecting dependable systems III, volume 3549 of LNCS. Springer, Heidelberg, pp 173–193

19. Małowidzki M, Dalecki T, Bereziński P, Mazur M, Skarżyński P (2016) Adapting standard tactical applications for a military disruption-tolerant network. In: International conference on military communications and information systems (ICMCIS), pp 1–5

20. Seifert H, Franke M, Diefenbach A, Sevenich P (2012) SOA in the CoNSIS coalition environment: extending the WS-I basic profile for using SOA in a tactical environment. In: Military communications and information systems conference (MCC), pp 1–6
21. Suri N, Morelli A, Kovach J, Sadler L, Winkler R (2015) Agile computing middleware support for service-oriented computing over tactical networks. In: IEEE 81st vehicular technology conference (VTC Spring), pp 1–5
22. Tahmoush D, Lofland C (2009) A prototype national emergency deployment system. In: IEEE conference on technologies for homeland security (HST), pp 331–338
23. Trainotti M, Pistore M, Calabrese G, Zacco G, Lucchese G, Barbon F, Bertoli P, Traverso P (2005) ASTRO: Supporting composition and execution of web services. In: Service-oriented computing - ICSOC 2005, volume 3826 of LNCS. Springer, Heidelberg, pp 495–501

Safety and Security in GNU/Linux Real Time Operating System Domain

Giuseppe Procopio[✉]

MBDA Italia S.p.A., via Monte Flavio, 45, 00131 Rome, Italy
giuseppe.procopio@mbda.it

Abstract. Historically, because of regulatory constraints systems focused only on hardware and software safety, and considered security independently as an add-on, if anything required.

But, it is widely recognized safety-critical systems today are quite certainly security-critical too, as well as safety and security functions may influence each other. It also happens that the system's usage context is not completely known nor understood at development time which means system maintenance will not just limited to bug-fixes and will involve continuous hazard analysis. Especially in Avionic and Automotive sectors, the growing awareness of conjoint safety and security pushed the research new paradigms for design, development, verification and validation, and the promotion of holistic methods and techniques for conducting safety and security co-engineering, co-assessment and certification/qualification. Finally, there is not a standard that provides conjoint guidelines for the safety and security domains so that compliance to multiple standards is currently the used approach.

This paper focuses briefly on the safety and security real-time operating systems, the architecture they are required to provide for addressing safety and security, and the applicable standards. It then highlights how a real-time GNU/Linux OS can be included in a formal certification package as demanded by SIL2 applications that meet the IEC 61508 requirements, and how such operating system should be improved for use into interconnected systems.

Keywords: Safety · Security · Co-engineering · GNU/Linux

1 Introduction

"As safety-critical systems become more and more complex, and more and more interconnected, they also become more and more vulnerable to cyber-attacks" [1]. It is today evident a safety-critical system can be compromised through cyber space so bypassing any physical barrier which means harms may be the result of breaches in the systems' security.

There is an increasing need to secure a safety-critical systems from cyber threats by including security requirements into so-called *Safety & Security Assessments* in order to better identify, analyze, evaluate and mitigate potential harms. But, as identified harms from a safety point-of-view can be different from a security point-of-view, it could also results that risk mitigations for safety and security aspects can be antagonists rather

© Springer Nature Switzerland AG 2020
P. Ciancarini et al. (Eds.): SEDA 2018, AISC 925, pp. 245–254, 2020.
https://doi.org/10.1007/978-3-030-14687-0_22

than mutually reinforced or independent. Then, for a cyber-targetable safety-critical system, a strategy to accept higher-level security risks in favor of lower-level safety risks does not apply.

Industry which were used to have separated safety and security software specifications addressed by separate engineering teams – which rarely talk each other and completely neglect their mutual dependence – are now conscious that safety requirements interfere with security requirements and vice versa. Let's have a look at few examples:

- Encryption is imposed by security requirements as method to preserve confidentiality of information stored or being in transfer. But the computational effort required by cryptographic algorithms can prevent the system from responding properly to safety events;
- Ongoing appraisal of the security state is required to trusted systems for gaining insights into their current state of availability, confidentiality, and integrity. But general resources consumption required by security functions can heavily reduce the reaction time against safety events;
- Security hardening imposes to disable or suspend an account after a definite number of its unsuccessful authentication attempts. But a brute-force attack may prevent legitimate users to perform safety operations;
- Security hardening forces protection against stack and heap overwrites as well as protection against PLT (Procedure Linkage Table) and GOT (Global Offset Table) overwrites. But management of the code added by the compiler may be unaffordable for the mandatory Safety Integrity Level;
- Frequent system patching is required to keep a cyber-resilience effective. But for a safety-critical system this imply running again a complete certification process;
- Virtualization could be an appropriate solution for enforcing the demanded information control via physical segregation. But performance's degradation may be not compatible with safety requirements;
- Security functions demand very high computational and memory resource under heavy cyber-attack. But such resource consumption should be limited in favor of safety functions.

From the standardization viewpoint, other problems have to be considered. There are currently widely accepted standards for safety and security separately, but the latter are accused of being sometimes completely incomprehensible to non-practitioners as they conflict with most important attributes like clearness, unambiguous, and non-interpretability. Anyway, no safety and security co-engineering standard is currently available.

2 Recommendations from the International Community

International Community is making a very large effort to fill the gap between system's safety and security. We will next mention only few of them.

During the W3C conference held in Paris in 2015 and addressing trustworthiness of Cyber-physical Systems-of-Systems, it was pointed out that new systems are more and more complex and interconnected, and operating in an unpredictable environment. In these new contexts, such systems *"must be adaptable, reconfigurable and extendable during their lifetime, since the classical predictability assumptions of safety and cyber-security assessment and certification no longer hold"* [2]. For such systems and services they provide, dependability must be evaluated in a holistic manner i.e. conducting safety and security co-engineering, co-assessment and certification/qualification, which imply the adoption of new paradigms for design, development, verification and validation. Usually, such systems are deployed into not well definite environment that introduces new elements of unknowns and uncertainties which in turn means current standards are not suitable anymore as they *"assume that a system's usage context is completely known and understood at development time"* [2]. This is proved by the fact that now the larger amount of effort is required in operation and maintenance phases for patching newly discovered vulnerabilities or adding new countermeasures to emerging threats which in turn invoke continuous hazard analysis. Even though with no flaws, the security of a system is required to evolve as threat evolves.

Also, research performed by the ARTEMIS-EMC^2 project [3] highlight that dependability (safety, reliability, availability, security, maintainability, etc.) must be evaluated in a holistic manner and provided important contribution for bridging the gap between safety and security assurance and certifications.

A study [4] from the Department of Digital Safety & Security AIT Austrian Institute of Technology recalls the need for holistic approaches for safety and security analysis and proposes improvements to address these current limitations for a combined safety and security analysis. On the same footsteps, a study from the Industrieanlagen-Betriebsgesellschaft mbH [5] for strengthening the overall air traffic system resilience against cyber threats recommend to *"Ensure consistency and enable synergies"* [5] for both safety and security and to align safety and security considerations under a common roof. Specifically:

- Safety and security models, use cases, considerations and documentation should be considered jointly;
- Identified hazards should come from *"integrated consideration of technical failures, acts of god, human error, organizational weaknesses and intentional acts"* [5];
- Safety and Security "language" should harmonize their expression and semantics.

3 Standards for Safety and Security

The need for a changing paradigm for the development of safety-critical system with enforced built-in security started from the 90's when defense programs in USA and Europe begun requiring Real Time Operating Systems (RTOS) which satisfied safety and security standards like ARINC 653 [6], DO-178B (now superseded by DO-178C) [7], and ISO/IEC 15408 [8]:

- ARINC 653 requires that each operating system must have dedicated time, space, and resources on the target hardware. Specifically, six functions need to be supported: (1) provision of secure and timely data flow; (2) controlled access to processing facilities; (3) provision of secure data storage and memory management; (4) provision of consistent execution state; (5) provision of health monitoring and controlled failures at application, system, and hardware level; (6) provision of general services;
- The DO-178B (C), originally introduced as software-safety standard for commercial planes, was also adopted by defense programs, sometimes in conjunction with ARINC 653 for enabling applications of different safety levels to share the same computing resources;
- The ISO/IEC 15408 at higher level of assurance (at least to Evaluation Assurance Level 4 (EAL4) augmented with flaw remediation) was adopted to provide assurance for security functions.

To meet such demand for OS in the Defense and Aerospace domain, many operating systems have then flooded the market but none of them addressed a safety and security co-engineering until the introduction of the Multiple Independent Levels of Safety/Security (MILS) concept at the beginning of the new century. MILS *"describes systems where different partitions hosting applications are either independent from each other or connected by communication channels without an explicit hierarchical ordering policy that would require attaching global security policy levels to each partition"* [9]. Originally proposed by Stanford Research Institute's John Rushby in the early 1980s, the MILS concept last received formal standardization thanks to EURO-MILS [9], a European founded program under grant agreement n° 318353 which aimed to provide directions for developing high-assurance safety and high-assurance security systems. The most important contributions of EURO-MILS have been the MILS Architecture [9] and the MILS Protection Profile for Operating Systems [10].

The Manufacturing Sector for safety-related systems is widely ruled by the IEC 61508 [11] standard which provides guidelines to achieve safety for an Equipment Under Control which gives rise to hazards. Its objective is to determine a system safety integrity level (ASIL or SIL accordingly) to be taken into account during the system-development in order to eliminate or properly reduce hazards. IEC 61508 is further adapted for Automotive (ISO 26262 [12]), Machinery (IEC 62061 [13]), Railways (IEC 62278 [14]), Robotic Devices (ISO 10218 [15]), Industrial Process (IEC 61511 [16]), Medical Device Software (IEC 62304 [17]), and others.

The IEC 61508 claims to cover functional safety of an equipment i.e. the function to be implemented by a safety-related system to achieve or maintain a safe state for the equipment under control, in respect of a specific hazardous event. It is also stated that *"Where Hazard analysis identifies that malevolent or unauthorized action, constituting a security threat, is reasonably foreseeable, a security threat analysis should be carried out"* [11] which means security is a component of functional safety so that malicious intrusion, either human or malware, have to be identified and managed when they can potentially induce hazards and/or inhibit the proper operation of safety functions. More than IEC 61508, cybersecurity is addressed by IEC 61511 which clauses explicitly calls for a security risk assessment, system security by design, secure maintenance, and secure operations.

4 Safety and Security Real-Time Operating Systems

An OS is certainly the most critical software component; providing a layer between the application programs and the hardware resources, any failure on it is a potential critical failure of applications running on the given hardware.

Safety's certification and security's certification became during the last years the dominant source of competitive differentiation for the OS's market, which is shared by few competitors. The Avionic sector is dominated by Green Hills, proposing the Integrity-178B as a successful examples of OS compliant to ARINC 653, DO-178C at Level A, ISO/IEC 15408 at EAL6, as well as MILS compliant OS. Other Operating Systems (OSs) are for instance VxWorks 653, provided by Wind River, and LynxOS-178 RTOS, provided by Lynx Software Technologies.

The industry of safety-critical systems is also well presided by Black Barry with its QNX OS, an RTOS certified IEC 61508 SIL3, ISO 26262 ASIL D, and IEC 15408 at EAL4+, and by SYSGO with its PikeOS, an RTOS including a hypervisor based separation microkernel designed for the highest levels of safety and security, and awarded with various certification standards as DO-178C, IEC 61508, EN 50128, IEC 62304, and ISO 26262.

Safety and security RTOSs have different functional specifications on automotive and aviation sectors. For the first one, we can take the AUTOSAR architectures [18] as reference, and EURO-MILS architecture for the second one.

- AUTOSAR is a standard which helps realizing safety and security requirements for safety-critical applications in automotive. Among its various specifications, we highlight the following protection mechanisms for application software:
  - Protection against illegal write accesses by applications, as well as protection against concurrent access;
  - Protection of non-volatile memory blocks;
  - Applications have their separate memory;
  - Objects are assigned to applications. A communication mechanism to transfer data between applications is provided but accesses to objects restricted;
  - Periodic system' self-check and self-test (including diagnostics), as well as ability to detect corrupted data;
  - Ensure confidentiality (including non-repudiation), integrity (including authenticity), and availability (including verification) of data;
  - Consistent access to shared data;
  - Consistent application' states and transitions;
  - Assigned execution time budget and related timeout action to tasks.
- EURO-MILS is a standard which helps realizing safety and security requirements for safety-critical applications in automotive and avionic domains. It addresses policies for resource allocation (in the space and time domains e.g.

memory, CPU, etc.), access control for exported resources, and information flow. Along with the Separation Kernel, it is central the concept of *partition*, i.e. a component that serves to encapsulate application(s) and/or data, and defined as a unit of separation with respect to resource allocation, access control, and information flow policies. Multiple partitions may coexist and co-operate via controlled information flows. Separation in space and in time is the mechanism used to manage interference between running applications in partitions, and it is provided by the *Separation Kernel* which is in charge to enforce a resource allocation policy, an information control policy, and an access control policy for partitions [9, 10]. EURO-MILS enhanced the classical tasks assigned to the MILS Separation Kernel (i.e. Data Isolation, Control of Information Flow, Resource Sanitization, Fault Isolation) as follow [9, 10]:

- **Separation in space:** Applications are hosted in different partitions with assigned memory resources. Applications of different partitions do not interfere each other, nor with the separation kernel.
- **Separation in time:** Partitions have their assigned CPU time budget and assigned priority to allow priority based scheduling within one time window. CPU registers and memory caches are zeroed on a partition switch CPUs.
- **Provision and management of communication objects:** Applications hosted in different partitions have their assigned communication objects allowing communication between partitions.
- **Separation kernel self-protection and accuracy of security functionality:** Error conditions trigger the transition of the Separation Kernel into a safe and secure state without system safety and security compromise.

We can observe large overlaps between the above mentioned EURO-MILS and AUTOSAR architectures.

4.1   Linux for Safety Related Systems

In 2002, a research report [19] from the CSE International Limited stated that *"Linux would be broadly acceptable for use in safety related applications of SIL 1 and SIL 2 integrity... It may also be feasible to certify Linux for use in SIL 3 applications by the provision of some further evidence from testing and analysis"*, but *"It is unlikely that Linux would be useful for SIL 4 applications and it would not be reasonably practicable to provide evidence that it meets a SIL 4 integrity requirement"* [19]. Few years later, a report by Nicholas Mc Guire of the Distributed and Embedded Systems Lab reported [20] the feasibility assessment of the development of a GNU/Linux which comply with the IEC 61508 requirements. It was clear the GNU/Linux OS project' status and its current development approach along with a more rigorous and structured documentation and testing were suitable for using such OS in IEC 61508 context [20]. This opportunity pushed the Open Source Automation Development Lab eG (OSADL) organization [21] to promote and coordinate the development of the SIL2LinuxMP project [22]. Officially started in 2015 and currently in progress, *"the SIL2LinuxMP project aims at the certification of the base components of an embedded GNU/Linux RTOS running on a single-core or multi-core industrial COTS computer board. Base components are boot loader, root filesystem, Linux kernel and C library bindings to access the Linux kernel.*" [22]. At the same time, it aims "*to create a framework that guides the safety engineer through the process of certifying a system based on Linux*" [23]. The SIL2LinuxMP uses state-of-the-art technologies as Linux Container (LXC) [24] and related resource allocators to implement the separation in time and in space. Specifically, the SIL2LinuxMP's architecture consider multiple isolated partitions (LXCs), each one with its allocated resources (CPU, memory, DRAM banks, etc.) and a resource control mechanism. LXC enforce multiple layers of protection which provide fault isolation and core system self-protection thus impeding that security compromises (also as potential source of harms) in a container can propagate to a second one, as well as impeding they can propagate to the core system. We can then observe a certain similarity to the MILS architectures.

But, the SIL2LinuxMP does not focus on security [23], at least in the sense of a claiming conformance to a security target as demanded by ISO/IEC 15408. It could be useful to provide such conformance, especially for interconnected systems. Leaving aside any security policy and any protection provided by the environment, we can attempt to formulate a minimal set of threats to be countered (of course this depend on assets to be protected. Not having defined these assets, we generically mention very common threats) i.e.:

- Unauthorized access to services, ineffective security mechanisms, or unintended circumvention of security mechanisms;
- Weak cryptography;
- Weak user authentication;
- Untrusted software updates;
- System integrity corruption;
- Resource exhaustion;
- Inability to associate an action to the requesting user;
- Inability to ensure data at-rest confidentiality and integrity;
- Inability to ensure data in-transit confidentiality and integrity.

Security objectives for countering the above mentioned threats are partially already provided by the SIL2LinuxMP's architecture. For the remaining part, it could be necessary the integration of new software components whose interference with safety-critical functions need to be investigated; of course, only independent or mutually reinforced safety and security functions can coexist. Anyway, security objectives are to be detailed into Security Requirements whose management, implementation, design, testing and delivery should comply with the IEC 15408 at EAL4 (with flaw remediation), at least.

As they demonstrated [25] significant overlaps between the DO-178B and the ISO/IEC 15408 at EAL5, we may assume overlaps of the latter with the IEC 61508 too i.e. we expect to see many artifacts reusable between a security certification process and a safety one, and vice versa, which in turn makes both safety and security certification more feasible.

4.2   Security-Only Certification for GNU/Linux OS

As security engineers do not need to take care of safety requirements (but we have seen that safety and security co-engineering is needed for safety engineers), security-only certification for GNU/Linux OS resulted to be less complex and more affordable to industries as demonstrated by numerous IEC 15408 certifications at EAL4 (with flaw remediation) for Linux-based distributions like FIN.X RTOS SE, Wind River Linux, Red Hat Enterprise Linux, etc. This is not surprising as safety standards [26]:

- Deals with the entire system while the IEC 15408 focuses almost exclusively on security requirements;
- Does not call for specific safety functions while security domain has well-defined functional specifications that need to be implemented;
- Requires (at higher levels) to demonstrate the absence of dead code and to perform structural testing to verify the absence of non-required functionality while IEC 15408's development and testing requirements apply only to security functions.

5   Conclusions

New generation of real-time operating systems are required to comply with demanding safety and security requirements, jointly. Especially in the Avionic and Automotive industry sectors, this OS's demand is captured only by few Commercial off-the-shelf (COTS) Real Time Operating Systems (RTOSs).

But this trend could not last forever. A GNU/Linux RTOS can be today considered high-quality software both in the specific project's organization, and in the software component integration, testing, and documentation. Of course, a formal compliance to a worldwide recognized standards requires appropriate financial budget, high technical skills, strong commitments, and strictly project's management, but the Open Source community already demonstrated the viability and profitability of such formal compliance as the IEC 15408 at EAL4, for instance.

The SIL2LinuxMP project is a new smart and promising initiative that is demonstrating a GNU/Linux OS can be successfully used in IEC 61508 contexts. But, to deploy it on interconnected and then accessible systems we suggested a formal compliance to the IEC 15408 too (at least to EAL4) so that safety and security are managed jointly. Though the SIL2LinuxMP project is still a work in progress, it is undoubted that it was a trailblazer towards safety market lands for a GNU/Linux OS and then we expect similar initiatives from industry companies or from the Open Source community, and this will reinforce the placement of the GNU/Linux on the OS's market for safety-critical and security-critical systems.

References

1. Report, MERgE Safety and Security, ITEA2 Project # 11011 Recommendations for security and safety co-engineering. Document version 1.0
2. Magazine, ERCIM News Trustworthy systems of systems safety & security co-engineering. Number 102, July 2015. https://ercim-news.ercim.eu/images/stories/EN102/EN102-web.pdf
3. Schoitsch E, Skavhaug A ERCIM/EWICS/ARTEMIS workshop on dependable embedded and cyberphysical systems and systems-of-systems, September 2014. https://www.researchgate.net/publication/289935138_Introduction_ERCIMEWICSARTEMIS_Workshop_on_Dependable_Embedded_and_Cyberphysical_Systems_and_Systems-of-Systems_DEC-SoS'14_at_SAFECOMP_2014
4. Schmittner C, Ma Z, Puschner P (2016) Limitation and improvement of STPA-Sec for safety and security co-analysis. In: SAFECOMP 2016 conference
5. Kiesling T, Kreuzer M (2017) ARIEL – Air Traffic Resilience Recommendations to strengthen the cyber resilience of the air traffic system, version 2.0
6. ARINC Specification 653P0-1 Avionics application software standard interface, part 0, overview of ARINC 653, 1 August 2015. https://www.aviation-ia.com/products/653p0-1-avionics-application-software-standard-interface-part-0-overview-arinc-653-2
7. RTCA Document DO-178C Software Considerations in Airborne Systems and Equipment Certification. https://www.rtca.org/content/standards-guidance-materials
8. ISO/IEC 15408-1:2009 Information technology – security techniques – evaluation criteria for IT security – part 1: introduction and general model. https://standards.iso.org/ittf/PubliclyAvailableStandards/c050341_ISO_IEC_15408-1_2009.zip
9. EURO-MILS Consortium Secure European virtualisation for trustworthy applications in critical domains, October 2012. http://euromils.eu/downloads/2014-EURO-MILS-MILS-Architecture-white-paper.pdf
10. EURO-MILS Consortium, Common Criteria Protection Profile (2016) Multiple independent levels of security: operating system, V2.03. http://www.euromils.eu/downloads/EURO-MILS-Protection-Profile-V2.03.pdf
11. IEC 61508 Functional safety of electrical/electronic/programmable electronic safety-related systems, parts 1 to 7, Edition 2010. https://www.iec.ch
12. ISO 26262 Road vehicles – functional safety, parts 1 to 12, Edition 2018. https://standards.iso.org
13. IEC 62061 Safety of machinery - functional safety of safety-related electrical, electronic and programmable electronic control systems, Edition 2005. https://www.iec.ch
14. IEC 62278 Railway applications - specification and demonstration of reliability, availability, maintainability and safety, Edition 2002. https://www.iec.ch
15. ISO 10218 Robots and robotic devices – safety requirements for industrial robots, parts 1 to 2, Edition 2011. https://standards.iso.org
16. IEC 61511 Functional safety - safety instrumented systems for the process industry sector - part 1: framework, definitions, system, hardware and application programming requirements, Edition 2016. https://www.iec.ch
17. IEC 62304 Medical device software - software life cycle processes, Edition 2006. https://www.iec.ch
18. AUTOSAR Development Partnership Requirements on AUTOSAR Features, Release 4.3.1. https://www.autosar.org/fileadmin/user_upload/standards/classic/4-3/AUTOSAR_RS_Features.pdf
19. Report, CSE International Limited for the Health and Safety Executive 2002 (2002) Preliminary assessment of Linux for safety related systems, Research Report 011

20. Mc Guire N (2007) Linux for safety critical systems in IEC 61508 context. https://www.osadl.org/fileadmin/dam/presentations/61508/61508_paper.pdf
21. Open Source Automation Development Lab eG (OSADL) Homepage. https://www.osadl.org/. Accessed 1 July 2019
22. OSADL-SIL2LinuxMP Homepage. http://www.osadl.org/SIL2LinuxMP.sil2-linux-project.0.html. Accessed 1 July 2019
23. Platschek A, Mc Guire N, Bulwahn L (2018) Certifying Linux: lessons learned in three years of SIL2LinuxMP
24. Linux Containers (LXC) Homepage. https://linuxcontainers.org/. Accessed 1 July 2019
25. Alves-Foss J, Rinker B, Taylor C (2002) Towards common criteria certification for DO-178B compliant airborne software systems. University of Idaho
26. Brosgol BM (2008) Safety and security: certification issues and technologies

Mapping Event-B Machines into Eiffel Programming Language

Victor Rivera$^{(\boxtimes)}$, JooYoung Lee, and Manuel Mazzara

Innopolis University, Innopolis, Russian Federation
{v.rivera,j.lee,m.mazzara}@innopolis.ru

Abstract. Formal modelling languages play a key role in the development of software since they enable users to prove correctness of system properties, in particular critical systems such as transportation systems. However, there is still not a clear understanding on how to map a formal model to a specific programming language. In order to propose a solution, this paper presents a source-to-source mapping between Event-B models and Eiffel programs, therefore enabling the proof of correctness of certain system properties via Design-by-Contract (natively supported by Eiffel), while still making use of all features of O-O programming.

1 Introduction

The importance of developing correct software systems has been increased in the past few years. Final users of systems trust systems and are not aware of the consequences of malfunctioning. Hence, the burden is on developers, engineers and researchers that have to pay close attention to the development of flawless systems. There are different approaches to tackle the problem, e.g. top-down and bottom-up approaches: using a top-down approach, one could think to start developing the system from a very abstract view point towards more concrete ones; in a bottom-up approach, on the other hand, one might think to start from a more concrete state of the system to then add more functionality to it. The key point on both approaches is to always prove that properties of the systems hold.

Event-B is a formal modelling language for reactive systems, introduced by Abrial [Abr10], which allows the modelling of complete systems. It follows the top-down approach by means of refinements. Event-B allows the creation of abstract systems and the expression of their properties. One can prove that the system indeed meets the properties to then create a refinement of the system: same system with more details. It has been applied with success in both research and industrial projects, and in integrated EU projects aiming at putting together the two dimensions.

On the other side of the spectrum, following a bottom-up approach, one can work with Eiffel programming language [Mey92]. In Eiffel, one can create classes that implement any system. The behaviour of such classes is specified in Eiffel using contracts: pre- and post-conditions and class invariants. These mechanisms

© Springer Nature Switzerland AG 2020
P. Ciancarini et al. (Eds.): SEDA 2018, AISC 925, pp. 255–264, 2020.
https://doi.org/10.1007/978-3-030-14687-0_23

are natively supported by the language. Having contracts, one can then verify that the implementation is indeed the intended, and also one can track the specifications against the implementation [NMR15]. After the implementation of the class, one can give more speciality or generalization by using inheritance. This paper gives a series of rules to generate Eiffel programs from Event-B model, bridging both top-down and bottom-up approaches. Rules take into account system specifications of the Event-B model and generate either Eiffel code or contracts. Thus, users will end up with an implementation of the system while they can prove it correct.

Several translations have been achieved that go in the same direction as the work presented on this paper. In [MS11], Mèry and Singh present the EB2ALL tool-set that includes a translation from Event-B models to C, C++ and Java. Unlike this translation, EB2ALL provides support for a small part of Event-B's syntax, and users are required to write a final Event-B implementation refinement in the syntax supported by the tool. The Code Generation tool [EB10] generates concurrent Java and Ada programs for a tasking extension of Event-B. Unlike these tools, the work presented here does not require user's intervention, while it works on the proper syntax of the Event-B model. In addition, these tools do not take full advantage of the elements present in the source language, e.g. invariants. The work presented in this paper, in addition to an implementation, generates contracts from the source language, making use of the Design-by-Contract approach. In [RCnWR17, RCn14, CR16], authors present a translation from Event-B to Java, annotating the code with JML (Java Modelling Language) specifications, and [RBC16] shows its application. The main difference with the work presented here is the target language. We are translating to Eiffel which natively supports Design-by-Contract. In addition, Eiffel comes with different tools to statically prove Eiffel code (e.g. Autoproof [TFNP15]) that fully supports the language. Another difference is the translation of carrier sets. EventB2Java translates them as a set of integers.

2   Preliminaries

2.1   Event-B

Event-B is a formal modelling language for reactive systems, introduced by Abrial [Abr10], which allows the modelling of complete systems. Figure 1 shows the general view of an Event-B machine and context. Event-B models are composed of contexts and machines. Contexts define constants (written after constant in context C), uninterpreted sets (written after set in context C) and their properties (written after axioms in context C). Machines define variables (written after variables in machine M) and their properties (expressed as invariants after invariant in machine M), and state transitions expressed as events (written between events and the last end). The initialisation event gives initial values to variables.

```
machine M sees C
variables v
invariants label_inv : I(s, c, v)
events
  event initialisation
    then A(s, c, v) end
  event evt
    any x
    where
      label_guard : G(s, c, v, x)
    then
      label_action : A(s, c, v, x)
    end
end
```

```
Context C
constant c
set S
axioms X(s, c)
end
```

Fig. 1. General view of an Event-B machine and its context.

An event is composed of guards and actions. The guard (written between keywords where and then) represents conditions that must hold for the event to be triggered. The action (written between keywords then and end) gives new values to variables.

In Event-B, systems are modelled via a sequence of refinements. First, an abstract machine is developed and verified to satisfy whatever correctness and safety properties are desired. Refinement machines are used to add more detail to the abstract machine until the model is sufficiently concrete for hand or automated translation to code. Refinement proof obligations are discharged to ensure that each refinement is a faithful model of the previous machine, so that all machines satisfy the correctness properties of the original.

2.2 Eiffel

Eiffel is an Object-Oriented programming language that natively supports the Design-by-Contract methodology. The behaviour of classes is specified by equipping them with contracts. Each routine of the class contains a pre- and post-condition: a client of a routine needs to guarantee the pre-condition on routine call. In return, the post-condition of the procedure, on routine exit, holds. The class is also equipped with class invariants. Invariants maintain the consistency of objects. Contracts in Eiffel follow a similar semantics of Hoare Triples.

Figure 2 depicts an Eiffel class that implements part of a Bank Account. The name of the class is ACCOUNT and it appears right after the keyword **class**. In Eiffel, implementers need to list creation procedures after the keyword **create**. In Fig. 2, make is a procedure of the class that can be used as a creation procedure. Class ACCOUNT structures its procedures in Initialisation, Access and Element change, by using the keyword **feature**. This structure can be use for information hiding (not discussed here). balance is a class attribute that contains the actual balance of the account. It is defined as an integer. Procedures in Eiffel are defined by given them a name (e.g. withdraw) and its respective arguments. It is followed by a head comment (which is optional). Procedures are equipped with pre- and post-conditions predicates. In Eiffel, a predicate is composed of a tag (optional) and a boolean expression. For instance, the pre-condition for withdraw (after the key work **require**) imposes the restriction on callers to provide and argument that is greater than or equal zero and less than or equal the balance of the account (amount_not_negative and amount_available are tags, identifiers, and are optionals). If the pre-condition of the procedure is met, the post-condition (after the key work **ensure**) holds on procedure exit. In a post-condition, the aid **old** refers to the value of an expression on procedure entry. The actions of the procedure are listed in between the key words **do** and **ensure**. The only action of withdraw procedure is to increase the value of balance by amount. Finally, The invariant is restricting the possible values for variables.

```
class ACCOUNT create make
feature —— Initalisation
  make
        —— Initialise an empty account.
    do
      balance := 0
    ensure
      balance_set: balance = 0
    end
feature —— Access
  balance: INTEGER
        —— Balance of this account.
feature —— Element change
  withdraw (amount: INTEGER)
        —— Withdraw 'amount' from this account.
    require
      amount_not_negative: amount >= 0
      amount_available: amount <= balance
    do
      balance := balance - amount
    ensure
      balance_set: balance = old balance - amount
    end
invariant
  balance_not_negative: balance >= 0
end
```

Fig. 2. Eiffel class

3   Translation

The translation is done by the aid δ : `Event-B` $\rightarrow$ `Eiffel`. δ takes an Event-B model and produces Eiffel classes. It is defined as a total function (i.e. $\rightarrow$) since any Event-B model can be translated to Eiffel. It uses two helpers: ξ translates Event-B Expressions or Predicates to Eiffel, and τ translates the type of Event-B variable to the corresponding type in Eiffel.

3.1   Translating Event-B Machines

Rule **machine** is a high level translation. It takes an Event-B machine M and produces an Eiffel class M.

$$\frac{\tau(v) = \text{Type } \xi(I(s,c,v)) = \text{Inv } \delta(\text{events } e) = \text{E}}{\delta(\text{event } \textit{initialisation } \text{then } A(s,c,v) \text{ end}) = \text{Init}} \text{(machine)}$$

δ(machine M **sees** C

variables v

invariants $label\_inv$: $I(s,c,v)$

event $initialisation$ then $A(s,c,v)$ end

events e

end) =

class M **create** initialisation

feature $--$ Initialisation

Init

feature $--$ Events

E

feature $--$ Access

ctx : CONSTANTS

v : Type

invariant

label_inv: Inv

end

Variables are translated as class attributes in class M. Event-B invariants are translated to Eiffel invariants. Both, Event-B and Eiffel, have similar semantics for invariants. Rule **context** generate an Eiffel class CONSTANT that contains the translation of Event-B constants and carrier sets defined by the user. Axioms, which restrict the possible values for constants are translated to invariants of this class. Constants in Event-B are entities that cannot change their values. They are naturally translated to Eiffel as **once** variables.

$$\frac{\delta(\text{axioms } X(s,c)) = \text{X} \quad \tau(c) = \text{Type}}{\begin{array}{l} \delta(\text{Context } C \\ \qquad \text{constant } c \\ \qquad \text{set } S \\ \qquad \text{Axioms } X(s,c) \\ \quad \text{end}) = \\ \textbf{class } \text{CONSTANTS} \\ \textbf{feature } -- \text{ Constants} \\ \qquad c: \text{ Type} \\ \qquad\qquad -- \text{ 'c' comment} \\ \qquad \textbf{once} \\ \qquad\qquad \textbf{create Type Result} \\ \qquad \textbf{end} \\ \textbf{invariant} \\ \qquad \text{X} \\ \textbf{end} \end{array}} \text{(context)}$$

Carrier sets represent a new type defined by the user. Each carrier set is translated as an afresh Eiffel class so users are able to use them as types. Rule `cset` shows the translation. Parts of the class are omitted due to space. Class EBSET [T] gives an implementation to sets of type T. Class S inherits EBSET [T] due to the nature of carrier sets in Event-B.

$$\frac{\tau(s) = \text{Type}}{\begin{array}{l} \delta(\text{Context } C \\ \qquad \text{constant } c \\ \qquad \text{set } S \\ \qquad \text{Axioms } X(s,c) \\ \quad \text{end}) = \\ \textbf{class } S \\ \textbf{inherit} \\ \qquad \text{EBSET [Type]} \\ \qquad \ldots \\ \textbf{end} \end{array}} \text{(cset)}$$

Rule `event` produces an Eiffel feature given an Event-B *event*. Parameters of the event are translated as arguments of the respective feature in Eiffel with its respective type. In Event-B, an event might be executed only if the guard is true. In Eiffel, the guard is translated as the precondition of the feature. Hence, the client is now in charge of meeting the specification before calling the feature. The semantics of the execution is handled now by the client who wants to execute the feature rather than the system deciding. The actual execution of the actions still preserve its semantics: execution of the actions is only possible if the guard is true. In Eiffel, for a client to execute a feature he needs to meet the guard otherwise a runtime exception will be raised: Contract violation.

Event-B event actions are translated directly to Eiffel statements. In Event-B, the before-after predicate contains primed and unprimed variables representing the before and after value of the variables. We translated the primed variable with the Eiffel keyword **old**. Representing old value of the variable. For simplicity, the rule only takes into account a single parameter, a single guard and a single action. However, this can be easily extended.

$$\frac{\xi(G(s,c,v,x)) = \text{G } \xi(A(s,c,v,x)) = \text{A}}{\tau(x) = \text{Type}} \text{ (event)}$$

δ(event *evt* any x
     where *label_guard* : $G(s,c,v,x)$
     then *label_action* : $A(s,c,v,x)$
     end) =
  evt(x : Type)
         -- 'evt' comment
     require
       label_guard: G
     do
       v.assigns(A)
     ensure
       label_action: v.equals(**old** A)
     end

Rule **init** below shows the translation of Event-B event *initialisation* to a creation procedure in Eiffel. The creation procedure initialises the object containing the constants definition. It also assigns initial values to variables taken from the initialisation in the *initialisation* event. In Eiffel, creation procedures are listed under the keyword **create**, as shown in rule **machine**. The **ensure** clause shows the translation of the before-after predicate of the assignment in Event-B.

$$\frac{\xi(A(s,c,v)) = \text{A}}{}\text{ (init)}$$

δ(event *initialisation*
     then
        label : $A(s,c,v)$
     end) =

initialisation
        -- evt comment
     do
       create ctx
       v.assigns(A)
     ensure
       label: v.is_equal(**old** A)
     end

3.2   Hand Translation

In this Section, we apply (manually) the translation rules to the Event-B model
in Fig. 3. The Event-B model is a well known model created by Abrial in [Abr10].
It models a system for controlling cars on an island and on a bridge. The model
depicted in Fig. 3 only shows the most abstract model of the system.

machine $m0$ **sees** $c0$
variables n
invariants
    inv1: $n \in \mathbb{N}$
    inv2: $n \leq d$
events
    event $INITIALISATION$
        then
            act1 $n := 0$
    end
    event ML_out
        where
            grd1 $n < d$
        then
            act1 $n := n + 1$
    end
    event ML_in
        where
            grd1 $n > 0$
        then
            act1 $n := n - 1$
    end
end

context $c0$
constants d
axioms
    axm1 $d \in \mathbb{N}$
    axm2 $d > 0$
end

Fig. 3. Controlling cars on a bridge: Event-B machine and its context.

Machine $m0$ sees context $c0$. $c0$ defines a constant d as a natural number
greater than 0. This constant models the maximum number of cars that can
be on the island and bridge. Machine $m0$ also defines a variable n as a natural
number (predicate **inv1**). Variable n is the actual number of cars in the island
and on the bridge. Predicate **inv2** imposes the restriction on the number of cars,
it must not be over d. Event *initialisation* gives an initial value to n: no cars
in the island or on the bridge. Event ML_out models the transition for a car
in the mainland to enter the island. The restriction is that the number of cars
already in the island is strictly less than d: there is room for at least another car.
Its action is to increase the number of cars in the island by one. Event ML_in
models the transition for a car in the island to enter the mainland. The only
restriction is that there is at least one car in the island. Its action is to decrease
the number of cars in the island. All these restrictions are ensured by the proof
obligations.

Figure 4 is the mapping to Eiffel programming language by applying the rules in Sect. 3.

```
class m0 create INITIALISATION
feature  --  Initalisation
    initialisation
    do
        create ctx
        n := 0
    ensure
        act1: n = 0
    end
feature  --  Events
    ml_out
    require
        grd1: n < d
    do
        n := n + 1
    ensure
        act1: n = old n + 1
    end

    ml_in
    require
        grd1: n > 0
    do
        n := n - 1
    ensure
        act1: n = old n - 1
    end
feature  --  Access
    ctx : CONSTANTS
    n : INTEGER
invariant
    inv1: n >= 0
    inv2: n <= d
end
```

Fig. 4. Excerpt of the Eiffel translation from the Event-B model depicted in Fig. 3.

4   Conclusion

We presented a series of rules to transform an Event-B model to an Eiffel program. The translation takes full advantage of all elements in the source by translating them as contracts in the target language. Thus, no information on the behaviour of the system is lost. These rules shows a methodology for software construction that makes use of two different approaches.

We plan on implementing these rules as an Event-B plug-in. We also plan of taking full advantage of the Proof Obligations generated by Event-B: translated them into a specification driven class so to help Eiffel provers in the process of proving the correctness of classes after any modification (extension) done by the implementer.

References

[Abr10] Abrial J-R (2010) Modeling in Event-B: system and software design. Cambridge University Press, New York

[CR16]   Cataño N, Rivera V (2016) EventB2Java: a code generator for Event-B. Springer International Publishing, Cham, pp 166–171

[EB10]   Edmunds A, Butler M (2010) Tool support for Event-B code generation, February 2010

[Mey92] Meyer B (1992) Eiffel: the language. Prentice-Hall Inc., Upper Saddle River

[MS11]   Méry D, Singh NK (2011) Automatic code generation from Event-B models. In: Proceedings of the second symposium on information and communication technology, SoICT 2011. ACM, New York, pp 179–188

[NMR15] Naumchev A, Meyer B, Rivera V (2015) Unifying requirements and code: an example. In: perspectives of system informatics - 10th International Andrei Ershov informatics conference, PSI 2015, in memory of Helmut Veith, Kazan and Innopolis, Russia, 24–27 August 2015, Revised Selected Papers, pp 233–244

[RBC16] Rivera V, Bhattacharya S, Cataño, N (2016) Undertaking the tokeneer challenge in Event-B. In: 2016 IEEE/ACM 4th FME workshop on formal methods in software engineering (FormaliSE), May 2016, pp 8–14

[RCn14] Rivera V, Cataño N (2014) Translating event-b to JML-specified java programs. In: Proceedings of the 29th annual ACM symposium on applied computing, SAC 2014. ACM, New York, pp 1264–1271

[RCnWR17] Rivera V, Cataño N, Wahls T, Rueda C (2017) Code generation for Event-B. Int J Softw Tools Technol Transf 19(1):31–52

[TFNP15] Tschannen J, Furia CA, Nordio M, Polikarpova N (2015) Autoproof: Auto-active functional verification of object-oriented programs. In: 21st international conference on tools and algorithms for the construction and analysis of systems. Lecture notes in computer science. Springer

A Tool-Supported Approach for Building the Architecture and Roadmap in MegaM@Rt2 Project

Andrey Sadovykh[1,2(✉)], Alessandra Bagnato[2], Dragos Truscan[3],
Pierluigi Pierini[4], Hugo Bruneliere[5], Abel Gómez[6], Jordi Cabot[7],
Orlando Avila-García[8], and Wasif Afzal[9]

[1] Innopolis University, 420500 Innopolis, Respublika Tatarstan, Russia
`a.sadovykh@innopolis.ru`
[2] Softeam, 21 avenue Victor Hugo, 75016 Paris, France
{`andrey.sadovykh,alessandra.bagnato`}`@softeam.fr`
[3] Åbo Akademi University, 20520 Turku, Finland
`dragos.truscan@abo.fi`
[4] Intecs S.p.A., Via U. Forti 5, 56121 Pisa, Italy
`pierluigi.pierini@intecs.it`
[5] IMT Atlantique, LS2N (CNRS) & ARMINES, 44000 Nantes, France
`hugo.bruneliere@imt-atlantique.fr`
[6] IN3, Universitat Oberta de Catalunya, Barcelona, Spain
`agomezlla@uoc.edu`
[7] ICREA, Barcelona, Spain
`jordi.cabot@icrea.cat`
[8] Atos, Subida al Mayorazgo, 24B, 38110 Tenerife, Spain
`orlando.avila@atos.net`
[9] Mälardalen University, Västerås, Sweden
`wasif.afzal@mdh.se`

Abstract. MegaM@Rt2 is a large European project dedicated to the provisioning of a model-based methodology and supporting tooling for system engineering at a wide scale. It notably targets the continuous development and runtime validation of such complex systems by developing the MegaM@Rt2 framework to address a large set of engineering processes and application domains. This collaborative project involves 27 partners from 6 different countries, 9 industrial case studies as well as over 30 different tools from project partners (and others). In the context of the project, we opted for a pragmatic model-driven approach in order to specify the case study requirements, design the high-level architecture of the MegaM@Rt2 framework, perform the gap analysis between the industrial needs and current state-of-the-art, and to plan a first framework development roadmap accordingly. The present paper concentrates on the concrete examples of the tooling approach for building the framework architecture. In particular, we discuss the collaborative modeling, requirements definition tooling, approach for components modeling, traceability and document generation. The paper also provides a brief discussion of the practical lessons we have learned from it so far.

© Springer Nature Switzerland AG 2020
P. Ciancarini et al. (Eds.): SEDA 2018, AISC 925, pp. 265–274, 2020.
https://doi.org/10.1007/978-3-030-14687-0_24

Keywords: Model-driven engineering · Requirement engineering · Architecture · UML · SysML · Traceability · Document generation · Modelio

1   Introduction

MegaM@Rt2 is a three-years project, funded by European Components and Systems for European Leadership Joint Undertaking (ECSEL JU) under the H2020 European program, that started in April 2017 [1]. The main goal is to create an integrated framework incorporating methods and tools for continuous system engineering and runtime V&V [2,8]. The underlying objective is to develop and apply scalable model-based methods and tools in order to provide improved productivity, quality, and predictability of large and complex industrial systems [9].

One of the main challenges is to cover the needs coming from diverse and heterogeneous industrial domains, going from transportation [3] and telecommunications to logistics. Among the partners providing use cases in the project, we can cite Thales, Volvo Construction Equipment, Bombardier Transportation and Nokia. These organizations have different product management and engineering practices, as well as regulations and legal constraints. This results in a large and complex catalog of requirements to be realized by the architecture building blocks at different levels of abstraction. Thus, the development of the MegaM@Rt2 framework is based on a feature-intensive architecture and a related implementation roadmap.

The MegaM@Rt2 framework plans to integrate more than 30 tools implementing the above mentioned methods and satisfying requirements of the case studies. The tools features are grouped in three conceptual tool sets:

1. MegaM@Rt Systems Engineering Tool Set regroups a variety of current engineering tools featuring AADL, EAST-ADL, Matlab/Simulink, AUTOSAR, Method B or Modelica, SysML and UML in order to precisely specify both functional and non-functional properties. Moreover, system level V&V and testing practices will also be supported by this tool set.
2. MegaM@Rt Runtime Analysis Tool Set seeks to extensively exploit system data obtained at runtime. Different methods for model-based V&V and model-based testing (MBT) will be rethought and/or extended for runtime analysis. Model-based monitoring will allow to observe executions of a system (in its environment) and to compare it against the executions of corresponding model(s). Monitoring will also allow a particular system to be observed under controlled conditions, in order to better understand its performance.
3. MegaM@Rt Model & Traceability Management Tool Set is a key part of the framework as it is dedicated to support traceability between models across all layers of the system design and execution (runtime). This can go from highly specialized engineering practices to low-level monitoring. Relying on the unification power of models, it should provide efficient means for describing, handling and keeping traceability/mapping between large-scale and/or heterogeneous software and system artifacts.

Model-based approaches for specification have been developed consistently during almost two decades [4]. Automated document generation was one of the first benefit offered by the Model-driven Architecture (MDA) [4,10]. Indeed, models as the first-class entities of the engineering process should contain all the necessary information for the design documentation. However, several challenges arise. First, the architect team should decide the right organization for the global architecture model. Second, it should be carefully planned which level of details is appropriate for the design of the individual contributions. Third, it should be considered that the architecture model will be used during 3 years of the project for numerous purposes, and thus needs to be prepared to accommodate for changes in methodology. Fourth, several documents need to be generated by extracting the relevant information from all over the architecture model.

In this paper, we present our experience on providing and utilizing model-based tool support for defining MegaM@Rt2 framework architecture, to identify the solution to be implemented in the context of the project and to obtain a corresponding roadmap for the development of architecture components throughout the project.

2   Architecture Specification Approach

We adopted a practical approach for the architecture specification that is particularly adapted to collaborative projects such as MegaM@Rt2, integrating tools coming from several parties [5,6]. As modeling language, we took a Systems Modeling Language (SysML) [11] subset for requirements specification and a Unified Modeling Language (UML) [12] subset for the high-level architecture specification. The approach to define the MegaM@Rt2 framework architecture is depicted in Fig. 1. We splitted the architecture model in several parts, dividing the responsibilities among: the Work Package (WP) leaders, the tool providers and the case study providers, as detailed in the following list:

- Requirements/Purposes are specified using SysML
  - Case Study Requirements are specified by case study providers,
  - Framework Requirements are specified by WP leaders, refining the Case Study Requirements
  - Tool Purposes are specified by tool providers with the aim to realise a specific subset of the Framework Requirements
- Architecture is specified in UML
  - Conceptual Tool Set is specified by the WP leaders and represents the basic architecture of the MegaM@Rt2 Framework
  - Tool Set Component is each tool instance, specified by tool providers, implementing part of the MegaM@Rt2 Framework functionalities
  - Common Interfaces - specified by tool providers,
  - Common deployment frameworks - specified by tool providers

In addition, to support the integration of each Tool Set Component into the MegaM@Rt2 Framework, the following additional elements has been identified:

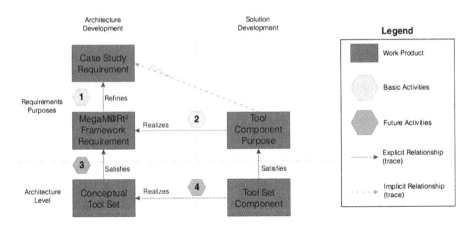

Fig. 1. Overview of the architecture and development process in MegaM@Rt2.

- Common Interfaces, specified by tool providers, to support data and model exchange between tools
- Common Deployment Frameworks, specified by tool providers, to highlight possible issues related to hardware and software platforms.

At the Requirements/Purposes level, the needs of industrial partners have been collected and classified by means of Case Study Requirements, from which we specified (Activity 1) the MegaM@Rt2 Framework Requirements. For the latter, we identify (Activity 2) a set of Tool Component Purposes that will realize the case study requirements. At the Architecture level, each Conceptual Tool Set component and the relevant interfaces are identified (Activity 3) to satisfy framework requirements. Then, for the Conceptual Tool Set we specify (Activity 4) concrete Tool Set Components to realize the desired functionality. Those Tool Set Components expose features (i.e. purposes) that are progressively available, during the project time frame, based on specific development plan. The roadmap is defined as the set of tools components purposes available at each project milestone.

3   Tooling Approach

Appropriate tooling support is important for the success of the model-driven engineering process shown in Fig. 1. In order to provide tool support for our architecture specification approach, we selected the Modelio and Constellation tools provided by one of the project participants: the SOFTEAM partner.

When collecting inputs of 50 users, it was important to provide guidelines and diagram templates. Otherwise, the integration work may become extremely challenging. As such, we defined a set of template diagrams both for specifying requirements and for collecting tool purposes. Users were able to clone these templates inside the model to describe their concrete tools.

Fig. 2. Requirements editing with Modelio.

In the next subsections we are providing more details on how different features of the tool were used to support our approach.

3.1 Architecture Specification

Modelio is an MDE workbench supporting standard modeling languages such as UML, SysML among others. All the modeling notations can be used in the same global model repository which is important for model traceability and management.

Requirements Modeling. In our approach, requirements originated from different sources, i.e. from 9 case study providers and 22 tool providers. In order to have an uniform approach for requirement specification that would facilitate gap analysis and roadmap identification, we defined requirement templates that were used to define the expected properties to be collected, such as criticality for the case study requirements and *planned release* date for tool purposes.

Modelio allowed us to edit requirements in both diagram view and tabular view (see Fig. 2). The requirements were manually edited or automatically imported from other documents, e.g. MS Excel.

Architecture Modeling. On the architecture level, we used Class and Deployment diagrams. We limited modeling to a subset of UML to enforce the common understanding of the architecture and simplify editing. In particular, we choose to use UML Components, Interfaces, Associations, Generalizations and Dependencies.

For tool components, we set a template for the architecture specification that included class diagram to specify functional interfaces, tool components subordinates and the relation to the conceptual tool set in the framework, and deployment diagrams to identify the execution environment of the tool component. In addition, Package diagrams have been used to define the high level structure of the MegaM@Rt2 framework architecture.

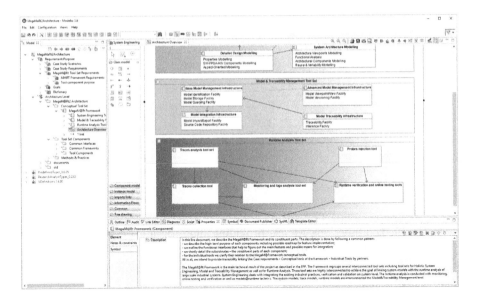

Fig. 3. Editing architecture and documentation with Modelio.

For instance, Fig. 3 shows that the MegaM@Rt2 framework architecture is composed of three parts corresponding to the three WPs of the project: System Engineering, Runtime Analysis, and Model and Traceability Management, respectively.

In Modelio, the documentation (Fig. 3) can be added in the textual notes or attached as separate documents. Both plain text and rich text notes are supported. In our work, we deliberately restricted editing to plain text notes to make sure that the generated documents are formatted correctly.

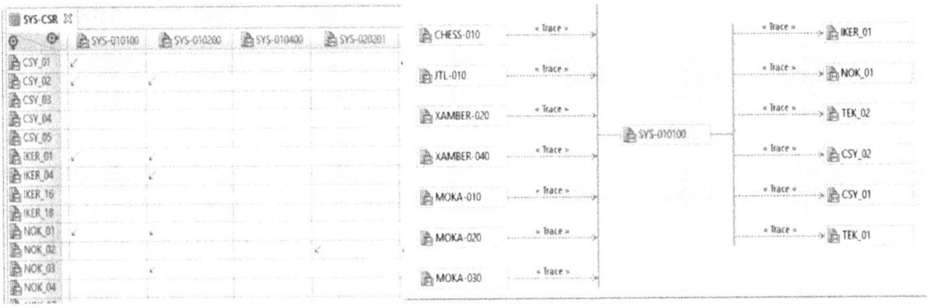

Fig. 4. Example: traceability links among the tool set, framework and case study requirements.

Requirements Traceability. Once the requirements have been specified, for each tool component we defined a traceability matrix to link case study requirements to framework requirements, and respectively framework requirements to tool purposes as depicted as described by Activities 1 and 2 of the modeling approach in Fig. 1. This allowed us to use instant traceability diagrams, as the one in Fig. 4, to visualise the whole set of dependencies for a given requirement. This proved beneficial not only for the requirement analysis and toolset integration planning, but also for identify common interfaces for tool components and visualise gaps for the requirements analysis.

Generating Documents. Modelio offers fairly flexible model query and document generation facilities that were used for editing and maintaining four specifications in the project. The template editor (Fig. 5) was particularly useful to implement custom extraction of model elements in order to create specific sections of the document.

In the example below, the template specifies that the generator will search for a Tool Components package, look at all the UML components to generate a tool section. This document section will include introductory paragraph, "Purpose" subsection, subsections for all class and deployment diagrams as well as section on the owned interfaces.

Fig. 5. Example: custom document generation template for individual tools section.

When editing the architecture model, it is quite useful to see the generation result. Thus, along with developing custom document templates, we integrated the document generation to the Modelio interface. That way regular users could call the document generation directly from the tool using a context menu (Fig. 6).

3.2 Collaborative Model Editing

Modelio Constellation [7] is the model sharing, collaborative editing, versioning and configuration management facility that allow hundreds of modellers to work together on the same common and shared model. Indeed, authoring an architecture deliverables in MegaM@Rt2 project required contributions of 27 partners. Thus, around 50 users worked together on the single model. On the regular basis, users connected to the Constellation server, synchronised the local model with the central repository, edited the architecture and generated the documents with alway updated templates. The documentation templates and user interfaces for document generation were developed continuously and had to be rolled out to the whole large team of modellers without interrupting the work process. It was important to provide versioning and conflict resolution when editing touched the common artifacts. Last but not least, several different deliverable were generated out of the same model. Therefore, the branching facility allowed to fix the state when the deliverable were released.

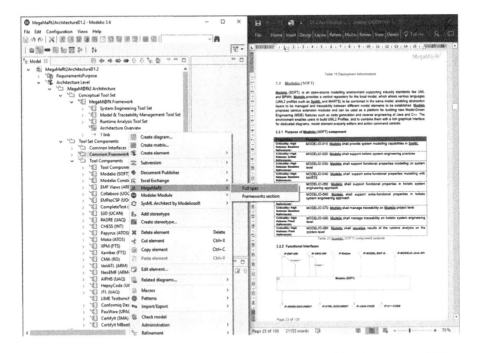

Fig. 6. Example: architecture document generated with Modelio document publisher.

4   Conclusions

In this paper, we presented the tool support we used in order to identify and specify the architecture of the MegaM@Rt2 framework using model-driven principles and practices. Our approach enforced the coordination and collaboration among many different stakeholders and thus manageability of this complex project. Indeed, the main benefits of our model-driven approach are that all information was collected from different stakeholders and stored using one single model using a single tool.

This model was used as a central repository, that every project partner can access and update using model versioning techniques. In addition, having all the information in one single place, allowed us not only to constantly monitor the status of the process and to trace the requirements of the framework components, but also to easily generate necessary artefacts (such as documents, tables, diagrams) from the model, whenever needed.

However, using a model-based approach also had some challenges and limitations.

The first of such challenge was that different project participants had different levels of familiarity with modeling tools in general and with Modelio and Constellation in particular. This issue has been addressed by providing several project-wide online webinars along with proper documentation on describing how the tools can be used to support the architecture specification approach. The Modelio support team was helpful in solving the licensing issues, helping with installation and resolving model versioning conflicts.

A second challenge came from the fact that 50 modellers worked collaboratively with the models which could trigger inconsistencies, conflicts and omissions in the collected information. Using the Constellation tool we were able to support model versioning and collaborative modeling. In addition, we splitted the model in several parts corresponding to each work package and we provided clear guidelines on how the work is organized.

A third challenge came from the limitations of the selected tools. For instance, there were different restrictions on how the styling of the documents generated from the models could be configured and how the information could be visualized using different types of diagrams. Manual effort is also required to create document templates and configure the document generators. However, once the generators were created they could be reused easily and effectively.

In addition to the above challenges, some industrial partners were already using an existing company-wide tool chain that is not part of our project consortium. In this case, they still gave their requirements, which were mapped to MegaM@Rt tool set capabilities. However, in such cases, the industrial partners had an additional validation of the acceptance of their requirements using both the MegaM@Rt tool set capabilities as well as the capabilities of their in-house tool set.

Overall, the experience with Modelio and Constellation was mostly positive and the tool will be further used in other architecture deliverables at later stages of the project and as the reference document.

Acknowledgement. This project has received funding from the Electronic Component Systems for European Leadership Joint Undertaking under grant agreement No. 737494. This Joint Undertaking receives support from the European Union's Horizon 2020 research and innovation program and from Sweden, France, Spain, Italy, Finland and Czech Republic.

References

1. ECSEL JU MegaM@Rt2 Project Website (2018). https://megamart2-ecsel.eu/
2. Fitzgerald B, Stol KJ (2017) Continuous software engineering: a roadmap and agenda. J. Syst. Softw. 123:176–189
3. Wallin P, Johnsson S, Axelsson J (2009) Issues related to development of E/E product line architectures in heavy vehicles. In: 42nd Hawaii international conference on system sciences (HICSS 2009), Big Island, HI, USA. IEEE
4. Di Ruscio D, Paige RF, Pierantonio A (2014) Guest editorial to the special issue on success stories in model driven engineering. J Sci Comput Program 89(PB):69–70. https://doi.org/10.1016/j.scico.2013.12.006
5. Afzal W, Bruneliere H, Di Ruscio D, Sadovykh A, Mazzini S, Cariou E, Truscan D, Cabot J, Field D, Pomante L, Smrz P (2017) The MegaM@Rt2 ECSEL project: MegaModelling at runtime - scalable model-based framework for continuous development and runtime validation of complex systems. In: 20th EUROMICRO conference on digital system design (DSD). IEEE
6. Bruneliere H, Mazzini S, Sadovykh A (2017) The MegaM@Rt2 approach and tool set. In: DeCPS workshop, 22nd international conference on reliable software technologies. ADA-Europe
7. Desfray P (2015) Model repositories at the enterprises and systems scale: the modelio constellation solution. In: 1st international conference on information systems security and privacy (ICISSP). IEEE
8. ISO/IEC/IEEE (2011) ISO/IEC/IEEE 29148: Systems and software engineering – life cycle processes – requirements engineering
9. ISO/IEC (2011) ISO/IEC 25010 systems and software engineering – systems and software quality requirements and evaluation (SquaRE) – system and software quality models
10. OMG: model driven architecture (MDA) Guide rev. 2.0. http://www.omg.org/cgi-bin/doc?ormsc/14-06-01
11. OMG: OMG systems modeling language (OMG SysML), Version 1.4. http://www.omg.org/spec/SysML/1.4/
12. OMG: unified modeling language (UML), Version 2.5. http://www.omg.org/spec/UML/2.5/

The Key Role of Memory in Next-Generation Embedded Systems for Military Applications

Ignacio Sañudo[1], Paolo Cortimiglia[2], Luca Miccio[1], Marco Solieri[1(✉)],
Paolo Burgio[1], Christian Di Biagio[2], Franco Felici[2], Giovanni Nuzzo[2],
and Marko Bertogna[1]

[1] Università di Modena e Reggio Emilia, Modena, Italy
{ignacio.sanudoolmedo,luca.miccio,marco.solieri,
paolo.burgio,marko.bertogna}@unimore.it
[2] MBDA SpA, Rome, Italy
{paolo.cortimiglia,christian.di-biagio,
franco.felici,giovanni.nuzzo}@mbda.it

Abstract. With the increasing use of multi-core platforms in safety-related domains, aircraft system integrators and authorities exhibit a concern about the impact of concurrent access to shared-resources in the Worst-Case Execution Time (WCET). This paper highlights the need for accurate memory-centric scheduling mechanisms for guaranteeing prioritized memory accesses to Real-Time safety-related components of the system. We implemented a software technique called cache coloring that demonstrates that isolation at timing and spatial level can be achieved by managing the lines that can be evicted in the cache. In order to show the effectiveness of this technique, the timing properties of a real application are considered as a use case, this application is made of parallel tasks that show different trade-offs between computation and memory loads.

Keywords: Real-time systems · Multi-core · Determinism · Memory interference

1 Introduction

The exponential increase in the embedded software's complexity and the integration of multiple functionalities in the same computing platform have completely changed the way in which vendors designed their solutions. Nowadays, many software manufacturers in the avionic domain are deploying solutions on top of multi-core processors whose cores are all disabled but one [1]. The main reason behind this decision is the lack of methods for guaranteeing core isolation, which in turn is mostly due, especially in the latest generation architectures, to the limited, or absent support of hardware mechanisms for the management of core-shared resources, e.g. the memory hierarchy or the I/O devices [2].

© Springer Nature Switzerland AG 2020
P. Ciancarini et al. (Eds.): SEDA 2018, AISC 925, pp. 275–287, 2020.
https://doi.org/10.1007/978-3-030-14687-0_25

In the most safety-critical systems, isolation is achieved either by integrating software components on a dedicated single-core System-on-Chip (SoC) which is used exclusively for one specific purpose, or by allocating all tasks (i.e. threads or processes) on the same core. As a matter of fact, the development of safety-related software components running on top of a heterogeneous multi-core system is still difficult, since achieving core-isolation and the consequent predictability is a challenging problem.

Indeed, on the certification level, the design, validation and integration of software components with different criticality levels on top of multi-core plat-forms is still hard. In order to ensure safeness in such process, different tight domain-specific standards have been proposed to provide a reference guidance— e.g. ISO-26262 [3] in the automotive domain, and DO-178C [4] in the avionic one. For the sake of simplicity, but without losing much generality[1] from now on we will focus on the considerations and terminology used in MCP CAST-32 [5] and MCP-CRI (still under development). The purpose of both documents is: to "*identify topics with Multi-Core Processors (MCP) with two active cores that could impact the safety, performance and integrity of the software for a single airborne system executing on MCPs*". Both documents also identify in the *temporal and spatial partitioning* the key requirement for the deployment of software components with different Development Assurance Level (DAL) in the same sub-system. In this way, a software component with a low DAL can be prevented from propagating it to a higher-DAL component, thus reducing design and production costs, by simply allocating the two in different partitions. Similarly, temporal and spatial partitioning mandates that variation in a task's computation time and memory usage, respectively, does not interfere with those another one's.

At system level, software components can be *interfered* in different ways. For example, one application can run for longer a time, thus increasing the execu-tion time of another application and potentially causing execution starvation. Also, an application can obtain and not release a shared resource, which may block another application that needs it, or cause even more severe pathologies like deadlocks or livelocks. Now, these interferences dramatically affects the esti-mation of the Worst-Case Execution time (WCET). Considering for instance the pervasiveness of image processing or neural networks components, it is easy to see that applications are becoming more memory-intensive. This accordingly shifted the concerns of authorities and industries towards the need of under-standing and taming the dominating part of WCET estimation—the access to shared resources, such as the memory subsystems.

Specifically, without applying *memory partitioning* techniques, multi-core systems feature DRAM banks and last-level caches that are extremely often accessed, but shared, hence prone to mutual inter-core contention. Safety-critical tasks can therefore be interfered both at time and space level. For instance, a cache line is evicted/replaced, the effective task execution time is impacted by

---

[1] Safety-critical software guidelines in different domains like ISO-26262 or IEC-61508 [6] are quite similar in many technical aspects.

the need of experiencing a miss event and a data fetch from central memory, which may even double the memory access latency. This leads to quite pessimistic assumptions in the WCET estimation, exacerbating the difference with respect to the average execution time (ACET) thus causing a reduction of the total CPU utilization.

At *cache* level, isolation or partitioning can be provided by using hardware mechanisms like cache locking, available on the old 7th generation of the ARM architecture, or cache partitioning, equipping a high-end server segment of Intel Xeon x86 architecture. The ARMv8 architecture, which powers most of the medium-to-high-performance modern system has instead no hardware assisting technology.[2] Now, the most prominent software technique that allows to implement cache partitioning is called *page coloring* [8] and consists in exploiting the cache mapping function to isolate non-contiguous memory partitions that injectively maps to cache indices partitions. This technique is founded on virtual address translation, which allows the striped allocation to be hidden at application level. An implementation at OS level would suffer both: great expensiveness, due to the sparseness in the OS choices by the embedded software companies; scarce applicability caused by the presence of custom OS. We advocate instead an easy-to-deploy, legacy-friendly *hypervisor* solution which additional enable seamless consolidation of single-core systems—cache coloring is placed below bare-metal applications and OSs. More precisely, we chose a novel extension [9], to the Jailhouse hypervisor [10], an open-source, minimal, partitioning hypervisor designed for real-time and/or safety-critical use cases. The solution has been developed in the context of the EU Horizon 2020 HERCULES research project [11], whose goal is to develop a system stack for a the next-generation, high-performance multi-core real-time systems. We consider a representative system where a real application from military domain runs on top of next-generation heterogeneous embedded platform, namely a Xilinx Zynq UltraScale+ MPSoC.

The remainder of the paper is organized as follows: Sect. 2.2 presents an analysis of interference patterns caused by memory contention. The reference application is presented in Sect. 3.1, showing some result of the capabilities of cache coloring to provide isolation at memory level in Sect. 3.3, finally a concluding discussion is presented in Sect. 4.

2   Platform Memory Profiling

In this section, we present the experiments conducted to determine the impact on memory latency caused by multiple sources of aggressive memory accesses, so to characterize the amount of interference.

2.1   Reference Architecture

We consider a widely adopted modern heterogeneous multiprocessor system on chip (MPSoC), the Xilinx Zynq UltraScale+, mounted on the ZCU102 board.

---

[2] The recently announced ARMv8.4-A will include the Memory System Resource Partitioning and Monitoring system, whose specification have been just released [7].

This platform is also used in the following to present the cache coloring mechanism. The hardware architecture considered in this work, the Xilinx Ultrascale+, has a 64-bit quad-core ARM Cortex-A53 CPU coupled with programmable logic in a single device, cache-coherent interconnect, and a shared memory system, among other peripheral interfaces. Figure 1 depicts potential contention points that can be found in the platform architecture. As can be observed in Fig. 1, the main source of contention between the processors and the programmable logic is found in the DDR memory controller when the DMA transfer is produced (1). Other sources of contention are the cache coherent interconnection (2) and the L2 cache (3).

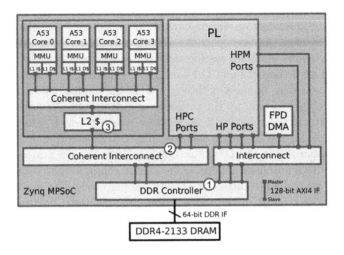

Fig. 1. Xilinx Zynq UltraScale and MPSoC.

2.2 Evaluation

Empirical activities have been performed within the HERCULES EU project [11] and are reported in a dedicated deliverable [12]. Images in this section are courtesy of respective authors and all rights belong to their respective owners. Along a first axis, experiments focused on three variants depending on where the measurement and interference run: CPU-CPU, CPU-FPGA and FPGA-CPU. On the CPU, it was employed the LMBench synthetic benchmark [13], which measures the average latency to perform a read memory operation; it was hosted on one of the CPU cores and configured to test both (i) sequential, and (ii) random stride reads. On the FPGA, a custom implementation has been realised and used.

CPU Impact on CPU. In this experiment is outlined the interference produced at intra core level (Fig. 2). In the 'x' axis the Working Set Size (WSS) in bits is shown, while in the 'y' axis is displayed the relative execution time in nanoseconds (ns). As can be observed from the charts, the impact is negligible whilst the WSS fits in the private L1 cache (32 KiB). The interference starts when we increase the working set, which increases by at most 2X. The latency starts to be significant when the WSS reaches the size of the shared L2 cache (1 MiB), in this case, the latency is greatly impacted by the presence of interference, 7X and 16X for sequential and random reads respectively. The latency slowdown gets serious when the WSS does not fit the L2 cache. In this case, the latency converges to a delay proportional to the number of interfering cores. The impact for random interference patterns is less pronounced, however it reaches a 2X and 1.5X for sequential and random reads, respectively.

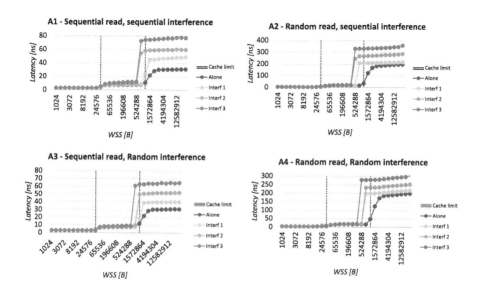

Fig. 2. CPU interference CPU.

FPGA Impact on CPU. In this experiment, authors characterized the interference experienced by the CPU when there is DMA activity with data required by the FPGA accelerator. The CPU cores perform sequential (D1) and random (D2) reads while FPGA accelerator leads to memory accesses with different bandwidth ratios. Figure 3 shows that, in both cases, memory interference degrades CPU performance. The 'X' axis plots the working set (WSS) in bits, that is, the amount of data accessed by the benchmark, while the 'Y' axis shows the corresponding execution time or latency to perform the operation of the target application in nanoseconds (ns). We see that, when the WSS becomes larger than the L2 cache, it is experienced a heavy performance degradation (almost 3×) if the FPD-DMA module transfers data at maximum bandwidth.

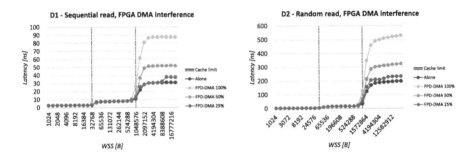

Fig. 3. FPGA interference to CPU.

CPU Impact on FPGA. The purpose of this experiment was to find out the interference caused by the CPU to the FPGA accelerator. The results of this experiment are presented in the Fig. 4. In (E1) the FPGA accesses memory via burst transfers (DMA), while the host cores perform sequential reads. In (E2) the CPU cores perform random interference. From these results, it is evident that the performance decreases proportionally to the number of interfering cores, as in the experiments shown above.

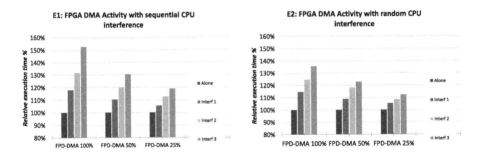

Fig. 4. CPU interference FPGA.

Discussion. The experiments presented in this section show that memory-intensive tasks can significantly influence the others task timing behavior. This issue is highly problematic for safety-critical applications and it cannot be mitigated in the discussed platform because the Xilinx Ultrascale+ does not present the capabilities to manage the memory accesses or infer the task priority when memory access are performed. To solve this problem, we will now briefly introduce an effective approach for resource partitioning, that mitigates this effect while providing higher bounds of predictability, this software technique is called cache coloring.

3   Software Solution for Seamless Cache Partitioning

Page Coloring. In order to enforce a deterministic cache hit rate on the most frequently accessed memory pages we leveraged on a software cache partitioning technique called *cache coloring*. By using cache coloring, virtual memory pages are mapped to contiguous memory sets into the physical cache by "coloring" the physical pages. By doing so, tasks are allocated a subset of the cache that cannot be evicted by other concurrent tasks. Page coloring and cache lockdown [14,15] mechanisms make easier the achievement of the goals established in the rationale defined in MCP_Determinism_8-11 (MCP-CAST 32) that are related to shared-resources.

Hypervisor Solution. Jailhouse [10] is an open-source, minimal, partitioning hypervisor designed for real-time and/or safety-critical use cases. It partitions hardware resources (e.g. CPUs, memory regions, PCI devices, interrupts) to cells, and assigns them to guests OS's or bare-metal applications called inmates. Jailhouse does not implement any form of resource sharing, scheduling or emulation. Jailhouse design's philosophy focuses on simplicity and openness – it features a small code base ranging between 7 and 9.7 KLoC depending on the architecture (ARMv8 and x86, respectively), implying much lower certification costs. It is licensed under GPLv2.

Coloring Support. The software mechanism proposed for the considered problem was implemented on top of the Jailhouse hypervisor [9] and Xilinx Zynq Ultra-Scale+. This approach has been successfully evaluated in a demanding real-time setup [16].

3.1   Use Case Architecture

Application software for military systems exposes high levels of functionality, integrity and dependability for company missile systems, support facilities and test applications. This software is typically multi-tasking and it is "hard real-time", i.e., the software tasks have to work within specific time constraints, and any minimal failure results in a complete loss of the missile and an unsuccessful engagement, hence it is essential that it is designed and verified to very exacting standards, and it exhibits very high levels of dependability. Dependability is a term that encompasses the needs associated with software systems that required levels of Safety, Security, Reliability, Integrity, Mission Criticality etc. In this context, most of the products provided by MBDA are "mission critical", i.e., a malfunction could result in the equipment under its control (a missile, radar, launcher, etc.) failing to operate correctly and thus failing to meet its operational need. However many of these products now contain features that require the development to meet the other aspects of Software Dependability. Embedded systems have very often real-time constraints dealing with predictability and determinism more than raw speed. Take for example flight software as a tactical application to support missile engagement mission: it runs guide, navigation and control algorithms in a real-time environment supported by multitasking O.S. in order to guarantee timing constraints with the objective to reach acquired target computing flight asset and adjusting its own route. Sensors management, data validity and actuators control must be very strictly related to time schedule and reaction performing the consequent task according to the real environment.

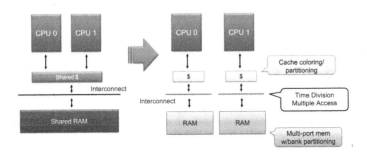

Fig. 5. Techniques to reduce memory interference.

Figure 6 shows the task decomposition of the application that we will consider in this work. The application is composed of different tasks and timers, the figure highlights the parallel execution of different threads ("Active nodes") and the data they exchange (junction nodes). We distinguish between threads executing algorithms, in circular boxes marked with "Algo", and the ones performing I/O activities (triangular boxes, marked as "Drivers") and interacting with external equipment, which are not shown in the diagram. Dashed lines represent movements of data, while solid, "stim" lines are events triggering node execution, that can optionally include also data movement. This application (developed by MBDA) was designed to execute on a single ARM core of the target board, clocked at 800 MHz, using a memory pool of 500 MiB.

3.2 Use Case Application

For the sake of simplicity we will consider only a subset of the taskset presented in the last section.

Test Application. The object under test is a single-core bare-metal application composed by two parts:

- *Test* A high-criticality, small-sized (512 B), either instruction- or data-bound, routine, whose periodic execution (from 200 us to 3 ms) is triggered through an AXI timer interrupt, in turn raised by the PL.
- *Internal interference* A lower-criticality, medium-sized (32 KiB) periodic (500 us) routine, independently executing on the same core as Test.

Observe that Internal interference is able to evict Test's instructions, or data, from L1 cache, forcing them to be retrieved in L2.

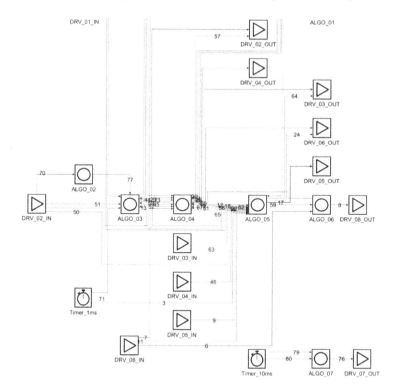

Fig. 6. Target application architecture.

Interference Application. A quite pessimistic model of memory activity is implemented in External Interference, a second single-core bare-metal application. It repeatedly and endlessly copies a 2 MiB memory segment into another one of equal size. To approximate saturation of the system memory bandwidth, and to maximise L2 cache pollution, we consider two running instances of External Interference.

Evaluation. We measured the execution time of Test under three different setup configurations:

- *Baseline* Only the test application is deployed on the original Jailhouse version 0.8.
- *Contention* We just add the interference application to the Baseline configuration.
- *Contention&coloring* We enable the coloring support to the Contention configuration. In particular, we assign:
  - half of the L2 cache to the test application, i.e. 8 colors, or 512 KiB;
  - quarter of the L2 cache to each inmate of the interference application, i.e. 4 colors, or 256 KiB.

In summary, we obtain the 48 configuration combinations that are reported in Table 1, each of which was run 10 K times. Results are shown in Fig. 7.

Table 1. Test configurations summary. a and b are assumed equal.

	Test	Internal	External
Core (id)	3	3	{none, 1–2}
Period (ms)	{0.2, 0.4, 0.6, 0.8, 1, 1.5, 2, 3}	0.5	0
Footprint size (B)	512	32 Ki	2 Mi
Footprint kind	{instr., data}[a]	{instr., data}[b]	Data

3.3   Discussion

We shall first comment on the bare-metal implementation, then compare it first to the hypervisor-hosted variation without coloring, and lastly discuss the colored version.

Baseline. Results for this configuration are plotted in the blue box plots of Fig. 7. This results represent our golden standard for attainable performances on this platform, and we notice it is indeed quite stable. Variations to the execution period produce effects on the execution time that are negligible both for the data-bound test, where no significant correlation is observable, and for the instruction-bound one, where even changes are hard to be measured. Average and maximum values are respectively around 0.13 and 0.17 mus for the instruction-bound routine, and around 0.23 and 0.61 mus for the data-bound routine, respectively.

Contention. The red box plots in Fig. 7 outline the dramatic detriment to the execution time predictability caused by the introduction of L2 cache stress activity. Both the average and the worst execution time, for both the instruction- and the data-bound configuration are now one order of magnitude greater than the baseline results. In the 2 ms period, instruction-bound case, for instance, we observe more than a 5.7 factor on the median value, and a 12.3 factor on the maximum one—0.75 and 2.1 mus, respectively.

Contention with Coloring. Effects of the activation of the cache coloring configuration are visualized in the green box plots of Fig. 7. It does not need to be much commented. The execution time is now only slightly higher than the baseline—in the worst case, the overhead amounts respectively at 0.11 and 0.22 μs for the instruction- and data-bound routines.

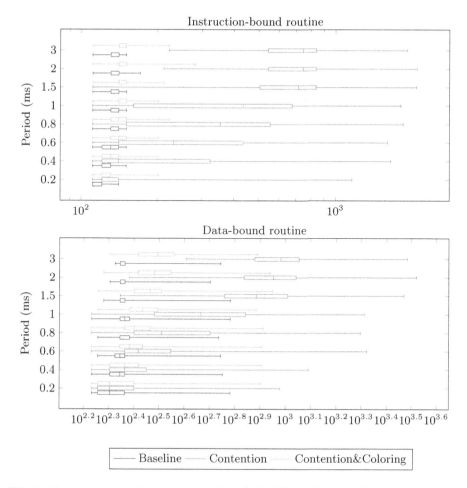

Fig. 7. Benchmark results: execution time (ns). Mixed linear and semi-logarithmic scales).

4   Conclusion

Discussion. In this paper, we proposed a system partitioning strategy to overcome timing scalability issues for real-time military application running on heterogeneous embedded SoCs based on multi- and many-core. We introduced an extensive analysis on a realistic application, aimed at pointing out the potential performance bottlenecks. Then, we presented our approach to mitigate their effect on the timing behavior of the application—a partitioning hypervisor enhanced with page coloring support, which avoids the problematic contention on shared caches. The evaluations shows that the solution brings substantial improvements for the system predictability, since the jitter of tasks' execution time is reduced up to one order of magnitude. We believe that the proposed approach represents not just a solid framework for both the system development and the deployment engineering in the real-time military domain, but also a starting point for the application of memory-aware task co-scheduling policies.

Further Works. We plan to control the interference caused by conflicting memory requests from the FPGA by using a memory-arbitration mechanism inspired by the one proposed by Capodieci et al. in [17]. SiGAMMA is a server-based mechanism that behaves as a memory arbiter between CPU and GPU, balancing the penalties of the concurrent memory accesses performed by the GPU engine. Figure 5 depicts the proposed approach that is based on splitting the system onto isolated partitions, preventing in this way unwanted interference between the multiple cores and the FPGA accelerator.

Acknowledgment. This work was supported by the I-MECH (Intelligent Motion Control Platform for Smart Mechatronic Systems), funded by European Union's Horizon 2020 ECSEL JA 2016 research and innovation program under grant agreement No. 737453.

References

1. Howard CE (2018) It's time: avionics need to move to multicore processors. https://www.intelligent-aerospace.com/articles/2018/01/it-s-time-avionics-needs-to-move-to-multicore-processors.html
2. Sañudo I, Cavicchioli R, Capodieci N, Valente P, Bertogna M (2018) A survey on shared disk I/O management in virtualized environments under real time constraints. SIGBED Rev. 15(1):57–63
3. ISO 26262-1:2011 (2011) Road vehicles – Functional safety
4. EUROCAE WG-12 RTCA SC-205 (2011) DO-178C, software considerations in airborne systems and equipment certification
5. CAST 32 Superceded by CAST 32A Multi-core Processors (2014) Standard
6. Functional safety of electrical/electronic/programmable electronic safety-related systems (2000) Standard, International Organization for Standardization, Geneva, CH
7. ARM. Reference Manual Supplement Memory System Resource Partitioning and Monitoring (MPAM), for Armv8-A Documentation. https://developer.arm.com/docs/ddi0598/latest
8. Mittal S (2017) A survey of techniques for cache partitioning in multicore processors. ACM Comput. Surv. 50(2):27:1–27:39
9. Kloda T, Solieri M, Mancuso R, Capodieci N, Valente P, Bertogna M (April 2019) Deterministic memory hierarchy and virtualization for modern multi-core embedded systems. In: 25th IEEE real-time and embedded technology and applications symposium, RTAS 2019
10. Kiszka J and Contributors (2018) Jailhouse
11. High-Performance Real time Architectures for Low-Power Embedded (HERCULES) Project Consortium (EU ID 688860 H2020 ICT 04-2015). The Hercules Project. https://hercules2020.eu/

12. Capodieci N, Cavicchioli R, Vogel P, Marongiu A, Scordino C, Gai P (2017) Deliverable 2.2 - detailed characterization of platforms. High-Performance Real-time Architectures for Low-Power Embedded (HERCULES) Project Consortium (EU ID 688860 - H2020 - ICT 04-2015). https://hercules2020.eu/wp-content/uploads/2017/03/D2.2_Detailed_Characterization_of_Platforms.pdf
13. McVoy L, Staelin C (2005) Lmbench - tools for performance analysis, version 2. http://www.bitmover.com/lmbench/
14. Caccamo M, Cesati M, Pellizzoni R, Betti E, Dudko R, Mancuso R (2013) Real-time cache management framework for multi-core architectures. In: Proceedings of the 2013 IEEE 19th real-time and embedded technology and applications symposium (RTAS), RTAS 2013. IEEE Computer Society
15. Tabish R, Mancuso R, Wasly S, Alhammad A, Phatak SS, Pellizzoni R, Caccamo M (April 2016) A real-time scratchpad-centric OS for multi-core embedded systems. In: 2016 IEEE real-time and embedded technology and applications symposium (RTAS)
16. Corradi G, Klingler B, Puillet E, Schillinger P, Bertogna M, Solieri M, Miccio L Delivering real time and determinism of ZYNQ Ultrascale+ A53 clusters with coloured lockdown and Jailhouse hypervisor. Xilinx Inc. whitepaper (November 2018, to appear)
17. Capodieci N, Cavicchioli R, Valente P, Bertogna M (2017) Sigamma: server based integrated GPU arbitration mechanism for memory accesses. In: RTNS

Vulnerabilities and Security Breaches in Cryptocurrencies

Gudmundur Sigurdsson[1], Alberto Giaretta[2(⊠)], and Nicola Dragoni[1,2]

[1] DTU Compute, Technical University of Denmark, Kongens Lyngby, Denmark
s172168@student.dtu.dk, ndra@dtu.dk
[2] Centre for Applied Autonomous Sensor Systems (AASS), Örebro University,
Örebro, Sweden
alberto.giaretta@oru.se

Abstract. Nowadays, 1375 different cryptocurrencies exist, and their market value totals up to $444.8 billion, at the time of writing. The interest revolving around cryptocurrencies is constantly growing, and this hype caused an increase of criminal attacks on various cryptocurrencies. In this paper, we cover the main aspects that concern cryptocurrencies vulnerabilities and related security breaches. Then, we propose possible solutions to prevent them, or to decrease the attackers' profit margins by increasing the costs they have to face, in order to strike some of these attacks. Alongside, we briefly describe a few attacks that have occurred in the past.

1 Introduction

In 2009, Satoshi Nakamoto introduced on the market the first cryptocurrency in history, Bitcoin [4]. Even though Bitcoin is still the most popular cryptocurrency, with a trading volume of $7.1 billion a day and 39.8% market share [7], at time of writing more than 1500 different cryptocurrencies have been created. One of the main alternatives is Ethereum, which holds a daily trading volume of $2 billion.

Thanks to its peer-to-peer protocol, Bitcoin is a fully distributed and decentralized currency that achieves trust without relying upon a trusted third party. Blockchain is the fundamental underlying technology, a linked list of information stored in blocks that, combined with a peer-to-peer (P2P) protocol, ensures distribution and decentralization. In the case of cryptocurrencies, currency transactions are the stored information. Apart from data, each block holds a hash of the previous one, as well as a timestamp, in order to enforce the chain validity. Indeed, any attempt to edit one of the blocks would invalidate hashes validity and, consequently, the whole blockchain [21].

It is possible to separate Bitcoin users into two different groups, regular users and miners. We describe hereby the two roles as separate entities, but a user can equally perform both of them. Firstly, regular users manage their bitcoins through a digital wallet, as like as they were using a traditional electronic home

P. Ciancarini et al. (Eds.): SEDA 2018, AISC 925, pp. 288–299, 2020.
https://doi.org/10.1007/978-3-030-14687-0_26

banking. In order to access the data, each user holds a private key, which ensures that no third party can access users' money. As we previously said, no central authority has power over Bitcoin, which means that if the private key is lost (and no backup was done) the bitcoins are lost forever. No third party can re-issue a password, reset an account or operate in any way over others' wallets.

Secondly, mining is the process where blockchain progresses and network generates new units of Bitcoin. Miners compete to solve complex mathematical problems by dedicating their computational power to this purpose. By solving these problems, they confirm the validity of Bitcoin transactions and they are rewarded with a fixed number of bitcoins [22], plus all the fees that users attached to their issued transactions as an incentive to choose their transaction. Whenever the amount of mined blocks reaches half of the number of the remaining minable blocks, the reward decreases by half. In the beginning, the fixed reward was 50 BTC (bitcoins) for each new block created, and the current reward equals 12.5 BTC per block. This halving mechanism will keep going on, until the upper limit of 21 million BTC is reached [18]. At that point, the whole Bitcoin network will rely exclusively upon the transaction fees.

When trying to mine a block, a miner must begin with collecting new transactions into the block. One the block has been created, a nonce is chosen and a pre-defined hash algorithm is applied to the block. The goal is to obtain a hash with a targeted number of zeros, so the miner keeps on increasing the nonce and performing the hashing over the block, until they hopefully find a good hash. In case this happens, the miner broadcasts the solution, which is checked by the other participants. In case the solution is correct and the block correctly formed, the miner receives the reward and the block is appended to the blockchain. Difficulty is adjusted automatically by the system, in order to achieve a ratio of 1 new block per 10 min, on average [22].

Blockchain is believed to be immutable and tamper-proof by design but, even though it is strongly resilient, this claim were proved to be incorrect. In this paper, we aim to explain the vulnerabilities that can be found in cryptocurrencies and propose solutions to prevent them from happening, along with some of the biggest security breaches that have occurred in the past. This paper revolves around Bitcoin, since it is the first cryptocurrency that appeared on the market, and even the most popular one.

2 Related Work

Bucko et al. [21] enlist various issues that can affect cryptocurrency trust. They suggest that cryptocurrencies would become more secure, if businesses that utilize this technology would follow the standards drafted by the CryptoCurrency Certification Consortium.

Latif et al. [30] state that the online environment could cause a security breach in cryptocurrencies. There is no security for the users' savings, since there is no third party to which users can turn to, in case problems with their money arise. They also state that cryptocurrencies depend on the power of speculations, investors can therefore over- or under-value the cryptocurrencies, which affects the market price. They also claim that the public nature of transaction records exposes users to de-anonymizing techniques.

Kaushal [29] talks about how selfish mining and malleability vulnerabilities affect Bitcoin. The author addresses the selfish mining problem and shows how a backward compatible approach was introduced, in order to solve it. Feder et al. [26] measured the impact of Distributed Denial of Service (DDoS) attacks on Bitcoin exchanges. They chose to use kurtosis (which measures whether the data is light- or heavy-tailed, with respect to a regular distribution) and skewness (which measures eventual lack of symmetry) for day-to-day transactions. They found out that when DDoS attacks occurred both kurtosis and skewness decreased. Therefore, the volume of day-to-day transactions changes and this means that less large transactions occur during such attacks.

Vasek et al. [31] studied 142 different DDoS attacks, mentioned on the website bitcointalk.org. The chosen timeline went from February 2011 to October 2013. From 2940 different pages, the team found out that 142 different DDoS attacks were performed on Bitcoin, during the chosen period. Vyas and Lunagaria [32] claim that storage, mining and transaction processes are the main vulnerable phases of Bitcoin. They state that, in order to increase security during the mining process, Bitcoin protocols need to change their framework. To operate this modification, roughly 80% of Bitcoin users need to agree on the change.

Conti et al. [22] thoroughly evaluated Bitcoin, with regards of both security and privacy issues. They give a detailed overview about the protocol, as well all the documented vulnerabilities together with possible countermeasures. Furthermore, the authors summarize the open challenges and suggest the aspects future work should focus on, in order to solve the current issues.

3   Vulnerabilities and Attacks

The heaviest loss in Bitcoin history happened in February 2014, when MT. Gox, a Bitcoin market in Tokyo, lost 850.000 BTC that totalled $474 million at that time. At the time of writing, taken into consideration the current market price, the losses would have totalled more than $7 billion. Later on, MT. Gox stated that the incident was caused by a transaction malleability, a known security issue [5] that we describe later in this section.

Another example is the attack performed on the Decentralized Autonomous Organization (DAO), a venture capital fund that enabled investors to directly fund proposals through smart contracts, based on the Ethereum platform. The attack led to a theft of 3.6 million ether coins, worth $3.1 billion accordingly to the current market price. The attackers exploited a loophole, which enabled them to recursively transfer funds from a parent account to a child, without updating the parent balance [2].

Apart from these two examples, there are many other vulnerabilities in cryptocurrencies. In this section, we describe the most important ones.

3.1   Selfish Mining

There are two types of miners: the honest ones, and the selfish ones. It does not matter if an honest miner works as an individual, in a single group or in a number of different groups. The main goal of selfish miners is to force the honest ones to waste their computational power on the stale public branch. Due to design reasons, it is possible (even though unlikely) that a blockchain splits in two concurrent branches. To solve this paradox, Bitcoin and other blockchain protocols enforce the longest chain as the correct one, thus invalidating all the blocks mined in the shortest parallel chain. The selfish miners leverage this characteristic, they keep private all the blocks and work on their secret branch. As soon as the honest miners are about to catch up, in terms of blockchain length, with the selfish miners, the latter group releases the secret chain. By doing so, the selfish miners nullify all the honest blocks and reclaim all the reward for themselves [25].

Not all cryptocurrencies are prone to selfish mining attacks. As an example, Ripple was created from the beginning with 100 billion XRP (the Ripple currency), which are not minable and are also wasted after the first usage. The company idea is to have possession of half of it and let the other half into circulation [1]. The only way to acquire this cryptocurrency is through exchange with other currencies, therefore Ripple is free from all attacks relating to the mining process.

3.2   Wallet

In order to manage their currency, whether it is to store it or issue transactions, users need a wallet. There are five different types of wallets, divided into three main categories: hardware, software (web, desktop, mobile), and paper wallet. Wallets need to be encrypted and an offline back-up is highly desirable: in case something happens, the facility that keeps the back-up can help its users to restore their wallets.

When it comes to choose a wallet solution, it is good to follow some criteria: the interface should to be user-friendly, backups have to be easy to create, security has to be taken into account, development team should actively support the wallet, and so on. As an example, if users want to be completely safe from attacks while the currency is in storage, they should get a paper wallet, which is also the cheapest wallet to choose [32]. Furthermore, the hardware wallet is a better choice than the software one, since it does not require to be continuously connected to the Internet. It does not matter what kind of cryptocurrency you are going to store in a wallet, wallets are all similar. Therefore, a user should always follow the same criteria when choosing their wallet [3].

Recently, a hacker stole 153.000 Ether (the Ethereum currency), worth almost $131.8 million dollars at the current price, by manipulating a vulnerability in Ethereum wallet. The attacker exploited a bug in Parity Ethereum client to withdraw currency from multi-signature wallets. Multi-signature wallets, also known as multi-sig, are accounts where numerous individuals democratically control the currency flow. A transaction is issued if, and only if, a pre-defined number of users sign it with their own private key [9]. Considering the cryptocurrencies value, alongside the general robustness of blockchains, hackers are focusing more and more on attacking directly users' wallets. As an example, as shown in Sect. 3.12, attackers use social engineering techniques to gain victims' trust and steal their wallets [12].

3.3   Distributed Denial of Service (DDoS) Attack

DDoS attacks aim to make a device (or a network) unreachable to users, for example through packet flooding techniques. These attacks are among the most common ones that affect cryptocurrencies, mainly because they are highly disruptive and relatively cheap to perform. The main targets of DDoS attacks, with respect to cryptocurrencies, are the currency exchange platforms and the mining pools.

As mentioned earlier, researchers [31] enlisted 142 different attacks in their 33 months timeline. Most attacks target mining pools, currency exchanges, electronic wallets, and financial services, but the most popular ones are performed on large mining pools and currency exchanges. These are more popular than others, since the attacker is likely to earn a larger amount of money than attacking small mining pools, or individuals. As an example, a trading and exchange platform called BTC-e experienced reiterated DDoS attack, during 2016 and 2017. These attacks overloaded the whole system, at such a point that it was offline for hours [6].

Moreover, mining pools are often target of DDoS attacks performed by other mining pools, which try to impede competitors from succeeding. Anti-DDoS services popularity varies between categories, but overall are only used by roughly 20% of Bitcoin pools. As expected, these protections are more popular among larger mining pools, which are more likely to experience attacks [31]. Another study [28] explored the trade-off between two different strategies to create bitcoins. These strategies are called construction and destruction. The construction paradigm is when mining pools devote more resources to increase their computing power, thus their chances of mining the next blocks. The destruction paradigm is when a mining pool chooses to invest resources into a DDoS attack, in order decrease the success of competing mining pools.

3.4   Malleability Attack

In malleability attacks, the attacker tries to modify the hash of transactions and pretend he never received the money, in order to deceive his victims and lead them to issue the same transaction a second time. First of all, the victim issues a legitimate transaction to the attacker (e.g., for a purchase). As soon as the attacker detects the transaction in the network, he modifies its signature, which creates a different transaction hash ID. After this, the attacker issues the altered transaction in the network. In this moment, both the legitimate transaction and the forged one are waiting for confirmation. If the forged transaction is confirmed, the legitimate is not acceptable anymore (therefore, it is discarded) and the attacker can try to persuade the victim to issue the transaction again [24]. The malleability attack should not be confused with the double spending attack. In the latter case, the coins are automatically double spent, whereas the former one can lead to a second spending phase if, and only if, the attacker succeeds in tricking the victim.

Malleability attacks can also lead to DoS attacks. If the attacker issues many forged transactions in the transaction pool, the miners have to spend a lot of time in verifying (and discarding) all these false transactions [22]. Bitcoin is prone to such attacks, which can be eliminated by slightly modifying the protocol [20]. Indeed, in 2015 Bitcoin experienced a malleability attack that succeeded in changing the hash of many transactions. The sums and the related recipients were not affected, but it took a long time before all the transactions were finally confirmed [13].

3.5 Double Spending Attack

Double spending happens when a malicious entity spends the same currency more than once. In a typical cryptocurrency like Bitcoin, when a transaction is issued, it goes in the transaction pool waiting to be confirmed as part of a block. As an example, if a merchant does not wait for the final confirmation of the incoming transaction, a malicious customer can try to spend the same coins by issuing another transaction to another merchant. When the confirmation happens, one of the two conflicting transactions is cancelled, since a coin can be spent only one time, but the attacker would have actually got more goods than he paid for.

Merchants are recommended to wait that, at least, six more blocks are mined after the block containing their transaction, before processing the order they were paid for. This time slot is enough to avoid the risk of a cancelled transaction [22].

3.6 Dropping Transactions

Transactions can be dropped when a miner deliberately does not pick up some transactions from the transaction pool. Someone suggested that this might become a problem, since that a miner, aiming to speedup its mining processes, could deliberately choose to drop all the transactions and mine empty blocks. As aforementioned in Sect. 1, in order to increase the chances that a miner picks up a transaction, the issuer attaches a fee. These fees seem profitable enough to avoid deliberate dropping.

Obviously, the higher the fee, the more probable a transaction is chosen. It is clear to see that a transaction has higher chances to be dropped if the attached fee is not profitable enough for the miner. As long as the transaction is not picked up from the transaction pool, it remains unconfirmed until a miner eventually chooses it [17].

3.7 51% Attack

The idea behind a decentralized cryptocurrency such as Bitcoin, is that no entity should have the power to take non-democratic decisions. A PoW protocol is perfect, under the initial assumption that miners work on they own, but mining pool can endanger this system. The more a mining pool grows, the more decisional weight it has over the blockchain it is mining. If, at any point, a pool achieves the 51% of the total computational power working on the blockchain, it can effectively take full control. The pool can allow its members to double spend their coins, prevent competing pools to mine the blockchain, and even impede the confirmation of transactions. Even without achieving the 51% of computational power, an important mining pool can still damage a cryptocurrency network. As an example, a mining pool that controls around 30% of computational power can still perform some of the aforementioned attacks [32].

The Proof of Stake (PoS) protocol makes these attacks more difficult than PoW-based cryptocurrencies. The core point of PoS is the coin age, measured by multiplying the number of coins per holding days. In PoW solutions, it is sufficient for the attacker to gain the 51% of resources, in order to control the network. Instead, with PoS ownership is not enough, as the attacker needs to hold the resources for a considerable amount of time. Furthermore, to avoid the selfish mining incentive in PoS networks, researchers [27] proposed to set an upper limit on coins age, after which the coin age is reset.

3.8 Timejacking Attack

Timestamps are essential to validate blocks in the Bitcoin protocol and, through this multi-step sophisticated attack, an attacker can increase the chances of performing a double spend attack [18]. First of all, an attacker can slow down the median time of its target by sending it wrong timestamps. The time can be skewed up to 70 min, accordingly to the protocol; over this threshold, the time reverts to system time. This can easily be done, since a few Tor nodes that wrongly report the time are enough to succeed in altering the victim time. On the other hand, to perform a timejacking attack the majority of the network has to be speeded up 70 min, which requires far greater resources than simply altering the time for a single node. At this point, if the attacker succeeds, the difference between the victim and the rest of the network is 140 min.

Keeping in mind that Bitcoin nodes are designed to reject blocks which differ more than 120 min from the local time, the attacker aims to create a "poison pill" block, with a timestamp 190 min ahead of the real time. By doing so, the block will be accepted from the speeded up network (since it is exactly within the 120 min threshold) but the victim will reject it, since the block time is 260 min far ahead the node time. At this point, the chains fork and the target is isolated from the normal Bitcoin network operations. Since the time difference between the new blocks and the local time is 140 min, all the legitimate blocks are automatically rejected without checking the contents.

Until the victim eventually catches up with the rest of the network (e.g., after a clock reset, an operation intervention or an unaffected node that pushes a correct timestamped block), the attacker can feed the node fake confirmations and double spend its own coins with the victim. If the victim is a miner, this results in a DoS attack, since it is unknowingly wasting computational power on a stale fork of the chain, while the majority of the network is normally progressing on another chain.

One of the possible solutions to this vulnerability, is to use the system time both when a new block is created and when the timestamps are compared. To further mitigate it, the acceptable time range could be tightened up from 70 min to 30 min leading to a restricted attack window, but this exposes the network to problems in case some nodes do not correctly handle daylight savings. Another suggestion would be to use the median blockchain time when validating blocks, as it already happens for the lower bound of the protocol. Indeed, Bitcoin protocol forces a node to reject any block which timestamp is earlier than the median of the previous 11 blocks. Adopting a similar strategy for the upper bound might resolve the timejacking vulnerability.

3.9 Sybil Attack

In a sybil attack, a malicious attacker fills the network with fake entities, and attempts to take over the regular network activities. Without adequate counter-measures, an attacker could try to confirm fake transactions or disconnect the victim from the network by not forwarding transactions and blocks. A malicious adversary could also forward only fake blocks, in order to attempt to double spend coins on the victim.

One of the main characteristics of cryptocurrencies is anonymity, therefore it is mandatory that anonymous validation algorithms are used (i.e., enforcing to link a miner to a physical person or organization is not an option). The Bitcoin Proof of Work (PoW) algorithm prevents the sybil attack for the mining processes. No matter how many virtual miners an attacker creates, the physical computational resources cannot be virtually multiplied, which entails that creating dozens of fake miners does not increase the chances to mine new blocks. PoW has some severe drawbacks, though, such as inefficiency and expensiveness [33].

To prevent sybil attacks and solve the Know-Your-Customer (KYC) problem, the Proof of Individuality (PoI) algorithm was proposed [14]. Based on Ethereum, the assumption is that a person can attend only one meeting at a time. PoI pairs users in random small groups of 5 people and a 10 min video conference starts for all the groups, at the same time. During this time the users have to actively engage, in order to prove they are attending to only one conference. After the, call the users within the same groups validate (or not) the others and each validated user receives an anonymous unique token. This proposal seems particular useful for voting scenarios. The operations happen at a specific time and, after the vote is cast, the token is destroyed [8].

3.10 Spam Transactions

In some cryptocurrencies, such as Bitcoin and Litecoin, the blocks have an upper size limit of 1MB [23] but it is not a rule for all cryptocurrencies. As an example, Ethereum do not have block size limits but a limit to the number of transactions [16], called gas limit. That being said, the results in both cases is that the number of maximum transactions per block is limited.

This inherent limit exposes the blockchain to DoS attacks, in the form of various flooding attacks. A malicious attacker can send multiple (and economically non-relevant) transactions, aiming to fill up the transactions pool and delay the legitimate transactions. In 2015, a set of unknown actors allegedly performed a flood attack on Bitcoin network [23], in order to convince the community that the 1MB size limit should be raised.

One way to counter flood attacks is to force a monetary commitment on the payer, whenever a transaction is issued. By doing so, the attacker eventually runs out of currency and resources to buy new coins. Bitcoin, Ripple, and many other cryptocurrencies, charge network transactions fees on their users [10, 19], which protects the network from this issue.

3.11 Segmentation

Segmented networks should perfectly work even if the connection between them is poor. If no means of communication exits, the network segments and the blockchain forks. When the connection is restored, accordingly to the Bitcoin protocol, the longest chain is automatically chosen as the correct one and all the transactions in the shortest chain are put back into the transaction pool. As long as a chain does not fork more than 120 blocks, all the transactions are still valid, even though they start again with a 0\unconfirmed status. If the fork is longer, such transactions are definitively invalid and unrecoverable.

A malicious attacker could take advantage of a segmented network to double spend coins on both blockchains, however if the attacker is able to connect to both blockchains, almost certainly even legitimate users can do the same. Even though it is unlikely that someone can leverage this characteristic to double spend coins [15], keeping track and evaluating a sudden drop, in numbers, of legitimate peers might help an honest user to avoid frauds.

3.12 Social Engineering

Social engineering attacks leverage the weaknesses of human psychology, which makes hard to counteract them through software and hardware solutions. This is the reason why social engineering attacks rank high, amongst the types of security issues. The only weapon against social engineering attacks is to educate the users to recognize and avoid attacks attempts. Social engineering techniques take advantage of human greed and curiosity, as well as many other emotions, which entails a wide diversity of possible attacks and related countermeasures.

Table 1. Comparison of cryptocurrencies attacks

Type of attack	Targets	Description	Effects	Solutions
Selfish mining	Mining process	Takes advantage of honest miners	Honest miners waste their resources on a stale chain	Backwards compatible modification [25]
Wallet	Businesses and regular users	Private key is lost or deleted	Loss of every content in the wallet	Have a backup of the wallet
DDoS attack	Miners, users, network and services	Makes devices or networks unreachable to their users	Isolate targets	Proof of activity (PoA)
Malleability Attack	Transaction process	Forge transactions, almost identical to the original copy	DDoS. Users can be tricked into re-issuing transactions	Modify the vulnerable protocols
Double spending attack	Transaction process	Spend the same currency on different transactions	Buying multiple things with the same coins	Confirm a transaction only after 6 valid blocks
Dropping transactions	Transaction process	Miners might never choose some transactions	Transactions are never issued	Attach reasonable fees for the miners
51% attack	Mining and currency exchange process	Mining pools have more than 50% of the power	Double spending and denial of service	Proof of Stake (PoS)
Timejacking attack	Mining and transaction process	Change the blocks timestamps	Miners work on a stale chain, users prone to fake transaction confirms	Use system time, and narrow time ranges
Sybil attack	Mining process	Fill the network with fake miners	DoS, double spending, and selfish mining	Adopt a resource-bound protocol (such as PoW, or PoI)
Spam transactions	Mining and transaction process	Flood the pool with multiple transactions	Delays and backlogs	Charge a fee for every transaction
Segmentation	Mining and transaction process	Segment the network and prevent communications	Double spending	Evaluate abrupt loss of peers
Social engineering	Miners, businesses, and regular users	Steal sensitive information	Money and identity theft	Educate users to identify potential frauds

Attackers can aim to steal login credentials of users' wallets, as well as infect third-party systems and use their resources for mining purposes.

As an example, in early December 2017 the Bitcoin mining marketplace Nice-Hash, accordingly to their head of marketing, experienced a sophisticated social engineering attack to their systems that led to 4700 BTC stolen (worth roughly $49.28 million) [12]. In another case, a malware called RETADUP infected a number of Israeli hospitals, aiming to spread a cryptocurrency mining software [11] and directly monetize from the infections.

4   Conclusion

The interest for cryptocurrencies, and consequently their economic value, has risen above all expectations in a few years. Bitcoin was the first cryptocurrency on the market and is still the most popular one, but many others have been proposed, with different degrees of success. It is clear to see that higher economic value, combined with the anonymous nature of the currency, entails higher interest from criminal parties. Cryptocurrencies are not only fertile soil for buying and selling illegal goods, but also an attractive target for currency thieves. In this paper, as shown in Table 1, we presented a number of different cryptocurrencies vulnerabilities, as well as some solutions and countermeasures. Moreover, we presented actual attacks perpetrated on different entities, and the severe economic impact that such attacks had on their victims.

New and intriguing ideas emerge every day from the cryptocurrency field, but hackers are quick to respond and adapt to new technology. On the one hand, cryptocurrencies foundations lay on the Internet and this makes them naturally prone to different vulnerabilities. On the other hand, blockchain is the promising core technology that could help to solve multiple trust problems in the present Internet architecture. One thing is sure, though: the exciting battle between attackers and defenders is unlikely to finish anytime soon.

References

1. 10 things you need to know about ripple. https://www.coindesk.com/10-things-you-need-to-know-about-ripple/
2. $55 million in digital currency stolen from investment fund. https://www.bankinfosecurity.com/55-million-in-digital-currency-stolen-from-investment-fund-a-9214
3. Best Ethereum wallets 2017: hardware vs software vs paper. https://blockonomi.com/best-ethereum-wallets/
4. Bitcoin: a peer-to-peer electronic cash system. https://coinmarketcap.com/all/views/all/
5. Bitcoin hack highlights cryptocurrency challenges. https://www.bankinfosecurity.com/bitcoin-hack-highlights-cryptocurrency-challenges-a-9305
6. Bitcoin, litecoin exchange platform under DDoS attack, security inadequate. https://cointelegraph.com/news/bitcoin-litecoin-exchange-platform-under-ddos-attack-security-inadequate
7. Cryptocurrency market capitalizations. https://coinmarketcap.com/all/views/all/
8. Ethereum based proof-of-individuality prevents sybil attacks. https://decentralize.today/ethereum-based-proof-of-individuality-prevents-sybil-attacks-9757864bbf61
9. Hacker uses parity wallet vulnerability to steal $30 million worth of Ethereum. https://www.bleepingcomputer.com/news/security/hacker-uses-parity-wallet-vulnerability-to-steal-30-million-worth-of-ethereum/
10. Introducing ripple. https://bitcoinmagazine.com/articles/introducing-ripple/
11. New RETADUP variants hit south America, turn to cryptocurrency mining. https://blog.trendmicro.com/trendlabs-security-intelligence/new-retadup-variants-hit-south-america-turn-cryptocurrency-mining/
12. Nicehash hacked at peak BTC prices, loses $64 million's worth of Bitcoin. https://cryptovest.com/news/nicehash-hacked-at-peak-btc-prices-loses-64-millions-worth-of-bitcoin/
13. The ongoing Bitcoin malleability attack. https://cointelegraph.com/news/the-ongoing-bitcoin-malleability-attack

14. Proof of identity on Ethereum (or the "KYC problem"). https://blog.oraclize.it/proof-of-identity-on-ethereum-or-the-kyc-problem-f4a9ee40af21
15. Re: Anonymity. Bitcoin forum. https://bitcointalk.org/index.php?topic=241.msg2071#msg2071
16. Re: Maximum block size? - Ethereum community forum. https://forum.ethereum.org/discussion/1757/maximum-block-size
17. Re: What is the incentive to collect transactions. Bitcoin forum thread. https://bitcointalk.org/index.php?topic=165.msg1595#msg1595. Accessed 26 Feb 2018
18. Timejacking & Bitcoin. http://culubas.blogspot.dk/
19. Transaction fee historical chart. https://bitinfocharts.com/comparison/transactionfees-btc-eth-bch-ltc-dash-xmr-vtc-aur.html
20. Andrychowicz M, Dziembowski S, Malinowski D, Mazurek Ł (2015) On the malleability of Bitcoin transactions. In: Brenner M, Christin N, Johnson B, Rohloff K (eds) Financial cryptographyand data security. Springer, Heidelberg, pp 1–18
21. Bucko J, Palova D, Vejacka M (2015) Security and trust in cryptocurrencies
22. Conti M, Kumar ES, Lal C, Ruj, S (June 2017) A survey on security and privacy issues of Bitcoin. ArXiv e-prints
23. McGinn D, Birch D, Akroyd D, Molina-Solana M, Guo Y, Knottenbelt WJ (2016) Visualizing dynamic Bitcoin transaction patterns. Big Data 4(2):109–119. https://doi.org/10.1089/big.2015.0056, pMID: 27441715
24. Decker C, Wattenhofer R (2014) Bitcoin transaction malleability and MtGox. In: Kutyłowski M, Vaidya J (eds) Computer security - ESORICS 2014. Springer, Cham, pp 313–326
25. Eyal I, Sirer EG (2014) Majority is not enough: Bitcoin mining is vulnerable. In: Christin N, Safavi-Naini R (eds) Financial cryptography and data security. Springer, Heidelberg, pp 436–454
26. Feder A, Gandal N, Hamrick JT, Moore T (2017) The impact of DDoS and other security shocks on Bitcoin currency exchanges: evidence from mt. Gox. J Cybersecur 3(2):137–144. https://doi.org/10.1093/cybsec/tyx012
27. Gao Y, Nobuhara H (2017) A proof of stake sharding protocol for scalable blockchains. In: Proceedings of the 14th APAN research workshop 2017
28. Johnson B, Laszka A, Grossklags J, Vasek M, Moore T (2014) Game-theoretic analysis of DDoS attacks against Bitcoin mining pools. In: Böhme R, Brenner M, Moore T, Smith M (eds) Financial cryptography and data security. Springer, Heidelberg, pp 72–86
29. Kaushal R (2016) Bitcoin: vulnerabilities and attacks. Imperial J Interdiscip Res 2(7) (2016). http://www.imperialjournals.com/index.php/IJIR/article/view/1238
30. Latif S, Mohd M, Mohd Amin M, Mohamad A (2017) Testing the weak form of efficient market in cryptocurrency. J Eng Appl Sci 12(9):2285–2288
31. Vasek M, Thornton M, Moore T (2014) Empirical analysis of denial-of-service attacks in the Bitcoin ecosystem. In: Böhme R, Brenner M, Moore T, Smith M (eds) Financial cryptographyand data security. Springer, Heidelberg, pp 57–71
32. Vyas CA, Lunagaria M (May 2014) Article: Security concerns and issues for Bitcoin. In: IJCA proceedings on national conference cum workshop on bioinformatics and computational biology NCWBCB, no 2, pp 10–12. Full text available
33. Xu X, Pautasso C, Zhu L, Gramoli V, Ponomarev A, Tran AB, Chen S (April 2016) The blockchain as a software connector. In: 2016 13th working IEEE/IFIP conference on software architecture (WICSA), pp 182–191. https://doi.org/10.1109/WICSA.2016.21

Question-Answer Nets Objectifying a Conceptual Space in the Design of a System with Software

Petr Sosnin^(✉)

Ulyanovsk State Technical University, Ulyanovsk 432027, Russia
sosnin@ulstu.ru

Abstract. Becoming of any system with software should be implemented especially carefully in a conceptual space (CS) where designers create the conceptual states of the system in parallel with its surrounding. In such work, they must build as useful conceptual models so regularities of the project space and its language, the controlled use of which is strictly obligatory. Thus, at the early stage of designing the system, the effectiveness and quality of work essentially depend on the form in which designers objectify the corresponding CS. This paper deals with the use of question-answer nets for objectifying the CS. In such conditions, providing the consistency between current states of the project and CS facilitate increasing the success of designing.

Keywords: Conceptual space · Designing · Question-answering · Software · Theory

1 Introduction

The term "conceptual spaces" has numerous definitions but in any case, it indicates on a set or a system of abstract objects bound by certain relations. Among such definitions, we mark the definition suggested by Gärdenfors, who tried to build the geometry of thought [1]. By his definition "A conceptual space is a geometric structure that represents some quality dimensions, which denote basic features by which concepts and objects can be compared."

The indicated function is not sufficient for the design. In our deep belief, CS oriented on the design must exist in two coordinated forms one of which has mental nature while the second is an artifact that supports the combinatorial, investigational and transformational activity of designers. After their use in the design processes, similar artifacts are better to include into the designed systems as components that also extend the interactive possibilities for the users of the systems.

Building and using the artifacts of the CS kind can lead to a positive impact on the increase of success in designing the software systems, for which more than 20 years, statistics register that the success rate of designing does not exceed 40%. Among basic reasons for such state of affairs, published reports indicate the negative manifestations of the human factor in the conceptual activity of designers [2]. The root reason for negatives is a lack of naturalness in Human-Computer Interactions (HCI) when a

© Springer Nature Switzerland AG 2020
P. Ciancarini et al. (Eds.): SEDA 2018, AISC 925, pp. 300–311, 2020.
https://doi.org/10.1007/978-3-030-14687-0_27

human (designer in the analyzed case) interacts with a computerized environment of designing.

The natural interactions of any human with physical surrounding have an empirical character, and they materialize in conditions of active use of senses. Any human has the richest experience of such behavior that is activated at the habit level.

HCI has an artificial nature materialized with the help of programmed interfaces that are oriented on very restricted use of senses in conditions of nonhabitual perception of the environment by the human. The controlled creation and use of the CS can increase the naturalness of HCI and correspondingly to reduce the negative manifestations of human factors, first of all, such as semantic errors and failures caused by misunderstanding.

In this paper, we present an approach for reifying the CS on the base of question-answer nets (QA-nets), different kinds of which were mastered in our research and practice of designing the systems with software (or shortly systems).

This kind of nets registers interactions of designers with applied experience that finds its expression in the CS representation, in which the regularities of the CS and the language of the project have the basics checked by experience. In such conditions, providing the consistency between current states of the project and CS facilitate increasing the success of designing.

2 Preliminary Bases

The last 15 years, our research and development have focused on the use of question-answering in the conceptual design of software intensive systems (SISs). The basis of studies was the reflection of operational design space onto the semantic memory (QA-memory) of the question-answer type that is shown in Fig. 1.

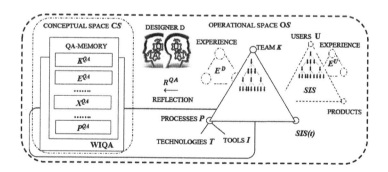

Fig. 1. Reflection of operational space on QA-memory

In the structure and content of the operational space (OS), we oriented on developing the systems with the use of methods and means that have similarity with components of Rational Unified Process [3] used at the conceptual stage of designing the SISs. Developed methods and means were integrated into the specialized toolkit WIQA (**W**orking **I**n **Q**uestions and **A**nswers) [4].

On the scheme, the reflection (R^{QA}) shows all that is involved in designing of the system S is found their expression as models in the QA-memory of the toolkit WIQA. Such reflection can be described by the following expression:

$$OS(P, K, E^D, E^U, S_i(t), \ldots, X, t) \xrightarrow{R^{QA}} CS(K^{QA}(t), E^{QA}(t), P^{QA}, (t), S_i^{QA}(t), \ldots, X^{QA}(t)),$$

where X indicates on any essence that is absent in the scheme, and other symbolic designations correspond to the names of essences in Fig. 1. Let us additionally note that results of the reflection R^{QA} are dynamic objects that complete the conceptual space in its current state CS(t).

This process of the reflection is based on real-time interactions of designers with accessible experience when they use workflows "Interactions with Experience" embedded into the toolkit WIQA. Such works leave their traces on current states of experience and models of its items.

Thus in the described case, the main feature of the CS(T) is its placement in the QA-memory. In the most general case, the memory cell must provide an information store that conceptually specifies the task to be solved in the project. Therefore, in specifying the cell, this requirement has led us to a set of basic attributes, one of which is intended for registering a textual part of a description of the task (its statement), while the necessary tables and graphics components of the task description are attached to the cell as files. Here we note, except the basic attributes designers can define additional attributes and references if it is useful for the object uploaded to the cell or their group.

In any case, cells of QA-memory are intended to upload the objects of the specialized types (QA-objects). Such objects can be simple if each of them is uploaded in one cell, for example, representations of "questions" of different types (Q-nodes), including questions like "task" (Z-nodes), and "answers" (A-nodes) to "questions." When an interactive object located in a set of the bound cells, it is a complicated QA-object. With each QA-object is associated not only its representation in the semantic memory but also the system of operations, including operations for visualizing the objects.

Cells of QA-memory can be used for uploading units of data that have appropriate structure content interpreted in the question-answer sense, for example, "name of objects – the question" and "some of its attributes – answer." Such simulating helped to develop the plug-ins "Organizational structure" that is intended for reflecting the features of the designers' team onto QA-memory in the form of the corresponding QA-net. The main function of this plug-ins is to provide the controlled distribution of the tasks among members of the team in designing a certain system.

The similar way was used for sets of designer practices, behavioral operations of which were simulated with using the following interpretations: "name of behavioral operation - the question" and "execution of this operation by the designer - answer." Such simulating has led us to behavioral programs that can be included in the work of designers in the WIQA environment. Such work is shown in Fig. 2 where one can see some QA-nets.

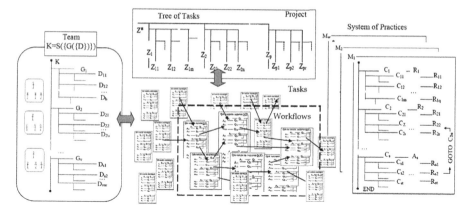

Fig. 2. QA-nets of the design process

The scheme shows:

1. QA-tree of the organizational structure (team K), which relates organizational units (the symbolic designation of groups G) onto any depth of their relations with the members of the collective (symbolic designation D). Due to additional attributes, the net nodes can obtain extended semantics, as well as relations, which makes it possible to present an organizational structure of any type.

2. QA-nets simulating workflows:

   • Technologies used in the process of conceptual design of the system being developed;
   • Business processes that will be implemented in the system after its commissioning.

3. Question-answer forms for encoding systems of methods S(WIQA, $\{M_n\}$) used in the technological workflows "Interaction with Experience," serving the application of the toolkit WIQA.

It should be noted, that in the encoding of techniques, QA-objects were used to represent commands $\{C_q\}$, which must be executed by designers (reactions $\{R_q\}$), who visually interact with steps of the method. In this interaction, the designers acted as a "processor," which also responds to the transition commands (GOTO C_{iu}).

At the end of the first step of our study, except QA-nets showed in Fig. 2, such artifacts were used for simulating project documents and for coding the tutorial practices and tests in automated learning.

At the second step of our research, we have developed an executable pseudo code language, traditional constructs of which are coded in the semantic memory as nodes of QA-nets or their fragments. Pseudo code possibilities helped us to muster ways of conceptual solutions of project tasks and their models prepared for the reuse as precedents.

The scheme in Fig. 3 demonstrates iterative building a set of specialized components of the precedent model in parallel with solving the corresponding task with the use of precedent-oriented approach [5]. In such a work, the designer interacts with necessary QA-nets. Moreover, any built component is also placed in QA-memory.

Stepwise Refinement in Solving the Task

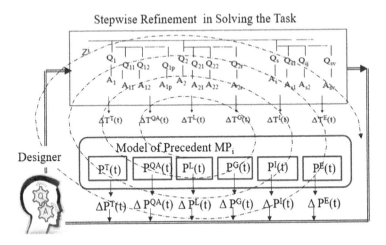

Fig. 3. Iterative creating the model of precedent

At the level of components, the precedent model includes: textual description $P^T(t)$ in the form of the statement of the corresponding task; QA-model of this task; the logical expression $P^L(t)$ of the regularity realized in the model; graphical presentation $P^G(t)$ oriented on understanding; the executable prototype $P^I(t)$ in the form of pseudo code program; and the program description $P^E(t)$ of the precedent in suitable language of programming.

In all enumerated applications of QA-nets, their use for a constructive specification of the CS was absent, and this essence was used metaphorically as a naturally artificial essence accompanied the human-computer activity of designers. However, such use of the CS(t) did not prevent to study and develop a number of findings in the experience-based human-computer interactions (conceptual experimenting, figuratively semantic support of automated imagination), conceptual solving the project tasks (question-answer and precedent-oriented approaches) and conceptually-algorithmic programming (automated interactions with experience, conceptual prototyping) [4].

3   Related Works

Among the works related to the research described in this paper, we distinguish the three areas of their interests. The first area is bound with versions of usage the phrase "conceptual space" in scientific publications. In this area, it should be noted the paper [6], where, in its beginning, the conceptual space is opposed to the perceptual space. After that, the author describes psychological space (space-as-experienced) from the viewpoint of "lived space." In this area, we also mark the paper [7] underlining the role of the CS in creative activity "someone who seeks to understand what creativity is, and how it is even possible, needs to consider the mental geography of conceptual spaces."

Studies of the second area are focused on "Conceptual space as a Geometry of Thought" suggested by Gärdenfors [1]. In this direction, spaces can be formalized with the use of metrics and useful ordering of their objects. Any object of the CS is presented as the certain domain with characteristics of object quality. These features are used for logical views on the CS [8], its theoretical descriptions [9] and thorough formalization [10].

The next area of interests is aimed at using the spaces in the design process. Among studies in this area, we mark the paper [11] where the design is understood and formalized with the use of three worlds (spaces) – the external world, the interpreted world, and the expected world, in which designers apply Function-Behavior-Structure framework (FBS-framework, FBS-model). In the paper [12], such a suggestion is applied for building the ontology of design. The useful modification of FBS-model that takes into account the important features of the design as a specific kind of human activity is disclosed in the paper [13].

All papers indicated in this section were used as sources of requirements in developing the set of instrumental means provided the creation and use of the proposed ontology.

4   Constructive Views on the Conceptual Space

4.1   Architectural Approach to Objectifying the Conceptual Space

As told in Sect. 2, any CS(t) is a naturally artificial essence be created for the certain project in reflections of the operational space of designing the corresponding SYS onto the semantic memory of the question-answer type. In any state, the CS(t) combines two sub-spaces, one of which is an artificial kernel that integrates QA-nets be generated at the time t.

It needs to remind that nodes of QA-nets keep the textual traces of interactions with the experience involved and applied in processes of solving the project tasks. When designers interact with textual traces, they activate processes in certain components of mental spaces located in their brain structures. In our version of understanding the CS (t), these mental processes come to life in the natural part of the CS(t). Activated components of the mental space we understand as an extension of the artificial part of the CS(t).

Described understanding has led us to an architectural approach of objectifying the CS(t). The offered approach is based on the standard ISO/IEC/IEEE 42010:2011 (Systems and software engineering—Architecture description) that recommends to define and construct the architecture of the SIS in the context of its environment. The standard orients the designers on the use of practices helped to build a system of diagrammatic views reflected concerns of persons (stakeholders) who are explicitly or implicitly involved in the project.

Thus, a system concern could pertain to any influence on a system in its environment including developmental, technological, business, operational, organizational, political, regulatory, or social influences. Consequently, the modeled concerns not only reflect the environment but also must be correspondingly materialized in the SIS to be designed. Concerns bind the SIS with its environment and continuously cross both of them on the course of the design process.

Such role of concerns leads to an interpretation of designing the SIS as its extraction from the CS(t) of the corresponding project. The essence of this interpretation is shown in Fig. 6 where the current state of the SIS(t) is understood as a systematically chosen part of the CS(t). In this sense, the CS(t) can play the role of a source of alternative versions of the SIS (Fig. 4).

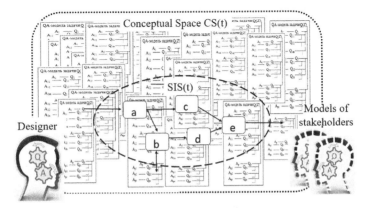

Fig. 4. Extraction of the SIS(t) from the CS(t)

The scheme underlines that designers are active in the environment of the system to be designed and they must model the stakeholders whose concerns must be considered in situational contexts. That is why it is useful to create the architectural views on the environment of designing and the CS(t) combining the coordinated models of the environment and the SIS in its current state. Thus, at the conceptual stage of designing, the CS(t) consists of objects that are conceptual models some of which are included in the SIS(t) while others present the environment.

It needs to note, in the general case, becoming of the CS(t) can start with the "empty" state that will accumulate the models to be created on the course of designing. Among these models will be a set of architectural models that can be expressed by the corresponding views as on the SIS(t) so on its environment. The composition of these views is a very informative representation of the CS(t), especially for its architectural description.

4.2   Basic Components of the Conceptual Space

In the described version of the CS(t), it is distinguished a number of levels of the structuring. The lowest level corresponds to the traces of interactions of designers (and other stakeholders) with natural experience. It is the level of expressing the questions and answers at the language of the project to be developed. Roles of similar traces were described above. Each unit of these kinds is uploaded in the corresponding cell or cells of the semantic memory of the question-answer type.

As told above, the processes of conceptual solving the project tasks are accompanied by creating and using the multifarious QA-nets some of which are accumulated for the future reuse in the specialized areas of QA-memory. The others are formed in the processes of reflecting the useful essences of the operational space on the CS(t). Such reflection must be based on a thorough analysis of the environment, typical steps of which are described in [14] where authors suggest a set of architectural views on the environment. Constructive considering the models of components chosen in the environment leads to the structure of the CS(t) shown in Fig. 5.

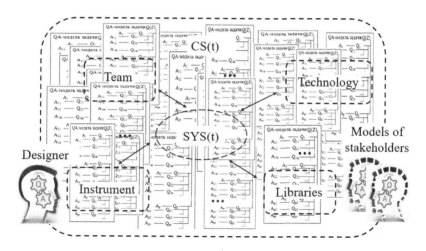

Fig. 5. Models of essences as components of CS(t)

The scheme demonstrates that a number of essences embedded in the conceptual environment are useful to interpret as systems with their borders in the CS(t). It needs to note that we include in the structure of the CS(t) only some models of essences indicated in Fig. 1.

The space CS(t) is a dynamic entity that is formed and applied on the course of designing. A typical reason for evolving the CS(t) is conceptual solving the next project task. In the described case for this, designers should use means of automated design thinking and precedent-oriented approach, which are embedded to the toolkit WIQA [4]. Applications of these means lead to generating the useful models that step by step fill the CS(t) by information in conditions that are shown in Fig. 6.

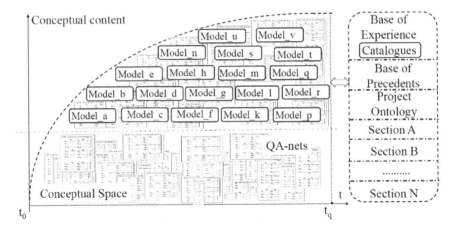

Fig. 6. Generating the results of conceptual solving the tasks

Marked conditions indicate that becoming of the CS(t) is accompanied by inter-actions of designers with the accessible experience, reusable models of which are accumulated in the specialized Base of Experience which as the subsystem is embedded to the toolkit WIQA. The informational content of this Base integrates the models of experience applied in related projects of SISs.

Additionally, in the toolkit WIQA, designers can use the executable pseudo-code language that is defined above QA-memory. Such possibility helps the designers to build and process the different objects placed (as corresponding QA-nets) in cells of QA-memory. These objects with relations among them compose the programmatic level of the CS(t). This level helps to automate the interactions of designers with the CS (t) and its components and system essences including the SIS(t) in its states in the conceptual stage of designing.

5   Theorizing the Conceptual Space

Conditions shown in Fig. 6 have led us to the following questions that defined a very important direction of our study:

- Which set of rules will facilitate the rational and consistent formation of the con-ceptual content indicated in the scheme?
- How can such set be implemented in the WIQA environment?

Answers these questions, we bound with using the scientific approaches based on experiments and theorizing in realizing the project of the SIS at the conceptual stage. The main role was laid on substantially evolutionary theories of the projects, each of which should be built as an applied Grounded Theory [15] for the project of the corresponding system.

Thus, in the course of mastering the processes of formation, transformation, integration, and use of QA-nets, questions of checking their correctness and systematization were repeatedly raised. All this led to a complex of methods and means of theorizing, allowing creating the substantially evolutionary theories of the project (below SE-theories), the typical structure and content of which is shown in Fig. 5 in the generalized form [16].

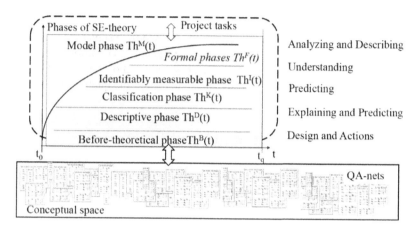

Fig. 7. Phase structure of the theory of the project

The scheme discloses that any applied SE-theory has a phase structure filling by the informational content in parallel with other activities of the design process. In this structure, we clarify only the following phases:

1. Before-theoretical phase $Th^{BT}(t)$ that accumulates facts of interactions of the designers with the accessible experiences. The informational content of this phase is identical to the content of the lowest level of structuring the CS(t).
2. Descriptive phase $Th^{D}(t)$, registered compositions of facts are used for formulating the theory constructs, for example, such as axioms, principles, hypothesis, thesis, argument, premise, theorems, causes, effects and regularities.
3. In building the theory, special attention focuses on vocabulary and especially on developing its part that presents a system of concepts. In Fig. 7, this part of the theory marked as the classification phase $Th^{K}(t)$. In the described version of SE-theory, this phase is built as a project ontology.
4. It should be noted; any theory is created for its uses, a very important kind of which is models M. Models are intermediary between theories and reality when people interact with them. Therefore, models form a useful area of theory applications that indicated in Fig. 5 as a model phase $Th^{M}(t)$.

If it is necessary or useful, designers can apply the current state of SE-theory for analyzing, describing, understanding and explaining of situations or for predicting the results of possible actions or for creating the useful models for implementing the next steps of the design process. Any application is constructive because it is accompanied by creating the necessary objects uploaded in the QA-memory.

In SE-theories, tasks are reasons of evolution. Any applied SE-theory begins its life cycle from the initial statement $St(t_0)$ of the project task $Z^*(t_0)$, QA-analysis of which leads to subordinated tasks combined in the tree of tasks. In our way of creating the SE-theories, it is used design thinking approach for the work with any new project tasks. In evolving applied SE-theory, any implementation of such approach fulfills a role of a "soft" rule of inference. Any such rule of inference corresponds to the certain cause-and-effect regularity in the conceptual space of the corresponding project.

Thus, creating the SE-theory is a very important line of designer activity. This line organically complements conceptual actions of designers in conditions when (in collaborative work) they build and apply the described version of the CS(t). Coordinated applying the SE-theory and CS(t) can facilitate enhancing the quality of designing the SISs.

6   Conclusion

In this paper, we described the constructive version of objectifying the conceptual space oriented on designing the systems with software. The essential feature of the suggested version is the use of QA-nets simulating the fundamental essences involved in the design process.

Our work with QA-nets in conceptual designing the systems, including the development of the toolkit WIQA and its extensions, confirmed that such kind of artifacts helps to build the CS that adequately reflects the operational space of designing.

The quality of such reflection was improved, following for evolving the potential of WIQA. When its current state is applied for conceptual designing the particular system, this toolkit provides the following possibilities:

1. To build conceptual solutions of the project task in the frame of the CS and parallel with its becoming;
2. To create such essential components of the CS as the Base of Experience including the Base of Precedents, corresponding the SE-theory including the Project Ontology;
3. To apply the current state of the CS for preventing semantic errors, their discovering and correcting, first of all, in controlled achieving the necessary degree of understanding and registering in the reusable form.

This space is opened for the combinatorial, investigational and transformational activity of designers, who creatively solve the necessary tasks. They can fulfill such work in the mode of behavioral programming.

Acknowledgements. The Russian Fund for Basic Research supported this work (RFBR), Grant # 18- 07-00989a, 18-47-730016 p_a, 18-47-732012 p_мк, and the State Contract №2.1534.2017/4.6.

References

1. Gärdenfors P (2000) Conceptual spaces – the geometry of thought. The MIT Press, Cambridge
2. Reports of standish group. www.standishgroup.com/outline. Accessed Oct 2018
3. IBM rational unified process. http://www-01.ibm.com/software/rational/rup/. Accessed Sept 2018
4. Sosnin P (2018) Experience-based human-computer interactions: emerging research and opportunities. IGI-Global, Pennsylvania
5. Sosnin P (2016) Precedent-oriented approach to conceptually experimental activity in designing the software intensive systems. Int J Ambient Comput Intell (IJACI) 7(1):69–93
6. Welwood J (1977) On psychological space. J Transpers Psychol 9(2):97–118
7. Boden MA (2009) Conceptual spaces. In: Meusburger P (ed) Milieus of creativity knowledge and space, vol 2. Springer, Heidelberg, pp 235–243
8. Gärdenfors P (2011) Semantics based on conceptual spaces. In: Banerjee M, Seth A (eds) Logic and its applications, vol 6521. LNCS. Springer, Heidelberg, pp 1–11
9. Rickard JT, Aisbett J, Gibbon G (2007) Reformulation of the theory of conceptual spaces. Inf Sci 177(21):4539–4565
10. Bechberger L, Kühnberger K-U (2017) A thorough formalization of conceptual spaces. In: Advances in artificial intelligence. LNCS, vol 10505, pp 58–71
11. Gero JS, Kannengiesser NN (2004) The situated function-behavior-structure framework. Des Stud 25(4):373–391
12. Gero JS, Kannengiesser U (2013) The function-behaviour-structure ontology of design. In: Chakrabarti A, Blessing L (eds) An anthology of theories and models of design. Springer, London, pp 263–283
13. Al-Fedaghi S (2016) Function-behavior-structure model of design: an alternative approach. Int J Adv Comput Sci Appl 7(7):133–139
14. Bedjeti A, Lago P, Lewis G, De Boer RD, Hilliard R (2017) Modeling context with an architecture viewpoint. In: Proceedings of IEEE international conference on software architecture, pp 117–120
15. Charmaz K (2006) Constructing grounded theory: a practical guide through qualitative analysis. Sage, London
16. Sosnin P (2017) A way for creating and using a theory of a project in designing of a software intensive system. In: Proceedings of the 17th international conference on computational science and its applications, pp 3–6

On the Time Performance of Automated Fixes

Jooyong Yi[✉]

Innopolis University, Innopolis, Russia
jooyongyi@acm.org

Abstract. Automated program repair has made major strides showing its exciting potential, but all efforts to turn the techniques into practical tools usable by software developers hit a crucial blocking factor: the timing issue. Today's techniques are slow. Too slow by an order of magnitude at least. The long response time implies that the currently available techniques cannot suit the actual needs of developers in the field. What developers want is a tool that can instantaneously propose a fix for a detected failure. A technique that can propose a patch instantaneously (or near-instantaneously) would provide the breakthrough that is required to turn automated program repair from an attractive research topic into a practical software engineering tool. Indeed, researchers have started to tackle this speed issue. In this paper, we survey recent approaches that were shown to be effective in speeding up automated program repair. In particular, we view automated program repair as a search problem—the ultimate goal of automated program repair is the search for a patch which is often preceded by other related searches such as a search for suspicious program locations, and a search for the specification for a patch. We describe how the problem of automated program repair has been decomposed into a series of search problems, and explain how these individual search problems have been solved. We expect that our paper would provide insight into how to speed up automated program repair by further optimizing the search for a patch.

1 Introduction

"To err is human", and software developers are not an exception. With the recent emergence of automated program repair (a.k.a., automated fixes), developers may finally receive necessary assistance from artificial intelligence. For example, Angelix—a technique we proposed earlier [16]—can automatically fix Heartbleed.

Automated program repair has made major strides showing its exciting potential, but all efforts to turn the techniques into practical tools usable by software developers hit a crucial blocking factor: the timing issue. Today's techniques are slow. Too slow by an order of magnitude at least. For example, it was reported that fixing the bugs of the PHP interpreter in the ManyBugs benchmark takes on average 1.84 h in GenProg [10], 2.22 h in SPR [11], and 1.03 h in Angelix [16], when experimented in commodity machines. The point here is not

© Springer Nature Switzerland AG 2020
P. Ciancarini et al. (Eds.): SEDA 2018, AISC 925, pp. 312–324, 2020.
https://doi.org/10.1007/978-3-030-14687-0_28

time differences in these tools, but the long response time commonly observed across these tools.

This long response time implies that the currently available techniques cannot suit the actual needs of developers in the field. What developers want is a tool that can *instantaneously* propose a fix for a detected failure. Such needs of developers were reported by Harman and O'Hearn [3], based on their recent experiences in deploying their tools to developers in Facebook. For instance, the bug fix rate for the faults reported by Infer[1] "rocketed" from close to 0% to over 70% when the deployment mode of Infer was switched from the batch mode to a more instantaneous mode whereby faults are reported when the developer submits code for review. Similarly, a technique that can propose a patch instantaneously (or near-instantaneously) would provide the breakthrough that is required to turn automated program repair from an attractive research topic into a practical software engineering tool.

Indeed, researchers have started to tackle this speed issue, and it is our proposition that this effort should be continued. In this paper, we survey recent approaches that were shown to be effective in speeding up automated program repair. In particular, we view automated program repair as a search problem— the ultimate goal of automated program repair is the search for a patch which is often preceded by other related searches such as a search for suspicious program locations, and a search for the specification for a patch. We describe how the problem of automated program repair has been decomposed into a series of search problems, and explain how these individual search problems have been solved. We expect that our paper would provide insight into how to speed up automated program repair by further optimizing the search for a patch.

2   Background: Automated Program Repair

The goal of automated program repair is to automatically change a buggy program into a correct one. This can be viewed as an extension of program verification, as shown in Fig. 1. As compared to the traditional program verification, automated program repair proactively fixes a buggy program that fails to be verified. After repair, verification should succeed.

Program verification typically requires formal specification. However, due to the lack of formal specification in industries, automated program repair, from early on, has used a test-suite instead of formal specification [24]. Accordingly, most of automated program repair techniques that are currently developed are test-driven, not verification-driven. However, it is also possible to use a verification technique as shown in DirectFix [15] where a test-suite is used as a weak specification of verification. There is also an approach such as AutoFix [18] that actively makes use of program contracts.

While the choice of using a test-suite can make automated program repair easily adopted in industries, the fact that a test suite is only an incomplete

---

[1] http://fbinfer.com/.

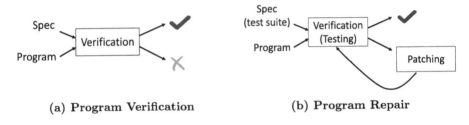

(a) **Program Verification**   (b) **Program Repair**

Fig. 1. Program repair can be viewed as an extension of program verification.

specification of the program inevitably poses a new challenge. In particular, a generated patch may merely pass available tests, without actually fixing the bug—which is often called a *plausible* patch, distinguishing it from a *correct* patch [12]. This phenomenon is called the overfitting problem of program repair, in the sense that a generated patch overfits to the available test suite [21]. A number of papers reported empirical evidences of the overfitting problem [9,12, 20,21,27].

Since overfitting occurs because the space of plausible patches is larger than that of correct patches, overfitting can be mitigated by tightening up the plausible patch space. One obvious way to tighten up the plausible patch space is to augment the test-suite—an overfitting patch p can be prevented from being generated with a test t that fails over p. However, it is not clear how to identify such t. Also, even if t is identified, it is not clear how to identify the expected output of t without human intervention.

The latter problem (the test oracle problem) can be worked around in some cases based on the similarity between execution paths [25]. Suppose that an existing test t_1 and a new test t_2 generate two similar execution paths π_1 and π_2, respectively. Since π_1 and π_2 are similar to each other, whether the original buggy program p passes a new test t_2 is likely to depend on whether p passes the existing test t_1. That is, if p passes (or fails) the existing test t_1, p is likely to pass (of fails) t_2 as well. Such educated guesses can be used to filter out patches that are likely to be incorrect. However, this approach only works for a new test which shows similar behavior to an existing test.

Given the difficulty of augmenting a test suite for program repair, various other approaches have been taken to overcome the overfitting problem. One widely-adopted approach is the minimality criterion [15]. That is, a patch that is as close to the original program (thus, requiring minimal changes) is more likely to be correct than those that further deviate from the original program. For example, DirectFix [15] generates a patch that preserves the structure of the original program as much as possible, utilizing a MaxSMT solver. As a result, a generated patch is syntactically close to the original program. While DirectFix applies the minimality criterion to the syntax of patches, Qlose [2] additionally minimizes semantic differences between the original buggy program and a patched program. Other approaches suggested to address the overfitting problem include anti-patterns (syntactic patterns that are commonly observed in overfitting patches) [22], and history-based patterns (syntactic patterns that often appear in the history of bug fixes) [8].

In the remaining part of the paper, we focus on the focal point of this paper: the time performance of automated program repair. We provide a survey on various techniques to improve time performance.

3   Program Repair as a Search Problem

We view automated program repair as a search problem—the ultimate goal of automated program repair is the search for a patch which is often preceded by other related searches such as a search for suspicious program locations, and a search for the specification for a patch. The reason we take this viewpoint is that formulating automated program repair as a search problem helps us discuss the efficiency of search, which makes direct influences on the time performance of automated program repair.

Broadly, automated program repair approaches are classified into two types: (1) generate-and-validate approach (a.k.a., search-based approach) and (2) symbolic approach (a.k.a., semantics-based approach or correct-by-construction approach). The generate-and-validate approach first generates a patch candidate and then validates whether the generated patch candidate passes all available tests. This process is repeated until either a patch is found or time budget is exhausted. Some tools, in particular early generation tools, most notably Gen-Prog [24], use search-based techniques such as genetic programming to generate patch candidates—hence, the generate-and-validate approach is also called the search-based repair. In addition to GenProg, many other tools (such as RSRe-pair [19], AE [23], SPR [11], PAR [6], SketchFix [4] to just name a few) take the generate-and-validate approach.

Meanwhile, the symbolic approach does not explicitly generate patch candidates. Unlike in the generate-and-validate approach, the symbolic approach extracts a repair constraint from the buggy program and the specification (typically, a test-suite as discussed in Sect. 2, though some symbolic approaches in principle can also be used with a formal specification, which is generally not possible in the generate-and-validate approach). This extracted repair constraint specifies the condition under which a modified program can pass all available tests. Thus, a patch can be synthesized by using the extracted repair constraint as the specification of synthesis. Tools implementing the symbolic approach include SemFix [17], DirectFix [15], Angelix [16], Nopol [26], and S3 [7].

Regardless of the types of the repair approaches, the time performance of automated program repair is governed by the following two factors: (1) the size of a search space, and (2) the velocity of search. We here use an empirical concept of velocity instead of a theoretical concept of complexity. Note that velocity is affected by various factors that include not only the complexity of a search algorithm, but also other various factors such as the overhead of a search procedure (e.g., compilation and input/output processing), computing power, and the degree of parallelization.

Recall that we view automated program repair as a series of search problems. In this paper, we explain how the aforementioned two performance influencing factors (i.e., the size of a search space and the velocity of search) change across different repair techniques. In particular, we provide an insight into how the performance influencing factors can be controlled by adequately addressing the following two closely related questions: (1) Which decomposition model to use to optimize the search for patches? (In other words, how to decompose the search problem for patches into several individual search problems?) (2) For each of an individual search, which search technique to use to optimally perform that search? Note that the answer to one question may affect the answer to the other question. In Sects. 3.1 and 3.2, we address these two questions, respectively, by comparing three repair approaches using the symbolic approach. Then, in Sect. 3.3, we also briefly discuss the same two questions with the generate-and-validate approach.

3.1 Search Problem Decomposition

As mentioned, automated program repair can be decomposed into a series of search problems, and how decomposition is done can affect the overall performance of repair. To discuss this point, we compare three closely related repair techniques, namely, Angelix [16], DirectFix [15], and SE-ESOC [14], all of which use the symbolic approach. The similarity between these three techniques help us address our two key questions mentioned earlier, by focusing on the differences between these techniques.

Angelix. We start with the description of Angelix. Angelix decomposes program repair into the following three search problems.

S1. A search for suspicious expressions (potentially buggy expressions),
S2. A search for the specification of the identified suspicious expressions, and
S3. A search for patch expressions that satisfies the extracted specification.

Figure 2(a) shows more concretely how these three searches take place in Angelix. Given a buggy program and a test suite that contains at least one failing test, Angelix first looks for suspicious expressions. Modifying these suspicious expressions properly may fix the bug. Angelix performs this first search (i.e., S1), using the standard statistical fault localization. In Fig. 2(a), expression $x > y$ is considered suspicious, and replaced with a placeholder expression E. Angelix also allows to modify multiple expressions by injecting multiple placeholder expressions.

At the second step, Angelix searches for the specification of the placeholder expression E. If E satisfies the identified specification, the program passes all tests. Given that the specification of the program is given in the form of a testsuite, the specification of E is essentially a table, as shown in Fig. 2(a). The table in the example shows the value that should be returned at E to pass the test (shown in the E column), for each case E is evaluated while executing the tests. The values of variables x and y when E is evaluated are maintained in the table.

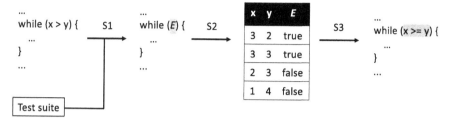

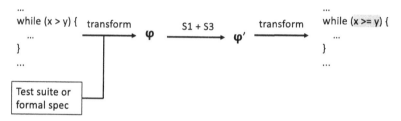

(a) Angelix decomposes program repair into three search problems.

(b) Directfix performs two searches S1 and S3 in one step (S2 is skipped for the reason described in the text). In principle, DirectFix can also work with formal specifications, as well as a test suite.

Fig. 2. Comparing the decomposition of Angelix and DirectFix

Given the extracted specification of E, Angelix performs its last search (S3), that is, a search for a patch. In this example, a patch expression x >= y satisfies the specification shown in the table. While not shown in the figure, Angelix also uses the structure of the original buggy expression (in this example, x > y) to find a patch that is as close to the original program as possible. In other words, Angelix *minimally* modifies the original buggy program. As discussed in Sect. 2, this minimality criterion is shown to be effective to improve the quality of automatically generated patches. To support the minimality criterion, Angelix transforms the original buggy expression e into a logical formula φ_e, and searches for φ'_e that (1) satisfies the extracted specification for e, and (2) preserves as many clauses of φ_e to look for a minimal patch. More technically, this search for a patch at the logical level is performed by customizing the component-based program synthesis [5]. While a detailed description is available in [15],[2] it is worth noting that the patch synthesizer of Angelix searches for a patch by feeding a transformed formula φ_e into an off-the-shelf SMT solver such as Z3 [1],[3] which implies that the search for a patch is performed inside an SMT solver.

Angelix performs the three searches one by one. It first selects suspicious expressions among a list of top N suspicious expressions (S1). For the chosen

---

[2] Angelix [16] reuses the synthesizer of DirectFix [15].
[3] More technically, a partial MaxMST solver built on top of Z3 is used.

suspicious expressions, Angelix performs S2 and only if S2 completes successfully, the last search S3 is started. If the search for a patch fails (either because S2 fails or S3 fails), Angelix selects another suspicious expressions, and repeat the procedure. This process continues until either a patch is found or resources such as time and a pool of suspicious expressions are exhausted.

DirectFix. The decomposition employed in DirectFix is simpler than in Angelix. As shown in Fig. 2(b), DirectFix performs a consolidated search after transforming a given buggy program and a test suite into a corresponding logical formula φ. In other words, DirectFix performs a search at the logical level using an off-the-shelf MaxSMT solver. While in Angelix, only the search for a patch (i.e., S3) is performed inside a MaxSMT solver, DirectFix performs all necessary searches inside a MaxSMT solver.

Initially, φ is unsatisfiable (i.e., there is no model that satisfies φ), reflecting the fact that the program fails at least one test in a given test suite (more technically, the program part of φ conflicts with the test part of φ). Thus, the goal of DirectFix is to search for a satisfiable formula φ', a variant of φ where the test part (or the specification part, if a formal specification is given, instead of a test suite) is preserved.

As mentioned, DirectFix performs a search in one step by feeding an unsatisfiable formula φ into a MaxSMT solver. We describe how a MaxSMT solver performs various searches for the purpose of discussion. We will later discuss how the implicit searches performed by a MaxSMT solver affects the time performance of program repair.

First, the search for suspicious expressions (i.e., S1) can be done by making use of the unsat-core feature supported by a MaxSMT solver. An unsat-core of an unsatisfiable formula φ is a group of clauses that conflict with each other. Conversely, the remaining clauses can be satisfied if the unsat-core is removed. Thus, only the clauses in a unsat-core are considered suspicious.

Meanwhile, the search for the specification of suspicious expressions does not take place, because the formula surrounding the suspicious clauses can be used as a specification. More specifically, given an unsatisfiable formula φ (e.g., $c_1 \wedge c_2 \wedge c_3 \wedge c_4$ where c_i is a clause of φ) and a set of suspicious clauses constituting an unsat-core (e.g., c_2 and c_3), the remaining part of the formula (e.g., $c_1 \wedge c_4$) can be considered the specification of the suspicious clauses. Finally, the search for a patch can be done in a similar way in Angelix, based on the available specification.

Angelix vs. DirectFix. As shown, Angelix generates a patch through a more fine-grained decomposition than DirectFix. DirectFix does not perform S2 of Angelix (the search for the specification of suspicious expressions). Also, though we described how DirectFix performs the two searches, S1 and S3, inside a MaxSMT solver, a MaxSMT solver does not have knowledge on different phases of searches. Thus, there is no guarantee that a MaxSMT solver first performs the first search, and then moves to the next search. Instead, different kinds of searches are performed in an intertwined way inside a MaxSMT solver.

This difference between the more fine-grained consecutive decomposition of Angelix and the lump-sum approach of DirectFix has performance consequences. To cut to the chase, Angelix outperforms DirectFix in terms of time performance. In fact, Angelix was introduced to overcome the scalability problem of DirectFix, while maintaining the ability to generate minimal patches [16].

The outperformance of Angelix over DirectFix can be explained in the following way. In Angelix, the search for a patch is started only when the previous two searches, S1 and S2, complete successfully. On the contrary, in DirectFix, a MaxSMT solver tries to find a patch for an expression e regardless of whether the modification of e can make all tests pass. Even when it is not possible to pass all tests by modifying e, DirectFix searches for a patch expression for e in vain. As a result, time is wasted to solve an impossible problem.

More generally, the clear distinction between S1 and S3, as done in Angelix, has an effect of reducing the search space. By performing S3 only when S1 completes successfully, the unreachable search space of S3 can be pruned out. However, as will be shown shortly, the search space is not always reduced whenever a search is decomposed into more steps.

Angelix vs. SE-ESOC. As shown in Fig. 3(b), SE-ESOC combines S2 and S3, resulting in a fewer search steps than in Angelix. On the contrary to DirectFix, merging two search steps leads to the reduction of the search space. The search space reduction takes place in the S2 step. In Angelix, S2 is performed via symbolic execution. Using symbolic execution, Angelix systematically explores symbolic execution paths, and collect those where all tests pass. Depending on a program, the search space of symbolic execution can be quite large.

However, not all symbolic execution paths that lead to test success leads to a patch. For example, Fig. 3(a) shows two tables corresponding to two passing symbolic execution paths, respectively. The second table can be considered a specification for an expression x % y == 0. However, if the synthesizer does not support the modulus operator, the mentioned patch expression cannot be synthesized by the synthesizer at hand.

SE-ESOC takes into account the expressive power of the synthesizer when performing symbolic execution. This essentially is equivalent to performing second-order symbolic execution; notice in Fig. 3(b) that a second-order symbol $E(x,y)$ is used instead of a first-order symbol E. During second-order symbolic execution, symbolic execution paths for which there is no model for $E(x,y)$—which means that a patch expression cannot be synthesized—are not executed, pruning out the search space.

In sum, decomposing S1 and S3 into two distinct steps reduces the search space. However, decomposing S2 and S3 into two distinct steps, on the contrary, enlarges the search space. This shows the importance of designing search-space decomposition.

3.2   Search Algorithms for Individual Search Problems

In Sect. 3.1, we showed how search problem decomposition affects the size of the search space. By adding or removing a search step properly, the search space

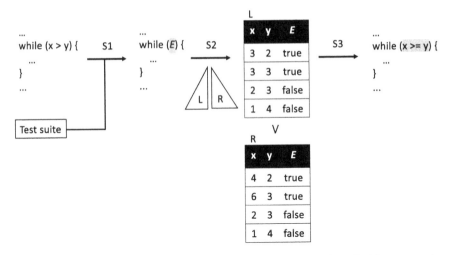

(a) The S2 of Angelix is performed via symbolic execution. In the example, the symbolic computation subtree **L** induces the first specification, whereas the symbolic computation subtree **R** induces the second specification. Not all specifications are supported by the synthesizer.

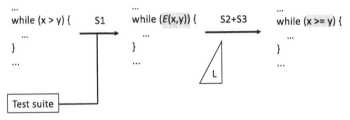

(b) SE-ESOC performs two searches S2 and S3 in one step. During the second-order symbolic execution, only the symbolic paths in which a patch can be synthesized are explored.

Fig. 3. Comparing the decomposition of Angelix and SE-ESOC

can be reduced. The search space of Angelix is smaller than that of DirectFix, but larger than SE-ESOC.

While SE-ESOC explores the smallest search space among the three approaches, this does not necessarily mean that SE-ESOC is fastest among the three. In fact, according to the experimental result shown in [14], the average running time of SE-ESOC (1 m 24 s) is longer than that of Angelix (1 m 5 s). For some subjects used in the experiment, SE-ESOC is more than two times slower than Angelix.

The fact that the space reduction of SE-ESOC does not always lead to faster running time implies that the size of the search space is not the only factor that affects the speed of automated program repair. Another important factor that affects the time performance of program repair is the *velocity* of search. As mentioned, SE-ESOC performs second-order symbolic execution, which typically takes more time than first-order symbolic execution used in Angelix. This is because performing second-order symbolic execution involves solving second-order constraints, which is typically more heavyweight and slower than solving first-order constraints. To truly speed up automated program repair, we need not only optimal decomposition that reduces the size of the search space, but also faster algorithms that can handle individual search problems more efficiently.

3.3   Generate-And-Validate Approaches

In the previous section, we show with *symbolic* repair approaches how two critical performance factors—the search space and the velocity of search—are influenced by search decomposition and the search algorithms of individual searches. In this section, we show that this observation also extends to generate-and-validate approaches.

Search Problem Decomposition. Similar to the previous section, we compare three different generate-and-validate approaches, namely, GenProg [10,24], SPR [11], and F1X [13], from the perspective of search problem decomposition. As shown in Fig. 4(a), SPR takes the same three steps as used in Angelix.[4] However, unlike in Angelix where a suspicious expression is replaced with a placeholder expression E, SPR transforms a suspicious expression into multiple different ways.

Meanwhile, GenProg takes simpler decomposition steps. As in DirectFix, GenProg performs two searches, S1 and S3, in an intertwined manner, which leads to a similar inefficiency to DirectFix. That is, regardless of whether it is possible to pass all tests by modifying a suspicious program location L, GenProg tries to modify L, wasting time in case patching is not possible at L. In comparison, the fact that SPR performs a search for a patch after completing a search for a specification of a suspicious expression as in Angelix reduces the search space, by pruning out the patches at infeasible program locations.

F1X further reduces the search space in a similar way to SE-ESOC. That is, F1X combines S2 and S3, as in SE-ESOC. While F1X enumerates over the finite patch space, patch candidates that are bound to fail a test are skipped over, which effectively prunes out the search space. Figure 4(b) shows an example where the first patch candidate $x < y$ fails a test T. Suppose that when the failure occurs, result $R1$ is observed, where the column x, y, z and E show the values of the three variables and the patch candidate, respectively, for each encounter with the patch candidate. Since variables y and z have the same values in all three instances, it can be inferred that the next patch candidate $x < z$ is certain to fail test T. Thus, F1X skips over this patch candidate without running it.

---

[4] Depending on the type of a suspicious program location, SPR also directly generates a patch candidate without searching for a specification. We ignore those cases in this discussion.

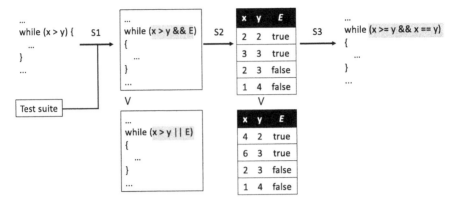

(a) **SPR decomposes program repair into three search problems.**

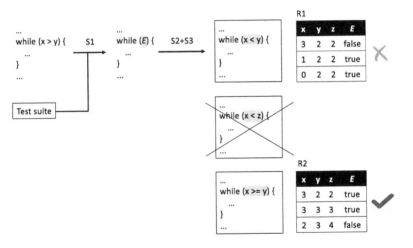

(b) **F1X performs two searches S2 and S3 in one step. During the enumeration of the patch space, patches that are certain to fail a test (denoted with the red cross mark) are skipped.**

Fig. 4. Comparing the decomposition of SPR and F1X

Search Algorithms. Repair tools using the generate-and-validate approach typically compiles each generated patch candidate, and runs tests over the compiled patch candidate. This implies that a cycle of generate and validate can be sped up by avoiding unnecessary compilation and test running. SketchFix [4] is a good example. Given a buggy program and a chosen suspicious program location L, SketchFix replaces L with a hole. The resultant program with such a hole is called a sketch. SketchFix compiles a sketch, instead of compiling each patch candidate. Since one sketch corresponds to many patch candidates (each patch candidate can be viewed as an instantiation of a hole with a specific expression or a statement), the number of compilation reduces in SketchFix. When a hole is executed in a compiled sketch, patch synthesis takes place on the fly.

While SketchFix increases the velocity of search by reducing the number of compilation, AE [23] speeds up the repair process by running a test that is likely to fail before a test that is likely to pass. This expedites test failure, and as a result AE can move more quickly to the next patch candidate.

4   Conclusion

We conclude this paper by sharing our long-term research goal. Given the low likelihood that automated fixes are always correct (due to the inherent overfitting problem), human developers would need to decide whether to accept a generated patch. Thus, a tool should provide a developer not only with bug fixes, but also with sufficient information based on which the developer can use for their decision. Also, it is equally important to have an effective mechanism in which a developer provides feedback for the tool, and in return quickly receives a refined patch. Developing a technique for high-speed automated fixes—the survey topic of this paper—is a prerequisite for such interactive fixes.

In this paper, we showed the time performance of automated program repair is affected by how the search for a patch is decomposed into smaller search problems, and how individual search is performed. We expect that new insights offered in this paper expedites speeding up automated program repair.

References

1. De Moura L, Bjørner N (2008) Z3: an efficient SMT solver. In: International conference on tools and algorithms for the construction and analysis of systems. Springer, Heidelberg, pp 337–340
2. D'Antoni L, Samanta R, Singh R (2016) Qlose: program repair with quantitative objectives. In: International conference on computer aided verification. Springer, Heidelberg, pp 383–401
3. Harman M, O'Hearn P (2018) From start-ups to scale-ups: opportunities and open problems for static and dynamic program analysis. In: SCAM
4. Hua J, Zhang M, Wang K, Khurshid S (2018) Towards practical program repair with on-demand candidate generation. In: ICSE, pp 12–23
5. Jha S, Gulwani S, Seshia SA, Tiwari A (2010) Oracle-guided component-based program synthesis. In: Proceedings of the 32nd ACM/IEEE international conference on software engineering, vol 1. ACM, pp 215–224
6. Kim D, Nam J, Song J, Kim S (2013) Automatic patch generation learned from human-written patches. In: Proceedings of the 2013 international conference on software engineering (ICSE). IEEE Press, pp 802–811
7. Le XBD, Chu DH, Lo D, Le Goues C, Visser W (2017) S3: syntax-and semantic-guided repair synthesis via programming by examples. In: ESEC/FSE, pp 593–604
8. Le XD, Lo D, Le Goues C (2016) History driven program repair. In: SANER, pp 213–224
9. Le XD, Thung F, Lo D, Le Goues C (2018) Overfitting in semantics-based automated program repair. Empir Softw Eng 23(5):3007–3033

10. Le Goues C, Dewey-Vogt M, Forrest S, Weimer W (2012) A systematic study of automated program repair: fixing 55 out of 105 bugs for $8 each. In: ICSE. IEEE, pp 3–13

11. Long F, Rinard M (2015) Staged program repair with condition synthesis. In: ESEC/FSE, pp 166–178

12. Long F, Rinard MC (2016) An analysis of the search spaces for generate and validate patch generation systems. In: Proceedings of the 38th international conference on software engineering (ICSE), pp 702–713

13. Mechtaev S, Gao X, Tan SH, Roychoudhury A (2018) Test-equivalence analysis for automatic patch generation. ACM Trans Softw Eng Methodol (TOSEM) 27(4):15

14. Mechtaev S, Griggio A, Cimatti A, Roychoudhury A (2018) Symbolic execution with existential second-order constraints. In: ACM joint meeting on European software engineering conference and symposium on the foundations of software engineering (FSE). ACM, pp 389–399

15. Mechtaev S, Yi J, Roychoudhury A (2015) DirectFix: looking for simple program repairs. In: ICSE, pp 448–458

16. Mechtaev S, Yi J, Roychoudhury A (2016) Angelix: scalable multiline program patch synthesis via symbolic analysis. In: ICSE, pp 691–701

17. Nguyen HDT, Qi D, Roychoudhury A, Chandra S (2013) SemFix: program repair via semantic analysis. In: International conference on software engineering (ICSE). IEEE, pp 772–781

18. Pei Y, Furia CA, Nordio M, Wei Y, Meyer B, Zeller A (2014) Automated fixing of programs with contracts. IEEE Trans Softw Eng (TSE) 40(5):427–449

19. Qi Y, Mao X, Lei Y, Dai Z, Wang C (2014) The strength of random search on automated program repair. In: Proceedings of the 36th international conference on software engineering (ICSE). ACM, pp. 254–265

20. Qi Z, Long F, Achour S, Rinard MC (2015) An analysis of patch plausibility and correctness for generate-and-validate patch generation systems. In: Proceedings of the 2015 international symposium on software testing and analysis (ISSTA), pp 24–36

21. Smith EK, Barr ET, Le Goues C, Brun Y (2015) Is the cure worse than the disease? Overfitting in automated program repair. In: Proceedings of the 2015 10th joint meeting on foundations of software engineering (FSE). ACM, pp 532–543

22. Tan SH, Yoshida H, Prasad MR, Roychoudhury A (2016) Anti-patterns in search-based program repair. In: Proceedings of the 2016 24th ACM SIGSOFT international symposium on foundations of software engineering. ACM, pp 727–738

23. Weimer W, Fry ZP, Forrest S (2013) Leveraging program equivalence for adaptive program repair: models and first results. In: Automated software engineering (ASE). IEEE, pp 356–366

24. Weimer W, Nguyen T, Le Goues C, Forrest S (2009) Automatically finding patches using genetic programming. In: ICSE, pp 364–374

25. Xiong Y, Liu X, Zeng M, Zhang L, Huang G (2018) Identifying patch correctness in test-based program repair. In: Proceedings of the 40th international conference on software engineering. ACM, pp 789–799

26. Xuan J, Martinez M, Demarco F, Clement M, Marcote SL, Durieux T, Le Berre D, Monperrus M (2017) Nopol: automatic repair of conditional statement bugs in Java programs. IEEE Trans Softw Eng (TSE) 43(1):34–55

27. Yi J, Tan SH, Mechtaev S, Böhme M, Roychoudhury A (2018) A correlation study between automated program repair and test-suite metrics. Empir Softw Eng 23(5):2948–2979

On the Parcellation of Functional Magnetic Resonance Images

Adam Folohunsho Zubair[1,2(✉)], Segun Benjamin Aribisala[1],
Marco Manca[3], and Manuel Mazzara[2]

[1] Department of Computer Science, Lagos State University,
Ojo PMB 0001, Nigeria
z.folohunsho@innopolis.university
[2] Institute of Technologies and Software Development, Innopolis University,
Innopolis 420500, Russia
[3] Conseil Européen pour la Recherche Nucléaire Openlab,
1211 Geneva, Switzerland

Abstract. Functional Magnetic Resonance Imaging (fMRI) is one of the techniques for measuring activities in the brain and it has been demonstrated to have a high potential in clinical application. However, fMRI is limited by some of the contradictory results reported by different studies. One of the possible reasons for this contradiction is the lack of standard and acceptable methods of analyzing fMRI data. Analysis of fMRI data in studies focusing on brain connectivity normally requires the definition of region of interest. This is normally done using regions of interest drawn on high resolution anatomical images. The use of anatomical images implies using structural information, thereby losing any functional information that could improve the analysis of fMRI data. In this article, we present the framework for a region of interest definition for fMRI using structural and functional information. Contrary to existing approaches, the proposed method will also consider the use of network properties. The method uses a bottom-up approach as it starts with structural information, then include functional information before it finally includes network properties. We hypothesize that the use of multiple information in defining the regions of interests in fMRI data will produce a more accurate, more reproducible and more trusted results than the use of structural information only. It is hoped that the use of the proposed model will lead to improved analysis of fMRI brain data, hence increasing its diagnostic potential.

Keywords: fMRI · Functional magnetic resonance images ·
Network properties · Connectivity · Graph theory ·
Region of interest and parcellation

1 Introduction

Functional magnetic resonance imaging (fMRI) has become an essential tool for investigating functional connectivity and brain networks in the human brain [1]. The approach uses task-based time course information obtained using a Blood Oxygenation Level Dependent contrast (BOLD) acquisition [2] to measure the temporal correlation

© Springer Nature Switzerland AG 2020
P. Ciancarini et al. (Eds.): SEDA 2018, AISC 925, pp. 325–332, 2020.
https://doi.org/10.1007/978-3-030-14687-0_29

between different regions of the brain within a single subject over time. Connectivity associations provide important diagnostic information for central nervous system related disorders. Due to its versatility, fMRI has been widely used in neuroscience, for example for investigating connectivity in brain states [3, 4], the relationship between connectivity and behavior [5], in schizophrenia [6], bipolar disorder [6] etc.

Several methods for processing fMRI data have been proposed, these include the seed methods [7], principal component analysis [8], independent component analysis [9] and clustering [10]. These methods have produced promising results, but findings vary between methods and there is no general consensus on the optimal strategy to data analysis [11, 12]. This lack of a standard method of analysis weakens the potential diagnostic or prognostic impact of fMRI data and could account for contrary findings in its clinical applications [13]. The development of a robust and standardized mathematical model for data analysis is essential to provide a framework for the use of fMRI in early and improved detection of brain related diseases.

Graph theory has been successfully used in many analysis of brain networks [14]. With this approach, the brain is modelled as a graph or network G(N, M), comprising of N "nodes" connected by M "edges". The nodes of the graph typically represent distinct anatomical or functional regions of the brain while its edges represent some measure of interactions between the regions. This representation allows the use of a rich set of mathematical tools and theoretical concepts in understanding brain network topology and dynamics [15–19]. Graph theory offers a range of measures for quantifying network properties [20], including small worldness [21], modularity [22], global efficiency [23], clustering coefficient [18] and hierarchical structure [18]. Previous research has demonstrated that these quantifiable network properties change during normal development [24], aging [22, 25, 26] and various neurological and neuropsychiatric diseases such as Alzheimer's disease (AD) [27] and late life depression (LLD) [7, 28].

The nodes of a fMRI brain graph can either be defined by the individual voxels of the brain imaging dataset (voxel-based representation) or the mean values computed from a group of voxels (region-based representation). Since analysis using the voxel based (VB) approach is conducted on a single voxel basis, the intrinsic Signal to Noise Ratio (SNR) is low whereas the region based (ROI) approach sums voxel values over a range of voxels giving higher SNR [29, 30]. Thus, the nodes of a fMRI brain graph are commonly represented using the mean values computed from a given region of interest [31].

One of the major methodological challenges of graph analysis of fMRI is appropriate definition of brain regions to represent the network nodes [13, 30, 32]. The universal method of defining regions of interest is to use an a priori anatomical template [10, 13, 16, 18, 19, 22, 24, 27, 34–39] which is the model-based method and an alternative method is the data-driven method which is independent of priori information. The performance of the model-based approach has been demonstrated to be better than the data-driven approach when used for detecting fMRI experimental effect [40]. Previous work has shown that the network's organizational parameters vary with template choice [32, 41]. The most commonly used templates are the Automatic

Nonlinear Image Matching and Automatic Labelling algorithm (ANIMAL) atlas [42], Automated Anatomical Labelling (AAL) toolbox [43] template and the Freesurfer software [44] template. These templates are used for parcellating each subject's fMRI brain data into multiple regions. Then, regional means and pairwise correlation are computed to form network nodes and edges respectively. This approach has proven very useful but suffers from the following shortcomings:

a. There is no acceptable gold standard for ROI definition because there are no clear macroscopic boundaries that can be used to delineate adjacent regions (The true boundary is instead defined by functional response). Thus, the criteria used are arbitrary and vary between templates. Additionally, there is large variation between individual brain anatomy, even when data is transformed into standard space.
b. There is no systematic method of defining the number of voxels within a region. Currently, this number varies between 10s and 1000s and this influences network organizational parameters [10, 41].
c. Regions are often chosen to be as large as possible (to maximize SNR). It is likely therefore that they include signals from several different functional sub-regions and this can complicate interpretation of results or even introduce partial volume errors [30].

2 Model Formulation

The interrelated problems in Sect. 1 are obstacles in fMRI brain data analysis which contributes to the irreproducibility of fMRI data analysis. This suggests that there is a need to develop an automatic and robust model for parcellating fMRI brain data into standard multiple regions.

Here, we propose the framework for a model for analysis of fMRI. The model will use structural, functional and network properties of the brain to group only related pixels into a region, hence eliminating the need to choose a fixed number of pixels within a region. The model will employ the use of many network properties thereby having a better performance than those measures based on single properties (Normalized Cut [45], Modularity [46] and Canonical correlation-based [37] measure). The network nodes will be the mean values from each ROI while the edges will be based on wavelet correlation coefficient as this has been demonstrated to perform better than Pearson correlation or other time dependent correlation coefficients [33].

2.1 Related Work

Golland et al. [53] proposed a parcellation method based on k-mean clustering. This method is subjective as it requires the user to choose the number of ROIs. Additionally, this method does not use any functional information which could lead to grouping functionally different pixels together.

Hagmann et al. [48, 49] proposed a method based on region growing. A major limitation of this approach is that all the ROIs are constrained to have the same number of pixels and there is no standard method of choosing an optimum pixel number. Poor choice of pixel number changes the coarseness of each region leading to either grouping of functionally different regions together or separating functionally similar pixels. Also, this approach does not use any functional information.

Delues [37] proposed a Canonical correlation (CC) based method of defining ROIs for fMRI brain data. This method defines ROI as a set of voxels with similar connectivity patterns to other ROIs. This method uses a network property (connectivity) but its major shortcoming is that a minimum of two ROIs is required to define (or redefine) one ROI making it difficult to define a single ROI.

Shen et al. [50] proposed and compared parcellation techniques based on the modularity detection algorithm [46], Normalised cut [45] (which analyses graph edge properties) and a Gaussian mixture model (GMM) [51] (a probabilistic model with the underlying assumption that the data has Gaussian density function). This study showed that both the modularity detection and NCUT algorithms perform better than GMM and that GMM is not good for parcellating fMRI data. This demonstrates that graph based algorithms have strong potential for parcellating fMRI brain data.

DonGiovanni et al. [47] proposed a method based on selection and clustering (S&C) of voxel to reproduce a consistent grouping of voxel belonging to the reference ROI based on spatiofunctional metric defined in a feature space given by spatial coordinates and functional weights of the low-dimensional representation. The limitation of this approach is that it does not consider the network properties of each voxel when grouping the voxels to the reference ROI.

2.2 Hypothesis

We expect the definition of regions of interest based on multiple network properties to produce more reproducible, consistent and stable brain networks than definition based on single property leading to improved methods of analysis of fMRI brain data.

We hypothesize that a region containing pixels which are structurally, functionally and graphically closely related will give high connectivity consistencies across ROIs.

2.3 Model Design

In the current graph theory analyses of brain fMRI data, the network nodes correspond to averages across groups of pixels representing each parcellated region. We will consider data at a lower level and apply graph theory within the parcellated pixels (i.e. consider each pixel in the parcel as a separate node) to identify the degree of similarity between pixels. Therefore, each pixel is grouped based on it level of interaction with other pixels, and the level of interaction is quantified using network properties to obtain its effect on the measure of integration, resilience and segregation of the network.

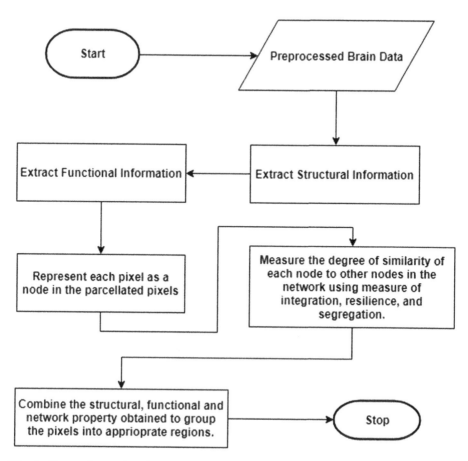

Fig. 1. Proposed procedure for region of interest definition based on combination of structural, functional and network properties

The proposed algorithm which is depicted in Fig. 1 will be implemented on fMRI brain data, and a performance measure based on connectivity consistencies between the ROIs of a single subject and across the control group will be investigated.

The regions obtained from the algorithm will be used to construct the whole brain network. All the existing ROI based analysis techniques of fMRI use a single template [10, 13, 16, 18, 19, 22, 24, 27, 34–39, 47], Previous MRI data analysis demonstrated that analysis based on single template is subjective and suffers from problems due to anatomical variability between individuals [52]. The study demonstrated that multiple template performs better than single template. In view of this the choice of multiple templates (e.g. AAL, ANIMAL and Freesurfer templates) will be investigated in this model, and the selected templates will be merged appropriately [52]. The final template produced from this approach will serve as the model's standard template.

3   Conclusion

The existing methods of analyzing brain fMRI data rely on the anatomical images which uses only structural information. Also, the existing methods do not have a systematic method of choosing the number of pixels in a region of interest. We have proposed the framework of an algorithm for improved analyzes of fMRI data. The algorithm will combine structural, functional and network properties of the brain to divide fMRI brain data into multiple regions of interest. It is hoped that when fully implemented, the proposed model will lead to improved analysis of fMRI brain data, hence increasing its diagnostic potential.

References

1. Chen JE, Glover GH (2015) Functional magnetic resonance imaging methods. Neuropsychol Rev 25(3):289–313
2. Ogawa S et al (1990) Brain magnetic resonance imaging with contrast dependent on blood oxygenation. Proc Natl Acad Sci 87(24):9868–9872
3. Elliott ML et al (2018) General functional connectivity: shared features of resting-state and task fMRI drive reliable individual differences in functional brain networks, p 330530
4. Martuzzi R et al (2010) Functional connectivity and alterations in baseline brain state in humans. Neuroimage 49(1):823–834
5. Hampson M et al (2006) Brain connectivity related to working memory performance. J Neurosci 26(51):13338–13343
6. Dezhina Z et al (2018) A systematic review of associations between functional MRI activity and polygenic risk for schizophrenia and bipolar disorder 1–16
7. Kenny ER et al (2010) Functional connectivity in late-life depression using resting-state functional magnetic resonance imaging. Am J Geriatr Psychiatry 18(7):643–651
8. Friston K et al (1993) Functional connectivity: the principal-component analysis of large (PET) data sets. J Cereb Blood Flow Metab 13(1):5–14
9. McKeown MJ et al (1998) Analysis of fMRI data by blind separation into independent spatial components. Hum Brain Mapp 6(3):160–188
10. Salvador R et al (2005) Neurophysiological architecture of functional magnetic resonance images of human brain. Cereb Cortex 15(9):1332–1342
11. Griffanti L et al (2016) Challenges in the reproducibility of clinical studies with resting state fMRI: an example in early Parkinson's disease. Neuroimage 124:704–713
12. Bennett CM, Miller MB (2010) How reliable are the results from functional magnetic resonance imaging? Ann N Y Acad Sci 1191(1):133–155
13. Cole DM, Smith SM, Beckmann CF (2010) Advances and pitfalls in the analysis and interpretation of resting-state FMRI data. Front Syst Neurosci 4:8
14. Islam M et al (2018) A survey of graph based complex brain network analysis using functional and diffusional MRI. Am J Appl Sci 14(12):1186–1208
15. Strogatz SH (2001) Exploring complex networks. Nature 410(6825):268
16. Albert R, Barabási AL (2002) Statistical mechanics of complex networks. Rev Mod Phys 74(1):47
17. Newman ME (2003) The structure and function of complex networks. SIAM Rev 45(2):167–256

18. Bullmore E, Sporns O (2009) Complex brain networks: graph theoretical analysis of structural and functional systems. Nat Rev Neurosci 10(3):186

19. Rubinov M et al (2009) Small-world properties of nonlinear brain activity in schizophrenia. Hum Brain Mapp 30(2):403–416

20. Finotellia P, Dulioa P (2015) Graph theoretical analysis of the brain. An overview. Scienze e Ricerche 9:89–96

21. Bassett DS, Bullmore ET (2017) Small-world brain networks revisited. Neuroscientist 23 (5):499–516

22. Meunier D et al (2009) Hierarchical modularity in human brain functional networks. Front Neuroinform 3:37

23. Di X et al (2013) Task vs. rest—different network configurations between the coactivation and the resting-state brain networks. Front Hum Neurosci 7:493

24. Supekar K, Musen M, Menon V (2009) Development of large-scale functional brain networks in children. PLoS Biol 7(7):e1000157

25. Wang L et al (2010) Age-related changes in topological patterns of large-scale brain functional networks during memory encoding and recognition. Neuroimage 50(3):862–872

26. Achard S, Bullmore E (2007) Efficiency and cost of economical brain functional networks. PLoS Comput Biol 3(2):e17

27. Buckner RL et al (2009) Cortical hubs revealed by intrinsic functional connectivity: mapping, assessment of stability, and relation to Alzheimer's disease. J Neurosci 29 (6):1860–1873

28. Cieri F et al (2017) Late-life depression: modifications of brain resting state activity. J Geriatr Psychiatry Neurol 30(3):140–150

29. Korhonen O et al (2017) Consistency of regions of interest as nodes of fMRI functional brain networks. Netw Neurosci 1(3):254–274

30. Fornito A, Zalesky A, Bullmore ET (2010) Network scaling effects in graph analytic studies of human resting-state fMRI data. Front Syst Neurosci 4:22

31. Stanley ML et al (2013) Defining nodes in complex brain networks. Front Comput Neurosci 7:169

32. Wang J, Zuo X, He Y (2010) Graph-based network analysis of resting-state functional MRI. Front Syst Neurosci 4:16

33. Zalesky A et al (2010) Whole-brain anatomical networks: does the choice of nodes matter? Neuroimage 50(3):970–983

34. Achard S et al (2006) A resilient, low-frequency, small-world human brain functional network with highly connected association cortical hubs. J Neurosci 26(1):63–72

35. Aubert-Broche B et al (2009) Clustering of atlas-defined cortical regions based on relaxation times and proton density. Neuroimage 47(2):523–532

36. Chen ZJ et al (2008) Revealing modular architecture of human brain structural networks by using cortical thickness from MRI. Cereb Cortex 18(10):2374–2381

37. Deleus F, Van Hulle MM (2009) A connectivity-based method for defining regions-of-interest in fMRI data. IEEE Trans Image Process 18(8):1760–1771

38. Estrada E, Hatano N (2008) Communicability in complex networks. Phys Rev E 77 (3):036111

39. Honey C et al (2009) Predicting human resting-state functional connectivity from structural connectivity. Proc Natl Acad Sci 106(6):2035–2040

40. Zhang J et al (2013) A manual, semi-automated and automated ROI study of fMRI hemodynamic response in the caudate. Nucleus 2(150):2

41. Wang J et al (2009) Parcellation-dependent small-world brain functional networks: a resting-state fMRI study. Hum Brain Mapp 30(5):1511–1523

42. Collins DL et al (1995) Automatic 3-D model-based neuroanatomical segmentation. Hum Brain Mapp 3(3):190–208
43. Tzourio-Mazoyer N et al (2002) Automated anatomical labeling of activations in SPM using a macroscopic anatomical parcellation of the MNI MRI single-subject brain. Neuroimage 15 (1):273–289
44. Fischl B et al (2004) Automatically parcellating the human cerebral cortex. Cereb Cortex 14 (1):11–22
45. Shi J, Malik J (2000) Normalized cuts and image segmentation. IEEE Trans Pattern Anal Mach Intell 22(8):888–905
46. Newman ME (2006) Modularity and community structure in networks. Proc Natl Acad Sci 103(23):8577–8582
47. DonGiovanni D, Vaina LM (2016) Select and cluster: a method for finding functional networksof clustered voxels in fMRI. Comput Intell Neurosci 2016
48. Hagmann P et al (2008) Mapping the structural core of human cerebral cortex. PLoS Biol 6 (7):e159
49. Hagmann P et al (2007) Mapping human whole-brain structural networks with diffusion MRI. PloS One 2(7):e597
50. Shen X, Papademetris X, Constable RT (2010) Graph-theory based parcellation of functional subunits in the brain from resting-state fMRI data. Neuroimage 50(3):1027–1035
51. McLachlan G, Peel D (2000) Finite mixture models, Willey series in probability and statistics. Wiley, New York
52. Heckemann RA et al (2006) Automatic anatomical brain MRI segmentation combining label propagation and decision fusion. Neuroimage 33(1):115–126
53. Golland Y, Golland P, Bentin S, Malach RJN (2008) Data-driven clustering reveals a fundamental subdivision of the human cortex into two global systems, 46(2):540–553

Author Index

P. Ciancarini et al. (Eds.): SEDA 2018, AISC 925, pp. 333–334, 2020.
https://doi.org/10.1007/978-3-030-14687-0

Printed in the United States
By Bookmasters